Bee Genetics
and Breeding

Bee Genetics
and Breeding

Edited by

Thomas E. Rinderer

Honey-Bee Breeding, Genetics,
and Physiology Laboratory
Agricultural Research Service
United States Department of Agriculture
Baton Rouge, Louisiana

Northern Bee Books

Northern Bee Books 2008

First published by Academic Press, Inc 1986

This re-print edition, issued under licence from Academic Press (London) July 2008 by:

Northern Bee Books
Scout Bottom Farm
Mytholmyrod
Hebden Bridge
HX7 5JS
UK

ISBN: 978-1-914934-03-2

A catalogue record for this book is available from the British Library

This edition 2010 printed by Lightning Source

Front cover illustration taken from
The Management of Bees: with a description of the "Ladies' Safety Hive"
Samuel Bagster 1834

To Walter C. Rothenbuhler
a pioneer in the field,
a mentor to many of us,
and a friend to all of us.

Contents

Contents

Contributors

Numbers in parentheses indicate the pages on which the authors' contributions begin.

Anita M. Collins (155, 283), Honey-Bee Breeding, Genetics, and Physiology Laboratory, Agricultural Research Service, United States Department of Agriculture, Baton Rouge, Louisiana 70820

Jean-Marie Cornuet (235), Station de Zoologie, Institut National de la Recherche Agronomique, 84140 Montfavet, France

Alfred Dietz (3), Department of Entomology, University of Georgia, Athens, Georgia 30602

Lionel Segui Gonçalves (345), Departamento de Biologia, Faculdade de Filosofia, Ciências e Letras de Ribeirão Preto, USP, Ribeirão Preto, São Paulo, Brasil

John R. Harbo (361), Honey-Bee Breeding, Genetics, and Physiology Laboratory, Agricultural Research Service, United States Department of Agriculture, Baton Rouge, Louisiana 70820

Gudrun Koeniger (255), Institüt für Bienenkunde der Polytechnishen Gesellschaft E. V., Universität Frankfurt am Main, 6370 Oberursel 1, Federal Republic of Germany

Jovan M. Kulinčević (391), Department of Biology, University of Belgrade and Beekeeping Combine, "Beograd," 11000 Beograd, Yugoslavia

Harry H. Laidlaw, Jr. (323), Department of Entomology, University of California, Davis, Davis, California 95616

Charles P. Milne, Jr. (205), Department of Genetics and Molecular Biology, University of Guelph, Guelph, Ontario, Canada N1G 2W1

Robin F. A. Moritz (121), Institüt für Bienenkunde der Polytechnishei Gesellschaft E. V., Universität Frankfurt am Main, 6370 Oberursel 1 Federal Republic of Germany

Robert E. Page, Jr. (323), Department of Entomology, The Ohio State Uni versity, Columbus, Ohio 43210

Thomas E. Rinderer (155, 305), Honey-Bee Breeding, Genetics, and Physi ology Laboratory, Agricultural Research Service, United States Depart ment of Agriculture, Baton Rouge, Louisiana 70820

Friedrich Ruttner (23), Institüt für Bienenkunde der Polytechnishen Ge sellschaft E. V., Universität Frankfurt am Main, 6370 Oberursel 1, Federa Republic of Germany

Antonio Carlos Stort (345), Departamento de Biologia, Instituto de Bio ciências, UNESP, Rio Claro, São Paulo, Brasil

H. Allen Sylvester (177), Honey-Bee Breeding, Genetics, and Physiolog) Laboratory, Agricultural Research Service, United States Department o: Agriculture, Baton Rouge, Louisiana 70820

Kenneth W. Tucker (57), Honey-Bee Breeding, Genetics, and Physiolog) Laboratory, Agricultural Research Service, United States Department o Agriculture, Baton Rouge, Louisiana 70820

Jerzy Woyke (91), Bee Culture Division, Agricultural University, Warsaw Poland

Preface

In a sense, this book represents the "coming of age" for bee genetics and breeding. Through the years, a few people have worked in these areas. Periodically, their contributions have been the subject of short reviews. In the last two decades many more people have studied the genetics and breeding of bees. Their combined efforts have resulted in important scientific advances that have been published in diverse journals and languages. Until now, no thorough review and synthesis of bee genetics and breeding have existed.

Collectively, the authors discuss the major subject areas of the field. Because of space limitations we have had to be selective. Nonetheless, while some publications have not been exhaustively reviewed, all the major research areas have received serious attention. Several authors have provided new, previously unpublished information and theory development.

The book has two parts. Part I deals with the scholarly issues of bee genetics. It is intended as a reference source for students of both bees and genetics. It could also serve as a text for university courses in bee genetics.

Part II deals more specifically with the practical issues of bee breeding. It contains sufficient guidance for bee breeders to initiate or improve breeding programs. Apiculturalists generally will find this part especially interesting since the quality of their own bee stock depends on the skills and knowledge of the breeders who produce their queens.

Several people have assisted in the preparation of this book. My wife, Vicki Lancaster, has been a constant source of encouragement. H. Allen Sylvester has read several edited chapters and his suggestions have led to the elimination of many errors. Jill Miranda, Lorraine Davis, and Sandra White have patiently done the bulk of the typing and retyping. Sandy Kleinpeter and Robert Spencer have produced all the illustrations. I thank them all.

Thomas E. Rinderer

Part I

Genetics

CHAPTER 1

Evolution

ALFRED DIETZ

I. INTRODUCTION

The evolution of eusocial insects involves two main issues: (1) the probable evolutionary steps, or phylogenetic origins, by which adaptive modifications accumulated over time, and (2) their adaptive significance, or how the modifications have been maintained in populations when they seem to decrease the fitness of the individuals endowed with them (Brockmann, 1984). The apparent contradiction and its importance was well recognized by Darwin (1859) when he described the complexity of cell-construction behavior in honey bees and the extent by which anything so close to perfection could have been favored by natural selection. In addition, he was concerned with the adaptiveness of sociality in general, because neuters or sterile females often possess behavioral and morphological characteristics which are distinctly different from those of reproductive colony members. How is it that such characteristics can be adaptive when the individuals endowed with them are unable to reproduce? These two issues have been debated extensively in the literature from the time they were formulated by Darwin (Starr, 1979; Brockmann, 1984).

II. PATHWAYS TO EUSOCIALITY

According to Wilson (1982), eusocial insects are characterized by three attributes: (1) cooperation among adults in brood care and nest construction, (2) overlapping of at least two generations, and (3) reproductive division of labor. Insects without these attributes are termed solitary, and those which lack either one or two of these attributes are known as presocial. In order to possess these attributes, it is essential that insects aggregate in some form of permanent grouping (Alexander, 1974), such as (1) groups of unrelated individuals which do or do not have social contact, (2) groups of uniformly related individuals which are not siblings, (3) groups of insects with various degrees of relatedness, including siblings, (4) groups of siblings where parents may or may not be present, and (5) groups consisting of identical individuals or clones. Grouping is generally obvious in social insects and has major importance in their defense. However, eusocial, or highly social, insects also show a considerable diversity in their behavioral patterns and life histories.

In eusocial Hymenoptera, colony founding occurs in two distinct patterns (Hölldobler and Wilson, 1977). A colony may be started by one or more reproductive females, who construct the nest, produce the eggs, and feed the larvae. The first brood is reared alone until they emerge and take over the work of the colony. The queen subsequently specializes in egg laying. This mechanism is known as independent founding. The second mechanism is known as swarming, whereby a new colony is founded by one or more queens and a group of workers from the original colony. The queen, in this situation, specializes in egg laying from the start.

In both mechanisms, the colony may be founded by a single queen (haplometrosis), or by several queens from the same generation (pleometrosis). A colony founded by several queens may either remain polygynous or become secondarily monogynous because of fighting among the queens or because workers eliminate all but one queen. Workers may also adopt newly mated sisters later which would result in secondary polygyny.

In honey bees, workers will not eliminate two queens introduced into a queenless colony (Dietz, 1968). They either coexist for an extended period of time if separated by queen excluders, or one queen becomes dominant, resulting in a discontinuation of egg laying and the ultimate disappearance of the other introduced queens (Dietz, 1985). However, Darchen and Lensky (1963) reported that the removal of stings of honey-bee queens sharing the same nest, to prevent the elimination of each other, resulted in the removal of surplus queens by workers. Dietz (1986) was able to maintain two queens separated by queen excluders for more than one year, but three or four queens could not be maintained together during the winter.

The phylogenetic of sociality in Hymenoptera can be investigated on the basis of the diversity of nesting patterns occurring in closely related species. Since one can find a continuous series of nests, starting with entirely solitary to highly eusocial, this sequence gives us clues to the evolutionary steps of eusociality (Brockmann, 1984). The two different sequences identified are known as the parasocial and subsocial (familial) routes (Fig. 1). The latter

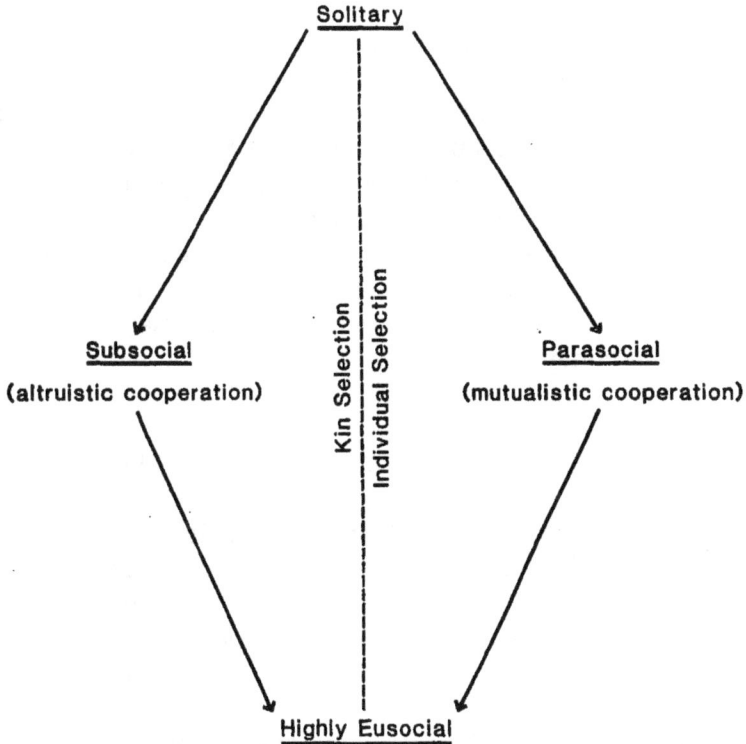

Fig. 1. Evolutionary pathways to eusociality. [After R. W. Matthews and J. R. Matthews (1978), copyright 1978 by John Wiley and Sons. Reprinted with permission.]

route is a continuum of nesting habits, including short encounters to long-term associations between mother and daughter (Wheeler, 1928; Michener, 1974).

In most solitary bees, the female provides her nest with an excess amount of food (mass provisioning), lays an egg into the cell, and after sealing it moves on to the next cell. Some solitary species, however, practice progressive feeding of the larva after is hatches from the egg. In this situation there is contact between the offspring and the mother, who provides partially or totally macerated food (Wilson, 1971; Michener, 1974).

In some instances, the offspring use the maternal nest and cooperate with the mother, if still alive, in the extension and defense of the common nest, even though each individual lays her own eggs. Most species, however, show a clear distinction in the reproductive abilities of mother and daughter. That is, the mother specializes in egg laying and the daughter in the care of the brood. This subsocial pathway to eusociality resembles the developmental pattern in an independently founded haplometric eusocial colony of ants, wasps, or bees (Brockmann, 1984).

The parasocial route to eusociality is a progression of social stages from communal to quasisocial to semisocial to highly eusocial (Michener, 1974; Matthews and Matthews, 1978). In general, the proposed pathway is from the solitary to the quasisocial state, with cooperative brood care by bees of the same generation. The next step would go from a semisocial stage, with its cooperative brood care and presence of a reproductive caste, to the eusocial stage. The familial, or subsocial, route has generally been considered as the universal pathway to eusociality.

Ribbands (1953) considers the development of the brood-food glands in honey bees as a distinguishing characteristic from their nonsocial relatives. Based on this assumption, he indicates that food has played a major role in the evolution of honey bees.

III. THEORIES OF INDIVIDUAL SELECTION

Four main theories on the evolution of eusociality have been advanced which are based on the natural selection of individuals: group selection, kin selection, mutualism, and parental manipulation. These theories attempt to offer an explanation of why the haplodiploid mode of sex determination in Hymenoptera appears to have been such a successful base for social evolution (Brian, 1983).

The restricted phylogenetic occurrence of eusociality in insects is clear since this condition has originated once in Isoptera (termites) and at least 11 times in the Hymenoptera. In Apidae, it has arisen at least twice (Winston and Michener, 1977; Sakagami, 1982). Remarkably, only about 15% of extant insect species are found in the Hymenoptera (Oster and Wilson, 1978).

A. Group Selection

In "The Origin of Species," Darwin introduced the concept of group selection to account for the evolution of worker castes in social insects in order to show that natural selection was operating at the colony level rather than on the individual organism. Group selection is theoretically possible because there appears to be no reason why selection above the level of the single organism might not, at times, affect gene frequencies and thus contribute to the extinction or persistence of a social trait (Matthews and Matthews, 1978).

In considering interdemic (interpopulation) selection, D. S. Wilson (1975) pointed out the importance of distinguishing the timing of an extinction event in the history of the population. That is, extinction is more likely to take place either early, when the population is struggling to become established, or when populations have declined to the point where they cannot be maintained. Extinction based on these criteria is called r extinction. The second form of extinction, K extinction, occurs after the population has increased above the environmental carrying capacity and thus is in danger of being eliminated due to starvation or habitat destruction. These forms of extinction have been named in appreciation of their close parallel with r and K selction (D. S. Wilson, 1975). E. O. Wilson (1973) considers r extinction to be more applicable to invertebrates, and K extinction more likely in vertebrates.

The following is a summary of the social characteristics which would be favored by the two types of group selection. In populations which are more susceptible to r selection, so called altruist traits formed by group selection would favor "pioneer" qualities, such as the clustering of small populations, cooperative foraging, mutual defense, and nest construction. In other words, r selection would favor qualities which will be beneficial in producing and maintaining a secure population level. The opposite is true in K selection. With it, the premium is on "urban qualities," such as resource conservation and reproductive restraint. Mutual aid is minimal.

Starr (1979) summarized two types of criticism of invoking group selection to account for altruism. First, the hypothesis advanced by Wynne-Edward (1962) that group selection influenced the development of altruism can be well explained by more conservative mechanisms, such as individual selection, parental manipulation, and kin selection. Also territoriality, as suggested by Williams (1966), can be readily accounted for by simple individual selection. Consequently, group selection is not necessary to explain altruism. Second, it may also be highly unlikely because of the restrictive conditions for its operation in nature. Starr (1979) cites several authors (e.g., Lewontin, 1970; Maynard Smith, 1964; Williams and Williams, 1957; Wilson, 1973; D. S. Wilson, 1975; and others) who presented various extraordinary conditions for extinction such as (1) extinction rates of selfish populations, (2) colonization rates of vacated habitats, (3) migration between populations and (4) population size.

In general, it can be concluded that group selection is theoretically possible but it is not very likely. Consequently, classical group selection does not appear to be a generally accepted mechanism for the evolution of sociality in insects, or for the formation of reproductive castes (Starr, 1979).

E. O. Wilson (1975) pointed out that even in colonial species there is no evidence to indicate that interdemic selection is superior to kin selection. The possibility also exists that populations of minimal size are decided indirectly by some still unknown form of individual selection. It is apparent that before group selection can take place, an allele still must first become established through selection at the level of the individual. Consequently, a better explanation of social behavior and apparent altruism at the individual level is required. Such a hypothesis has not only been advanced by Hamilton (1964, 1972, 1974) with his kin-selection theory, but it has also been advanced by others (Alexander, 1974; Alexander and Sherman, 1977; Michener and Brothers, 1974; Trivers and Hare, 1976; West-Eberhard, 1975).

B. Kin Selection

Dzierzon (1845) was the first to report that the drones of honey bees develop from unfertilized eggs, and females, queens or workers, from fertilized eggs. Although the original adaptive significance of haplodiploidy remains unresolved, it had a profound influence on the social evolution of Hymenoptera (Oster and Wilson, 1978).

Hamilton (1964) was first to provide a genetic explanation as to why the haplodiploid mode of sex determination in Hymenoptera has been such a successful starting point in the evolution of eusociality. The concept of altruism and inclusive fitness, based on kin selection, has had a profound impact on social evolutionary thought. Hamilton (1964, 1972, 1974) suggested that because of male haploidy, a female has $\frac{3}{4}$ of her genes in common with her full sister, since the half of the genotype of each individual which comes from their father is identical. However, she is related to her mother or daughter by only $\frac{1}{2}$. Consequently, assisting a sister would be better from a genetical point of view than producing her own female offspring, i.e., her inclusive fitness would be greater than direct (or indirect) fitness. This advantage in caring for siblings rather than reproducing, however, is cancelled in the production of males, which are only $\frac{1}{4}$ related to their sisters, unless this asymmetry in genetic relatedness is put to use as suggested by Trivers and Hare (1976), in which the investment in females is three times as great as that in males.

Bertram (1982) pointed out that Hamilton (1964) was able to show that altruistic behavior among close relatives can be selected for by natural selection, since the genes responsible for the altruistic behavior are possibly present not only in the offspring of the altruist, but also in her close relatives. Selection for altruistic genes is obvious if the altruism they produce results in there being more of them in future generations, regardless of whether they are in the bodies of the offspring of the altruist or in the bodies of the offspring of her relatives. They are less likely to be present in the body of a relative the less closely related he or she is to the altruist. Consequently more distant relatives must be reared to compensate for each of her own offspring which the altruist failed to produce.

Bertram (1982) indicated that the algebra of cost and benefit is not com plicated (Hamilton, 1964; E. O. Wilson, 1975). That is, altruistic genes will spread if $K > 1/r$, where K is the ratio of the recipient's benefit to the altruist's cost and where r is the coefficient of relatedness between the altruist and the recipient. The cost and benefit are measured in terms of individual fitness.

E. O. Wilson (1975) presents a model that includes all relatives affected by the altruism. Thus, to benefit first cousins ($r = \frac{1}{8}$), an altruist without offspring would have to multiply a cousin's fitness eightfold. An uncle ($r = \frac{1}{4}$), on the other hand, would have to be advanced fourfold; and so on. In the extreme brother-to-brother situation, $1/r = 2$, the loss of fitness for an altruist without offspring is considered to be total (that is, $= 1.0$). Consequently, to bring about an increase in the shared altruistic genes, K (the benefit to cost ratio) must exceed 2, i.e., the brother's fitness must increase more than double.

The evolution of selfishness can be explained by the same model (E. O. Wilson, 1975). It could appear that selfishness in any degree pays off so long as the outcome is the increase of one's genes in the next generation. However, this is not the case if related individuals are being harmed to the extent of losing too many of their genes shared with the selfish individual by common descent (among siblings, between parents and offspring, or between other relatives). In this situation, the inclusive fitness must be in excess of 1; however, the result of increasing that threshold is now the spread of selfish genes. In general, as pointed out by Lin and Michener (1972), selfishness is an activity which helps in the perpetuation of an individual's own genes in a direct manner. Altruism, however, is the promotion of the same genes by means of another individual.

Hamilton's (1964) concept of inclusive fitness was a major step in explaining helping behavior as the evolutionary product of reproductive competition between individuals. His concept of fitness was divided into two components: (1) the kinship (or indirect) component, expressed as the effect of the individual on the fitness of his neighbors multiplied by his fractional relatedness to them, and (2) a direct (or classical) fitness component realized through his own offspring. The sum of these two components comprise the individual's inclusive fitness, that is, the total genetic contribution of the individual to future generations, which tends to be maximized by natural selection. Consequently, strategies of reproduction should arise to emphasize direct or indirect components based on the likelihood of genetic pay-off by each, and variations in individual strategies should reflect individual differences in expectation of success by direct or indirect routes (Matthews and Matthews, 1978; Noonan, 1981).

The theory of kin selection has received considerable attention and wide acceptance, even though there are some basic problems. Perhaps the major

problem is based on the concept that single matings by female Hymenoptera are the rule. If a female mates with more than one male, then there will
be several types of male gametes available for the fertilization of eggs. As a
result, the coefficient of relationship between sisters will decrease below $\frac{1}{2}$,
and the intrinsic advantage in helping sisters rather than offspring will be
lost (Wilson, 1971). It is known that some species of social Hymenoptera
mate only once (Wilson, 1971), but this condition is frequently not found in
highly social species. Honey bees are an excellent example of social insects
in which multiple insemination is the rule. Taber (1954) determined that
queens generally mate six or seven times, and Taber and Wendel (1958)
reported some sperm transfer from seven to 10 matings.

Wilson (1971) suggested that these difficulties could be explained away if
the males responsible for multiple matings were closely related to one another, and thus their sperm would be genetically very similar or even identical. Another solution to this apparent problem was advanced by Alexander
(1974), Hamilton (1964, 1972, 1974) and Lin and Michener (1972), who
suggested that multiple insemination is a desired condition in each highly
social phyletic line. The basic hypothesis is that monogamy was originally
essential when the reproductive differences between females were very
small. However, an increasingly social mode of reproduction resulted in
increased productivity, a situation which could facilitate the evolution of
sterile castes.

C. Mutualism

Lin and Michener (1972) pointed out that "Hamilton and others have
placed too much emphasis on the evolution of altruistic behavior through
kin selection," especially since there is sufficient evidence to indicate that
social behavior in insects is in part mutualistic. Trivers (1971) presented a
model on "reciprocally altruistic" behavior to show that certain classes of
behavior can be selected for even when the recipient is so distantly related to
the individual performing the altruistic act that kin selection can be ruled
out. West-Eberhard (1981) also showed that social colonies of insects without altruism are highly probable. Thus, mutualism takes place when organisms act in ways that enhance their own classical fitness while fortuitously
also contributing to that of their neighbors.

According to Lin and Michener (1972), there are several factors, in addition to kin selection, which appear to be responsible for the numerous
origins of sociality in Hymenoptera. The need for defense is an important
factor in favor of mutual tolerance and initial colony formation (Chao and
Hermann, 1986). This applies not only to primitively social Halictinae,
where two or more females are in the nest to protect it by constricting the

nest entrance, biting at or driving off intruders, and blocking the nest entrance with the abdomen, but also for large colonies with many workers that protect the nest against intruders at all times. Guarding is uncommon in halictines when the nest is occupied by a single individual.

Lin and Michener (1972) suggest that regardless of the evolutionary route taken by certain hymenopteran insects to become eusocial, the evolution of joining and accepting behavior and the start of semisocial colonies can be accounted for if the following assumptions are made: (1) the productivity of females, even in solitary form, varies considerably; (2) individuals which are highly productive increase their advantage over less productive ones by constructing their nest first; and (3) the percentage of mortality among the offspring of all individuals in a colony is equal to or lower than that of the offspring of individuals housed alone in their own nests. The reduced mortality rate for offspring of females residing in colonies is mainly due to improved defense. A significant factor in the origin of joining behavior in some groups could be that joiners are in excellent position to become queens and to take over the nest if the established queen disappears (Hermann, 1979).

Once a joining class originated, selection would appear to favor colonies that produce their own joiners (Lin and Michener, 1972). Such colonies ordinarily would be eusocial. A eusocial society could also become established secondarily from a semisocial society with or without kin selection acting as a predominant factor in the origin of the semisocial organization. However, after the origin of such eusocial colonies (family groups), evolution of special queen and worker features is a possibility through kin selection.

Mutualism points to the parasocial route (incorporating communal, quasisocial, and semisocial levels of organization) to eusociality in two ways: (1) mutual benefits such as common defense against parasites and predators is provided by group living, and (2) nonreproductive helpers are afforded some probability that they will in fact be able to reproduce even if they are unmated and thus able to produce only haploid eggs. Thus, mutualism is considered as the basis of the familial route to eusociality, but since the associating females are related, their fitness is enhanced by the inclusion of a kinship component (Fletcher and Ross, 1985).

D. Parental Manipulation

The kin-selection hypothesis of the origin of social behavior has been criticized by several authors, including Alexander (1974), Lin and Michener (1972), Michener and Brothers (1974), and West-Eberhard (1975). They have suggested various additional advantages of early eusociality as being

equally important in Hymenoptera, especially defense, superiority of mutual nest construction, oviposition by workers in the presence of the queen
or taking over the nest when the queen dies or leaves, and finally the
advantage the queen obtains from controlling and exploiting the workers at
the expense of their own genetic fitness. Lin and Michener (1972), Michener
and Brothers (1974), and West-Eberhard (1975) consider these factors as
being supplementary to kin selection. However, Alexander (1974) takes the
position that exploitation, or parental manipulation, is of primary importance while kin selection plays a negligible role.

The basic concept in Alexander's (1974) development of the parental
manipulation theory is that the parent will manipulate investment in such a
manner as to maximize its own fitness. The parent is expected to prevail if
there is a conflict over investment distribution between parent and offspring. The parent will concentrate, or disperse, investment based on the
plan which supplies the highest yield (Starr, 1979). Parental investment is a
commitment of resources or energy by the parent to the fitness of an offspring and is accompanied by some reduction in expense to the ability of the
parent to invest in other offspring (Trivers, 1972).

The concept of parental manipulation has received support by Michener
and Brothers (1974), who suggested that eusocial behavior in halictine bees
evolved by the successful control and domination of some female bees over
others, in contrast to unforced submission of the dominated bees as a result
of kin selection. They were able to show that queens of the primitively
eusocial bee *Lasioglossum zephyrum* controlled the activities and oviposition
of other adult females, and they concluded that sterile castes and division of
labor did not evolve primarily through selection of the workers to maximize
their inclusive fitness, but rather of the queens to maximize their classical
fitness (Starr, 1979).

In their discussion of the implications of haplodiploidy for inclusive fitness, Trivers and Hare (1976) pointed out that there is a fundamental
conflict of interest between the queen and her worker daughter over the
kinds and ratios of reproductives that should be produced by the colony.
Their suggestion starts from the observation that full sisters, related to each
other by $\frac{3}{4}$, are females which are related to their brothers by only $\frac{1}{4}$. Thus,
the average relatedness will be $\frac{1}{2}$ in a balanced sex ratio. Under such conditions, a worker can increase her inclusive fitness by lowering the fitness of
the mother in either of the following two ways: (1) she can lay some of the
male-producing eggs herself or let her sister do it, which would result in
some of the new males being her sons ($r = \frac{1}{2}$), or nephews ($r = \frac{3}{8}$), rather
than brothers; or (2) she can invest differentially in brood of the sex with
which she is more closely related.

The fact that fertilized eggs develop into females, queens or workers,
while unfertilized eggs become males, means that males inherit their entire

genome from their mother, but females inherit only half of their genome from each parent. Trivers and Hare (1976) have pointed out that this circumstance results in asymmetrics in the degrees of relationship. Based on the use of this asymmetry in genetic relatedness, they suggest an investment ratio of 1 for queens and $\frac{1}{3}$ for nonlaying workers. The underlying assumptions are that (1) the amount of inbreeding is negligible, (2) the queen is inseminated by a single male, and (3) the colony is monogynous.

Using sex-ratio data and dry weights for samples of 21 ant species as a measure of relative investment, Trivers and Hare (1976) found that the ratio of investment is generally lower than 1 and frequently in the vicinity of $\frac{1}{3}$. They concluded that there is a queen–worker conflict over the sex ratio and energetic investments in the new queens and males. This interpretation is in support of the kin-selection theory. However, Alexander and Sherman (1977) suggest that the lowered investment ratio cannot be interpreted as any specific number such as $\frac{1}{3}$. They are of the opinion that these data can just as well be explained as the result of competition between related males in populations of small effective size, whereby queens and workers could find it beneficial to invest in a reduced number of males as opposed to females. Information presently available on these ant species is not in support of this alternative hypothesis (Oster and Wilson, 1978).

It has been pointed out by Starr (1979) that the parental manipulation theory has attractive features with regard to the Hymenoptera. Most importantly, multiple insemination, in contrast to kin selection, causes no problem for parental manipulation, especially since mother–daughter relatedness is independent of relatedness between daughters. The situation is complicated if parental manipulation and kin selection function together. Parental manipulation, however, cannot provide an explanation for the near monopoly of eusociality in the Hymenoptera.

E. Concluding Remarks

After a review of the effects of group selection, kin selection, mutualism, and parental manipulation and their influence on steps by which societies began and evolved, the question remains as to why all Hymenoptera are not social insects. The assumption that the same haplodiploid reproductive system may confer an advantage for the Hymenoptera is unlikely since there are thousands of solitary species. Also, there are many other animal classes with a haplodiploid system, such as the mites and rotifers, which are not social. However, there are some advantages present in the haplodiploid sex system, i.e., diploid females such as honey-bee queens, have the ability to determine the sex of the offspring by regulating sperm entry into the ovum (Nedel, 1960). Thus, males can be excluded from the society and used in gene dispersal.

The answer to the rise in sociality may lie in the concept of preadaptation (Matthews and Matthews, 1978). In other words, a preadaptive trait must already be present in an animal prior to the occurrence of an event which provides for a selective advantage, i.e., "for a trait to be a selective advantage, the animal must already be equipped to use it in advance of its occurrence." New adaptations are generally built on such stepping stones as prior behavior patterns, morphological characters, and physiological conditions.

Matthews and Matthews (1978) present a listing of various other preadaptions of importance in the evolution of hymenopteran eusocial behavior. It can also be argued that the haploidiploid sex system may have been initiated by parental manipulation or group selection. However, Levin and Kilmer (1974), in their computer simulation study, concluded that group selection appears not to be the primary force in the evolution of altruism. They concluded that its major role is perhaps in synergistic association with kin selection. Hamilton's theory of kin selection is an excellent explanation of the subsocial pathway to eusociality; however, it fails to explain the semisocial, mutualistic behavior observed by Lin and Michener (1972) for halictid bees. Thus, once group living became established through mutualism, reproductive competition became a major factor in the evolution of social behavior (West-Eberhard, 1981).

Hermann (1986) pointed out that altruism has been defined in a number of different ways and thus the theories on the evolution of altruism have resulted in confusion. Wilson (1971) defined it as self-destructive behavior performed for the benefit of others. Originally, altruism was considered to be the opposite of selfishness; however, it may also have its roots as a mechanism which is intrinsically selfish, as shown by Lin and Michener (1972), Orlove (1974), and West-Eberhard (1975). The term "altruism" has been called a misnomer by Lin and Michener (1972). Thus, activities that directly contribute to the perpetuation of an individual's own genes are known as "selfish," but the advancement of the same genes in another individual is known as "altruistic." Selfish behavior is suitably explained by classical evolutionary theory and does not require a close relationship of the partner. Finally, as pointed out by Brian (1983), these theories are not mutually exclusive, and the starting point for bees and wasps may well have been parental manipulation and kin selection.

IV. GEOLOGICAL HISTORY OF BEES

The geological record of the earliest insects is still vague. Fragments of small arthopods, discovered in Devonian chert in Scotland, have been classified as Collembola, even though their identity will remain in doubt until more is known about them (Carpenter, 1952). The oldest unquestionable insects were discovered in rocks extending back to the Upper Carboniferous (or Pennsylvanian) Period (Carpenter and Hermann, 1979). Their age dates back about 350 million years. Orders of insects which still exist today were already present at the end of the Paleozoic Era.

Marked changes in insects occurred not only at the beginning of the Permian period, or 50 million years after the appearance of the first insect, but also during the Mesozoic era. The difference between archaic fauna of the Permian and relatively modern fauna of the Triassic period is as large as that between the Triassic and Recent periods. There is essentially no difference between the Triassic and Jurassic insect faunas, except for the occurrence of more existing families and the absence of flower insects (Carpenter, 1952).

Flowering plants had become established by the beginning of the next period, the Cretaceous, and with it the types of insects associated with these plants. Unfortunately, there is very little information available about the Cretaceous insects due to the lack of adequate specimens (Carpenter, 1952). Insects found in Baltic amber, now considered as of early Tertiary age, are an important link to extant genera and species. Ant fauna of the Baltic amber, for example, includes 43 genera, of which 24, or 55%, are still present, whereas all but one of the genera of bees in such amber are extinct (Carpenter, 1952).

Wilson and Taylor (1964) described the only fossil colonial assemblage of any kind so far discovered as belonging to the ant *Oecophylla leakeyi*. Members of this colony, composed of 360 specimens including larvae, pupae, and worker subcastes, resemble currently extant species that have not undergone any major change in morphology and caste pattern for a period of over 10 million years. This find is highly significant because it represents the first occurrence of an insect colony and thus dates the existence of a formicid society in the Miocene period. It also indicates that the social organization of these ants, and perhaps bees, was as highly developed as those of the present forms in regard to caste differentiation and polymorphic workers.

The superfamily Apoidea includes all bees which on morphological grounds are similar to the sphecoid wasps, even though the absence of an adequate fossil record has made it impossible to determine the exact ancestral phyletic line (Wilson, 1971). However, Evans (1969), in his description of sphecoid wasps from the Cretaceous Period, reported on one specimen which appeared to be generalized enough to be considered as a possible ancestor of the Apoidea. This find could indicate that the present bee fauna probably originated more than 70 million years ago. The first appearance of bees is thus closely tied in with a change in food from insect prey to pollen and nectar obtained from the flowers of angiosperms. In some eusocial species, including honey bees, the larval food is derived from glandular secretion which ultimately is obtained from the pollen and nectar of flowering plants.

The geologically oldest and most completely preserved honey bees were found in Baltic amber in East Prussia and date to the upper Eocene period, or roughly 50 million years ago (Zander and Weiss, 1964). These bees have morphological characters which partially point to the present-day Meliponini and partially to the Apini. Haskins (1970) reported on a find of a worker apparently of Meliponid affinities. This eusocial bee, complete with well developed pollen baskets, was discovered in the Mexican amber of Chiapas. It is believed to date from the Oligocene period and thus is roughly 30–40 million years old. Other bees found in Baltic amber (lower Oligocene) include representations of the Andrenidae, Apidae, and Megachilidae (Carpenter and Hermann, 1979).

Fossil bees which are morphologically very similar to the present-day Apini have been discovered in the lower Oligocene beds at Rott (Siebengebirge), Germany (Zander and Weiss, 1964). Some of these bees, such as Synapsis henshawi (Cockerell, 1907), originally described under the genus Apis subgenus Synapsis, have been estimated as 30 million years old and are clearly species belonging to the present-day genus Apis. Apis oligocenia (Meunier) is its synonym (Maa, 1953). Other finds, dating to the upper Miocene period (about 15 million years ago), came from the Randecker Mar near Göttingen, Germany. Based on their wing venation, Armbruster (1938, cited by Bischoff, 1960; Zander and Weiss, 1964) considers them to be very close to the present-day Apis. Their pollen-collecting structures, however, are morphologically more primitive. Another find, consisting of 17 individuals tightly enclosed in a piece of red marble sinter (stalactite), dates to the same period, and was made in Böttingen (Schwaben), Germany. These bees, described by Zeuner (1931, cited by Bischoff, 1960) as Apis armbrusteri, are also very similar to present-day honey bees. A thorough review of all fossil honey-bee species is presented by Maa (1953). A honey-bee find in East Africa, from the Upper Pleistocene period (100,000 years ago), has been reported by Bischoff (1960). This bee cannot be differentiated from the contemporary African honey-bee subspecies.

While fossil records will be useful in interpreting the social evolution of bees on a morphological basis, they provide no evidence about the evolution of social behavior. Nevertheless, the present-day assemblage of ancient forms of ants and bees provides a living record of evolution, including near-facsimiles of types that were dominant on our planet 50 million or more years ago. In such "living fossils" we can observe in detail the specific evolution of many of the behavioral and physiological patterns that have perfected the sociality of honey bees over an extremely long period of time.

V. THE RISE OF HONEY BEES

The evolution of insect-pollinated plants and nectar- and pollen-feeding insects were complementary developments that commenced in the Jurassic period (approximately 180 million years ago). Currently, more than 65% of all flowering plants are insect-pollinated, and 20% of all insects depend on flowers for food during their developmental period (Dietz, 1982). The abundance of nectar and pollen as a readily available source of larval food was a contributing factor in the change of some wasps from a predatory existence (Sphecoidea) to that of collecting nectar and pollen.

Most species of bees are solitary. Each solitary female, after courtship, constructs her own nest either in the hollow stem of a plant, in a burrow in the ground, or in another protected place. After completing the first nest cell, the female collects nectar and pollen and mixes this material into a paste. She subsequently places a pellet of this food in the bottom of the cell and deposits an egg on it. The pellet provides all the food necessary for the larva to complete its development. After the cell is sealed, another cell is constructed by the bee, more food is collected, and another egg is deposited. The female subsequently dies without ever seeing her offspring (Dietz, 1982).

Several species of small bees of the genus *Halictus* provide the clue for the next stages in the evolution of sociality. In this case, females live long enough to provide the growing larvae with food when necessary (Butler, 1975). This type of provisioning is known as progressive feeding, which differs from mass-provisioning behavior where the larva receives food in excess of its requirement. Mass feeding is practiced by honey bees during the early portion of the larval period, and they switch to progressive provisioning after the larvae are 2–3 days old. An exception to this sequential feeding arrangement is found in queen honey-bee larvae, which are mass fed throughout their larval period (Dietz, 1972).

Females in several other species of *Halictus* have already advanced to a moderately eusocial stage. Even though these bees are mass provisioners, each female establishes her own nest in early spring and continues to live for several months. The developing workers, which are slightly smaller than their mothers, assist with nest construction, food collection, and larval feeding. In the summer, queens and males are produced and, after mating, females hibernate until the next spring, when each attempts to start her own nest. The old queen, males, and workers die in the fall, and thus each nest lasts for less than a year (Dietz, 1982).

The next stage in the evolution of bee sociality is evident in colonies of bumble bees. Here again, the overwintering bumble-bee queen begins searching for a nesting site, a period which lasts several days to 2 weeks or more (Heinrich, 1979). Near the entrance to her nest, she constructs a thimble-sized honey pot out of wax scales produced from glands located between the segmental plates on both the top and bottom of her abdomen (Heinrich, 1979). The foraging queen deposits collected nectar in the honey pot and drops the collected pollen loads onto the floor of the nest. The pollen is subsequently moistened with nectar and formed into a pollen clump, upon which about eight to 10 eggs are deposited. The progressively fed larvae derive all their nutrients from the nectar-moistened pollen. After 16–25 days, the first workers emerge from their cocoons. They require pollen as a protein source to provide for muscle and other tissue development for only a few days; they then feed exclusively on nectar (Heinrich, 1979). A similar situation is found in honey bees (Dietz, 1975).

The longevity of bumble-bee colonies is correlated with the length of the season. The colony lasts until late summer or fall in temperate regions and concludes when larvae develop into males and potential queens. The new queens seek underground shelter for hibernation after insemination, and the old queen, drones, and workers die before the onset of winter. Bumble bees are characterized by distinct differences in the queen and worker castes; once the colony is well established, the workers perform all the duties initially carried out by the queen, with the exception of egg laying (Heinrich, 1979).

The eusocial Meliponidae, or stingless bees, occupy the social position between that of the bumble bees and honey bees. The sting is reduced and cannot be used for nest defense. However, the workers of most species can bite and eject a burning liquid from the preoral area (Wilson, 1971).

The meliponines are most abundant in South and Central America; however, they are also found in Africa, Asia, Australia, and some Pacific islands. These are separated into two major groups: species of the genus *Melipona*, which are restricted to Central and South America, and species of the genus *Trigona*, which are present in the Old World as well as in Central and South America (Richards, 1953; Dietz, 1962).

The workers range in size from about 2 to 20 mm and are sometimes, therefore, slightly larger than a honey bee. There are distinct behavioral and morphological differences between queens and workers. Intermediate forms are most often absent.

Reproductive swarms in meliponines differ from those of honey bees in that the queen does not leave the old nest until the new one is completely established (Wilson, 1971). However, the social organization of these bees is similar to that of *Apis mellifera* (Linne), except that mass provisioning is employed for nursing larvae (Butler, 1975). A detailed comparison of the social organization of stingless bees with that of honey bees has been presented by Sakagami (1971, 1982).

Honey bees are the best known eusocial insects. With the exception of the wag-tail dance, which is not found in other highly social insects, the honey bee has a number of features that are also present in stingless bees, such as chemical communication, colony size, division of labor, reproductive castes, and thermoregulation (Wilson, 1971).

REFERENCES

Alexander, R. D. (1974). The evolution of social behavior. *Annu. Rev. Syst. Ecol. 5,* 325–383.

Alexander, R. D., and P. W., Sherman. (1977). Local mate competition and parental investment in social insects. *Science 196,* 494–500.

Armbruster, L. (1938). Versteinerte Honigbienen aus dem obermiocanen Randecker Marr. *Arch. Bienenkd. 19,* (1) 1–48; (2) 73–93; (3/4) 97–133.

Bertram, B. C. R. (1982). Evolutionary conflicts of interest. *In* "Current Problems in Sociobiology" (King's College Sociobiology Group, eds.), pp. 251–267. Cambridge University Press, New York.

Bischoff, H. (1960). Stammesgeschichte der Biene. *In* "Biene und Bienenzucht" (A. Büdel, and E. Herold, eds.), pp. 1–4. Ehrenwirth Verlag, Müchen.

Brian, M. V. (1983). "Social Insects—Ecology and Behavioral Biology." Chapman and Hall, New York.

Brockmann, J. H. (1984). Evolution of social behavior in insects. *In* "Behavioral Ecology: An Evolutionary Approach" (J. R. Krebs and N. B. Davies, eds.), pp. 340–361. Sinauer Assoc. Inc., Sunderland, Mass.

Butler, C. G. (1975). The honey bee colony—life history. *In* "The Hive and The Honey Bee" (R. Grout, ed.), pp. 39–74. Dadant and Sons, Hamilton, Ill.

Carpenter, F. M. (1952). Fossil insects. *In* "Insects, The Yearbook of Agriculture," pp. 14–19. U.S. Department of Agriculture, Washington, D.C.

Carpenter, F. M., and Hermann, H. R. (1979). Antiquity of sociality in insects. *In* "Social Insects" (H. R. Hermann, ed.), Vol. 1, pp. 81–89, Academic Press, New York.

Chao, J. T., and Hermann, H. R. (1986). Ant predation on broods of *Polistes annularis* L. (Hymenoptera: Vespidae) in northeastern Georgia. *J. Kansas Entomol. Soc.* In press.

Cockerell, T. D. A. (1907). A fossil honey bee. *Entomologist 40,* 227–229.

Darchen, R., and Lensky, J. (1963). Quelques problèmes soulevés par la création de sociétés polygynes d'abeilles. *Insectes Soc. 10,* 337–357.

Darwin, C. R. (1859). "On the Origin of Species by Means of Natural Selection, or the Preservation of Favoured Races in the Struggle for Life." 1st Ed. John Murray, London.

Dietz, A. (1962). A short natural history of the honey bee family Apidae (Leach, 1817). *Australas. Beekeep. 63,* 187–188.

Dietz, A. (1968). "Beekeeping in Maryland." Ext. Bull. 223. University of Maryland, College Park.

Dietz, A. (1972). The nutritional basis of caste determination in honey bees. *In* "Insect and Mite Nutrition" (J. E. Rodriguez, ed.), pp. 271–279, North-Holland Publ., Amsterdam.

Dietz, A. (1975). Nutrition of the adult honey bee. *In* "The Hive and The Honey Bee," pp. 125–156. Dadant and Sons, Hamilton, Ill.

Dietz, A. (1982). Honey bees. *In* "Social Insects" (H. R. Hermann, ed.), Vol. 3, pp. 323–360. Academic Press, New York.

Dietz, A. (1985). Problems and prospects of maintaining a two queen colony system in honey bees throughout the year. *Am. Bee J. 125,* 451–453.

20 Alfred Dietz

Dietz, A. (1986). Monogyny and induced polygyny in honey bee colonies. (Manuscript in preparation).

Dzierzon, J. (1845). Gutachten über die von Hrn. Direktor Stöhr im ersten und zweiten Kapitel des Generalgutachtens aufgestellten Fragen. *Eichstadt. Bienenzeitung. 1,* (11) 109–113; (12) 119–121.

Evans, H. E. (1969). Three new *Cretaceous aculeate* wasps (Hymenoptera). *Psyche (Cambridge)* 76, 251–261.

Fletcher, D. J. C., and Ross, K. G. (1985). Regulation of reproduction in eusocial Hymenoptera. *Annu. Rev. Entomol. 30,* 319–43.

Hamilton, W. D. (1964). The genetic evolution of social behavior. I and II. *J. Theor. Biol. 7,* 1–52.

Hamilton, W. D. (1972). Altruism and related phenomena, mainly in social insects. *Annu. Rev. Ecol. Syst. 3,* 193–232.

Hamilton, W. D. (1974). Evolution sozialer Verhaltensweisen bei sozialen Insekten. *In* "Sozial-polymorphismus bei Insekten" (G. H. Schmidt, ed.), pp. 60–93. Wiss. Verlagsges. MBH Stuttgart.

Haskins, C. P. (1970). Researches in the biology and social behavior of primitive ants. *In* "Development and Evolution of Behavior" (L. R. Aronson, E. Tobach, D. S. Lehrman, and J. S. Rosenblatt, eds.), pp. 355–388. W. H. Freeman and Co., San Francisco.

Heinrich, B. (1979). "Bumblebee Economics." Belknap Press of Harvard Univ. Press, Cambridge, Mass.

Hermann, H. R. (1979). Insect sociality—an introduction. *In* "Social Insects" (H. R. Hermann, ed.), Vol. 1, pp. 1–33. Academic Press, New York.

Hermann, H. R. (1986). Social organization in insects. *In* "Self Organization" (S. Fox, ed.), Liberty Press, Indianapolis. In press.

Hölldobler, B., and Wilson, E. O. (1977). The number of queens: an important trait in ant evolution. *Naturwissenschaften 64,* 8–15.

Levin, B. R., and Kilmer, W. L. (1974). Interdemic selection and the evolution of altruism: a computer study. *Evolution 28,* 527–545.

Lewontin, R. C. (1970). The units of selection. *Annu. Rev. Ecol. Syst. 1,* 1–18.

Lin, N., and Michener, C. D. (1972). Evolution of sociality in insects. *Q. Rev. Bio. 47,* 131–159.

Maa, T. (1953). An inquiry into the systematics of the tribus Apidini or honey bees (Hym.). *Treubia 21,* 525–640.

Matthews, R. W., and Matthews, J. R. (1978). "Insect Behavior." John Wiley and Sons, New York.

Maynard Smith, J. (1964). Group selection and kin selection: a rejoinder. *Nature (London) 201,* 1145–1147.

Michener, C. D. (1974). "The Social Behavior of the Bees: A Comparative Study." Belknap Press of Harvard University Press, Cambridge, Mass.

Michener, C. D., and Brothers, D. J. (1974). Were workers of eusocial Hymenoptera initially altruistic or oppressed? *Proc. Natl. Acad. Sci. USA 71,* 671–674.

Nedel, O. J. (1960). Morphologie und Physiologie der Mandibeldrüse einiger Bienen-Arten. *Z. Morphol. u. Okol. Tiere 49,* 139–183.

Noonan, K. M. (1981). Individual strategies of inclusive-fitness-maximizing in *Polistes fuscatus* foundresses. *In* "Natural Selection and Social Behavior: Recent Research and New Theory" (R. D. Alexander and D. W. Tinkle, eds.), pp. 18–44. Chiron Press, New York.

Orlove, M. J. (1974). A model of kin selection not invoking coefficients of relationship. *J. Theor. Biol. 49,* 289–310.

Oster, G. F., and Wilson, E. O. (1978). "Caste and Ecology in Social Insects." Princeton University Press, Princeton, N.J.

Ribbands, R. (1953). "The Behavior and Social Life of Honey Bees." Bee Research Association, Ltd., London.

Richards, O. W. (1953). "The Social Insect." MacDonald, London.

Sakagami, S. F. (1971). Ethosoziologischer Vergleich zwischen Honigbiene und stachellosen Bienen. Z. Tierpsychol. 28, 337–350.

Sakagami, S. F. (1982). Stingless bees. In "Social Insects" (H. R. Hermann, ed.), Vol. 3, pp. 361–423. Academic Press, New York.

Starr, C. K. (1979). Origin and evolution of insect sociality: a review of modern theory. In "Social Insects" (H. R. Hermann, ed.), Vol. 1, pp. 35–79. Academic Press, New York.

Taber, S. (1954). The frequency of multiple mating of queen honey bees. J. Econ. Entomol. 47, 995–998.

Taber, S., and Wendel, J. (1958). Concerning the number of times queen bees mate. J. Econ. Entomol. 51, 786–789.

Trivers, R. L. (1971). The evolution of reciprocal altruism. Q. Rev. Biol. 46, 37–57.

Trivers, R. L. (1972). Parental investment and sexual selection. In "Sexual Selection and the Descent of Man" (B. Campbell, ed.), pp. 363–378. Aldine, Chicago.

Trivers, R. L., and Hare, H. (1976). Haplodiploidy and the evolution of social insects. Science 191, 249–263.

West-Eberhard, M. J. (1975). The evolution of social behavior by kin selection. Q. Rev. Biol. 50, 1–33.

West-Eberhard, M. J. (1981). Intragroup selection and the evolution of insect societies. In "Natural Selection and Social Behavior: Recent Research and New Theory" (R. D. Alexander and D. W. Tinkle, eds.), pp. 3–17. Chiron Press, New York.

Wheeler, W. M. (1928). "The Social Insects, their Origin and Evolution." Harcourt, Brace and Co., New York.

Williams, G. C. (1966). "Adaptation and Natural Selection." Princeton Univ. Press, Princeton, N.J.

Williams, G. C., and Williams, D. C. (1957). Natural selection of individually harmful social adaptations among sibs with special reference to social insects. Evolution 11, 32–39.

Wilson, D. S. (1975). A theory of group selection. Proc. Natl. Acad. Sci. USA 72, 143–146.

Wilson, E. O. (1971). "The Insect Societies." Belknap Press of Harvard University Press, Cambridge, Mass.

Wilson, E. O. (1973). Group selection and its significance for ecology. Biosciences 23, 415–419.

Wilson, E. O. (1975). "Sociobiology, The New Synthesis." Belknap Press of Harvard Univ. Press, Cambridge, Mass.

Wilson, E. O. (1982). Of insects and man. In "The Biology of Social Insects" (M. D. Breed, C. D. Michener and H. E. Evans, eds.), pp. 1–9. Westview Press, Boulder, Col.

Wilson, E. O., and Taylor, R. W. (1964). A fossil ant colony: new evidence of social antiquity. Psyche 71, 93–103.

Winston, M. L., and Michener, C. D. (1977). Dual origin of highly eusocial behavior among bees (Hymenoptera: Apidae). Proc. Natl. Acad. Sci. USA 74, 1135–1137.

Wynne-Edward, V. C. (1962). "Animal Dispersion in Relation to Social Behavior." Oliver and Boyd, Edinburgh.

Zander, E., and Weiss, K. (1964). "Das Leben der Biene." Ulmer, Stuttgart.

Zeuner, F. (1931). Die Insektenfauna des Böttinger Marmors. Fortschr. Geol. u. Paläontol. 9, 292.

CHAPTER 2

Geographical Variability and Classification

FRIEDRICH RUTTNER

I. INTRODUCTION

Variability in honey bees, even within one colony, has been observed since antiquity (Fraser, 1951). The most obvious variability within a colony is in body coloration; the most obvious variability between populations is in size and pilosity. In the first century of taxonomy after Linnaeus, almost every honey bee found was given a separate name because of differences in their characteristics. In 1953 Maa presented for members of the genus *Apis* a list of 146 names which he selected from 600 names used up to that date.

Buttel-Reepen (1906) was the first to systematize the taxa of the genus *Apis*. He introduced a trinary nomenclature to distinguish taxa below the level of species. Present taxonomy of honey bees is based on this scheme, although it is now somewhat modified. Alpatov (1929, 1948) and Goetze (1930, 1940, 1964) introduced biometric methods into the microsystematics of honey bees. Using measurements of parts of the body as well as quantified characteristics of hairs, wing venation and color, they developed the first clear descriptions and classifications of honey bees.

II. TAXONOMY OF THE GENUS *APIS*

Honey bees offer taxonomists patterns of biological variation rarely found in other genera. The majority of this variation is within the species *mellifera*. This species has an unusually large distribution which encom-

Copyright © 1986 by Academic Press Inc.

passes widely different environments. Consequently, the species has many
special local types. Honey bees have been cultured for several thousand
years and have been intensively studied by scientists and apiculturalists.
These studies include the hybridization of even the most divergent types
and have produced an enormous treasure of knowledge rarely available for
other species. It is now clear that the whole complex of *Apis mellifera* constitutes one single species.

By integrating the abundant data on *mellifera* with information concerning the other less known species, the taxonomic structure of the genus is
found.

The first question in this connection is: How many species of honey bees
are we able to discriminate, since the "species is the basic unit in taxonomy
and evolution" (Mayr, 1963)? This question is answered without difficulty,
if the main and generally accepted questions of evolutionary genetics and
taxonomy are considered: (1) do groups have allopatric or sympatric occur-
rences; (2) do groups freely hybridize or not; and (3) do groups have species-specific characters?

The most important species-specific characters of the four generally recognized *Apis* species are listed in Table 1. They include two quantitative
characters (length of fore wing as a measure of body size, and cubital index)
because of their large predominantly discontinuous variation (Fig. 1).

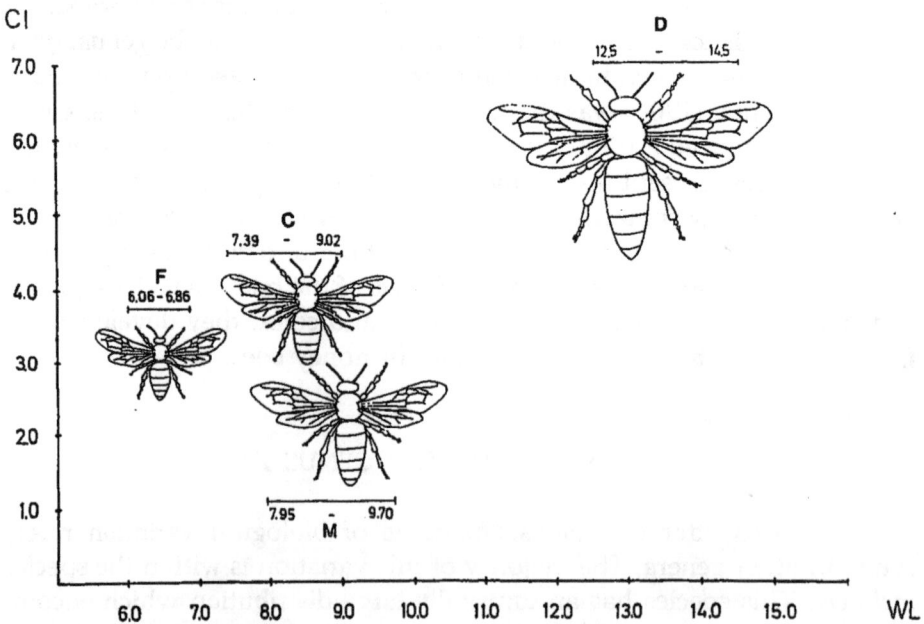

Fig. 1. Total variation of two quantitative characters (length of fore wing, WL, and cubital
index, CI) of the four *Apis* species. F, *A. florea*; C, *A. cerana*; M, *A. mellifera*; D, *A. dorsata*. Bar
with figures: minimum and maximum values of wing length, as measured thus far. Position of
bee on vertical line: mean value of cubital index.

TABLE 1. Some Species-Specific Characters of the Genus *Apis*

Character	Species			
	mellifera	*cerana*	*dorsata*	*florea*
Workers				
fore wing, longitudinal (mm)	8.0–9.7	7.4–9.0	12.5–14.5	6.0–6.9
cubital index	1.65–2.95	3.1–5.1	6.1–9.8	2.8–3.7
tomenta	Tergites 3–5	Tergites 3–6 +	Tergites 3–6 +	terg. 3–6 Variable
hind wing; extension of radialis vein	−			
sequence of melittin amino acids (deviation from *mellifera*-type)	0	0	3 Amino acids changed	5 Amino acids changed
Drones				
endophallus	One pair of cornua; bulb with chitin plates	One pair of cornua; initials of 3 others; no chitin plates	4 Pairs of very long thin cornua	One pair of long cornua, distal part of endophallus elongated
metatarsus 3	Thin pad of fine branched hairs on median surface	As *mellifera*	Thick pad of sturdy strongly branched hairs	Deep incision forming a "thumb," inside coated with branched hairs
Behavior				
capping of drone cells	Solid	Perforated	Solid	Solid
nest	Several combs in cavity	As *mellifera*	Single large comb fixed at bottom of branch or projecting rock	Single comb on twig encircling the branch and forming a "dancing platform"
communication	Sun-oriented dance on vertical comb in the dark	As mellifera	Sun-oriented dance on vertical comb with free sight to sky	Sun-oriented dance on platform with free sight to the sky
Interspecific				
hybridization	None	None	None	None
Distribution	Allopatric	Sympatric	Sympatric	Sympatric

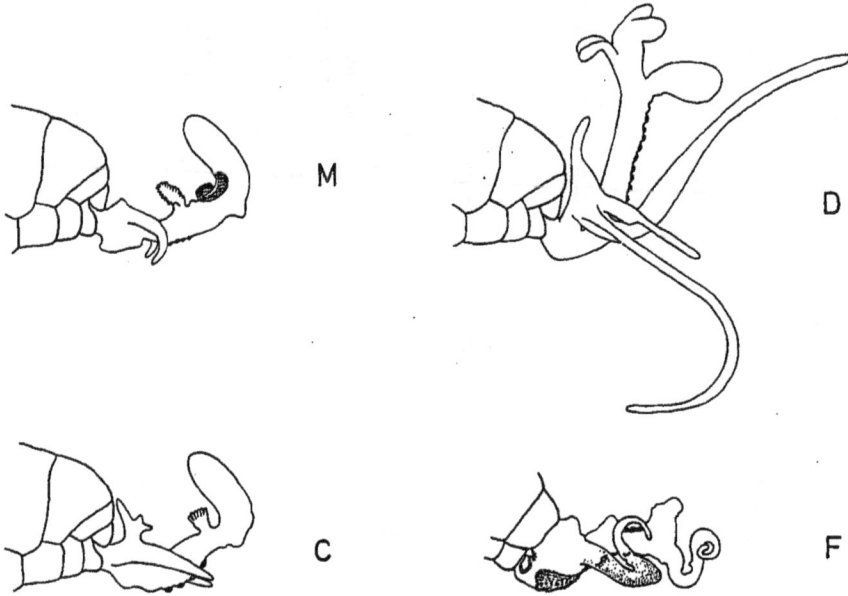

Fig. 2. Endophallus of four *Apis* species, everted. Symbols as for Fig. 1.

The male copulatory organ is one of the most important taxonomic characters, although it has rarely been studied by honey-bee taxonomists. The
bizarre endophallus of *Apis* species is a very specific development in the
genus (Fig. 2) and for a long while was completely neglected even by
anatomists.

In modern taxonomy, the characteristics of both sexes and of all castes
must be considered. If new species are established—as Maa (1953) has
done with 20 taxa of this rank—without even mentioning the males and
without considering the available data on *mellifera*, they cannot be regarded
as valid.

Usually, the *Apis* species are classified according to their nesting behavior
in two groups: (1) those that nest with several combs in cavities (*Apis
mellifera* L. and *A. cerana* F.), and (2) those that nest with one single comb in
the open (*A. dorsata* F. and *A. florea* F.). Which group is primary remains a
point of discussion.

III. METHODS OF MORPHOMETRIC ANALYSIS

The discrimination of the four species of *Apis* is an easy task. The classification of intraspecific geographic variation requires much more effort. Variation is only qualitative, and differences frequently occur within narrow
limits. This requires that exact measurements of a number of characters be

taken from groups of individuals. The values for such measures used in this chapter are based on the measurements of 40 characters in samples of 20 bees originating from single colonies (Ruttner *et al.*, 1978) and analyzed at the Institute of Apicultural Science in Oberursel. The 40 characters were selected from a greater number of characters measured. Several of them are new in bee taxonomy, although angles of wing venation were first used by DuPraw (1965). As "operational taxonomic units" (Sneath and Sokal, 1973), the means of the standardized individual data were used. Statistical analyses were done by one of the standard multivariate methods (principal component analysis, discriminant analysis, or cluster analysis).

IV. SURVEY OF THE GENUS *APIS*

A. *Apis mellifera* (Linn., 1758: 576)

Apis mellifera is one of the most successful species in the animal kingdom. It has colonized a vast area within a relatively short period — short enough that no separate species evolved as they have in the closely related *Meliponini*. *Apis mellifera* became independent from environmental conditions to a great extent: one and the same species is able to survive in semidesert tropical regions as well as in cold-temperate zones.

The phenotypic variability of the species involves behavior as well as morphology. This includes such essential features as reaction to cold, susceptibility to diseases, rhythm during communication, and the specificity of learning (von Frisch, 1967; Lauer and Lindauer, 1973; Koltermann, 1973).

In multivariate analysis (principal components), the clusters of the geographic varieties formed by representing the factors 1 and 2 in a coordinate system take the shape of a lying "Y" with one long and two short branches (Fig. 3): A-branch with the African races south of the Sahara; M-branch with the races of the West Mediterranean up to the race *A. m. mellifera;* and C-branch with the races of the East including *A. m. carnica* and *A. m. ligustica.*

The overlapping clusters in the global analyses of the entire species (Fig. 3) are easily separated by detailed analysis of the bees from a given region.

A description of the many geographic races evolved within the species is given later in the chapter.

Autochthonous distribution: North–south from southern Scandinavia to the Cape of Good Hope, west–east from Africa's west coast to the Ural and East Iran (Fig. 4).

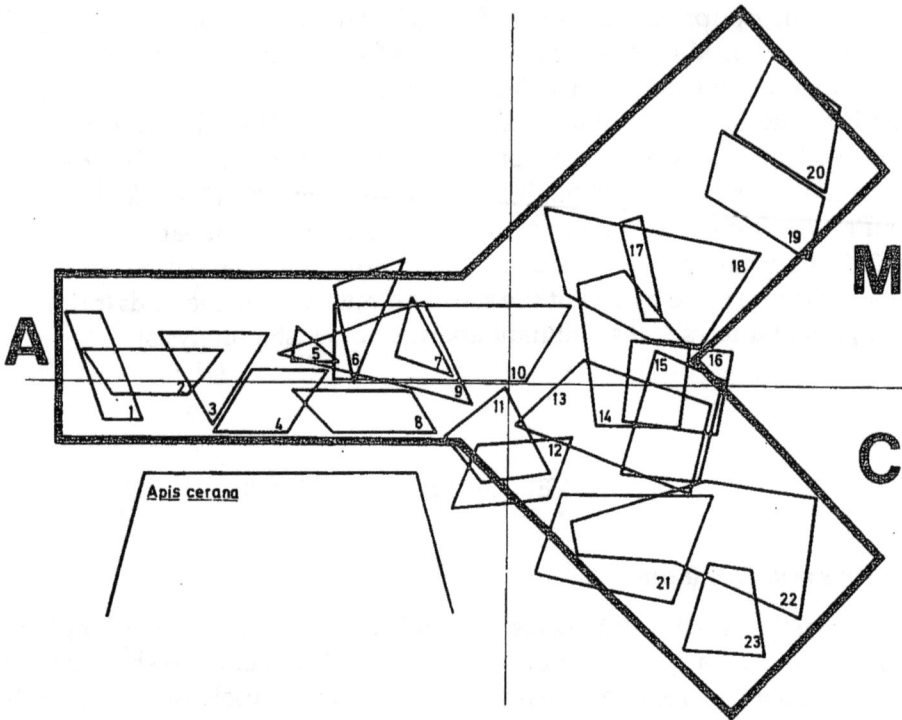

Fig. 3. Structure of the geographical races of *A. mellifera*, resulting from principal component analysis. Small races are to the left; large ones to the right. A, C, M: African, carnica, and
mellifera branches, respectively. List of geographic races: (1) *A. m. yemenitica*, (2) *A. m. litorea*,
(3) *A. m. lamarckii*, (4) *A. m. adansonii*, (5) *A. m. capensis*, (6) *A. m. unicolor*, (7) *A. m. sahariensis*,
(8) *A. m. scutellata*, (9) *A. m. syriaca*, (10) *A. m. monticola*, (11) *A. m. meda*, (12) *A. m. cypria*, (13) *A.
m. iran*, (14) *A. m. anatoliaca*, (15) *A. m. sicula*, (16) *A. m. adami*, (17) *A. m. caucasica*, (18) *A. m.
intermissa*, (19) *A. m. iberica*, (20) *A. m. mellifera*, (21) *A. m. ligustica*, (22) *A. m. carnica*, (23) *A. m.
cecropia*.

B. *Apis cerana* (Fabr., 1793: 327)

Apis cerana is the East Asiatic counterpart of *A. mellifera* (see Butler, 1954;
Koeniger, 1976a). Its morphology and behavior are so similar to *A. mellifera*
that for a long time it was considered as an *A. mellifera* subspecies (Buttel-
Reepen, 1906). However, it has several qualitative species-specific characters in addition to those listed in Table 1 (e.g., fanning with the abdomen
toward the entrance). Moreover, it is genetically separated from *A. mellifera*
(Ruttner and Maul, 1969, 1983).

It is not true that *A. cerana* is smaller than *A. mellifera*. These species
greatly overlap in size (Fig. 1). The northern types of both are generally
larger than southern types.

Few morphometric data are available for this species. Thus, no quantitative description can be given for its many geographic races [or even "species" as listed by Maa (1953)].

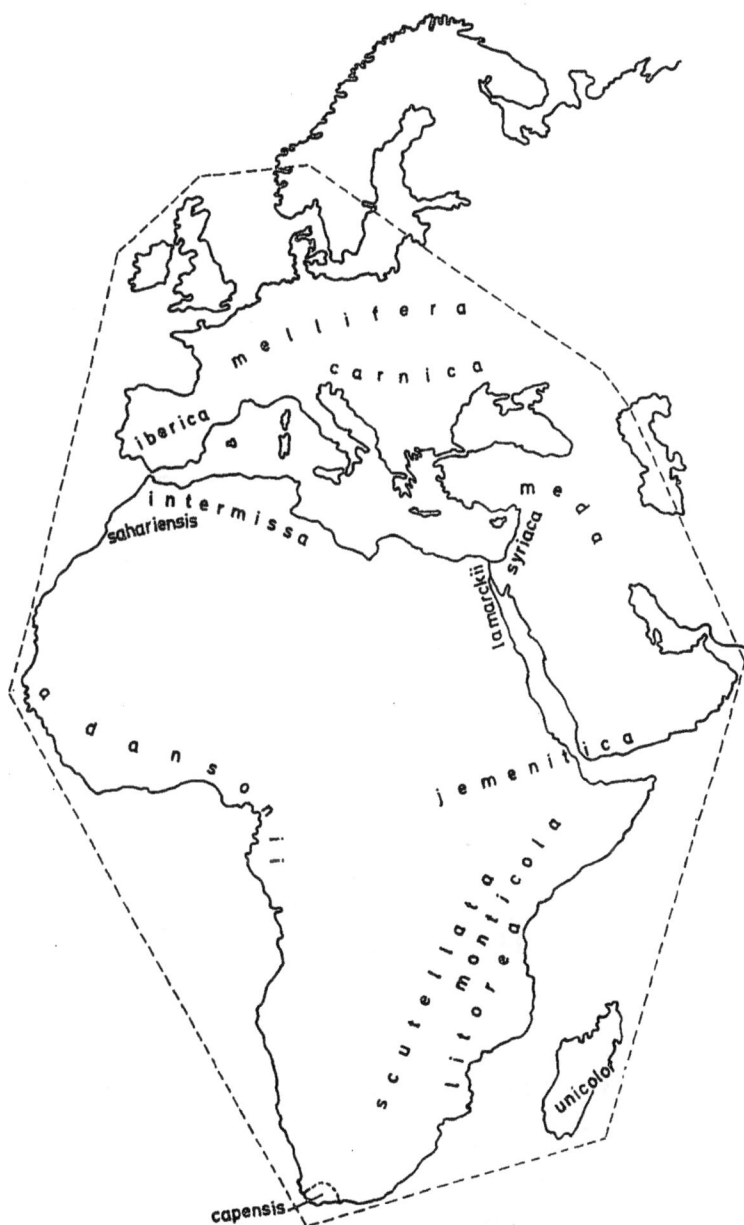

Fig. 4. Geographic distribution and limits of the species *Apis mellifera* and its major races.

Ecological requirements of *A. cerana* are about the same as those of *A. mellifera*. This species also succeeded in colonizing forested areas in the cool temperate zone (northern China to Ussuria in East Siberia). Thus, its area of distribution is very large; it extends from West Afghanistan to Japan (Fig. 5).

When kept sympatrically, *A. cerana* and *A. mellifera* colonies frequently rob each other (Koeniger, 1982). In Japan, *A. cerana* (originally the only

Fig. 5. Geographic distribution of *Apis cerana* (————), *A. florea* (—•—•) and *A. dorsata* (— — —).

honey bee there) is now largely replaced by imported *A. mellifera* colonies. Another cause of failing coexistence of the two species is attempted inter-mating, which produces lethal offspring (Ruttner and Maul, 1983).

A new problem is the shifting of parasites from one species to another since the geographic isolation had been broken by humans. A parasitic mite of brood and adults (*Varroa jacobsoni* Oud.) is coadapted to *A. cerana* and causes no serious damage since it is limited to reproduction on drone brood (Koeniger *et al.*, 1981). In Siberia and in Japan, where both bee species now are kept together, the parasite infested *A. mellifera* colonies and became a severe pest to the unadapted host.

Apis cerana is kept in hives. Colonies are reported to be smaller than *A. mellifera* colonies as are honey yields (Sharma *et al.*, 1980).

C. *Apis dorsata* (Fabr., 1793: 328)

Apis dorsata has an area of distribution largely overlapping but somewhat shifted to the east to that of the distribution of *A. florea* (Fig. 5). It includes South China, Celebes, and Timor, but not Iran or the Arabian Peninsula. It is found in altitudes up to 2000 m; the mountain type named *laboriosa*, possibly a separate species, is found even higher in Nepal (Sakagami *et al.*, 1980). More than 90% of Indian honey and wax are from this bee (Ghatge, 1956).

The morphology of *A. dorsata* is described by Morse and Laigo (1969) and Sakagami *et al.* (1980). *Apis dorsata* is the largest of the honey bees: its external characters are in a size range far beyond those of the other three species (Fig. 5). Since descriptions of the varieties (as subspecies or even species) do not include data on males, the taxonomic status of subgroupings remains uncertain.

The large combs (up to 1 m²) are fixed on the undersides of thick horizontal branches of large trees. Sixty or more nests may be found on such "bee trees." In treeless regions the bees also accept small trees, buildings, or projecting rocks as nesting sites. Data on honey yield vary within wide limits; a realistic figure seems to be 5–10 kg per colony per season. The nest has a certain functional structure; different activities are conducted at different places on the comb. Young bees hang in a curtain at a bee-space distance over both sides of the comb.

Two behavioral characteristics of *A. dorsata* are remarkable. First, they have a well organized mass defense reaction. When a colony is disturbed, clusters of alerted bees form, which hang on the lower edge of the comb and soon fall. This entire group of excited bees is activated for defense. An intruder, once marked by the odor of a specific pheromone (2-decen-1-yl-acetate) by being stung, is followed for kilometers (Koeniger *et al.*, 1979). "The giant honey bee *A. dorsata* is one of the most dangerous animals of Asia" (Koeniger, 1975). However, *A. dorsata* can be habituated to the presence of humans and observed from the closest distance for months (Roepke, 1930; Koeniger *et al.*, 1975). Colonies routinely nest on buildings in the center of crowded cities without showing any defensive activities.

Second, *A. dorsata* seasonally migrates (Koeniger and Koeniger, 1979; Koeniger, 1980; Roepke, 1930; Hadorn, 1948) to locations 100–200 km distant. The routes of migration seem to be constant through the years. Migrating swarms make stops on their route for 2–3 days. The timing of migration is correlated with the change of the season (rainy and dry period).

D. *Apis florea* (Fabr., 1787: 305)

In general, *A. florea* is a bee of the plains up to 500 m. However, seasonal migrations occur up to 1500 m and even higher (Muttoo, 1956). The migrations may be an adaptation to survive a cool or hot climate. *A. florea* colonies nest in the protection of house walls during winter; in summer, they nest in the shade of trees and bushes [Iran: Tirgari, (1971); Pakistan and Sri Lanka: Koeniger (1976b)]. Rings of resin at its points of attachment protect the comb against ants. Such behavior essentially enlarges the range of adaptation. In spite of its small size it competes well with the other *Apis* species (Koeniger, 1976b). The reproductive isolation of the three sympatric species in southeast Asia is favored by different flight times of drones (Koeniger and Wijayagunasekara, 1979).

Distribution (Fig. 5): Coasts of the Persian Gulf, Pakistan, India, Sri Lanka, Thailand, Malaysia, Indonesia, Philippines (Palawan). Museum material at Oberursel is not sufficient to permit a full evaluation of *A. florea's* phenotypic variation. However, it shows the same trend as the other species: larger types in the north and smaller ones in the south. *Apis florea* is a typical member of its genus with the identical number of chromosomes (Fahrenhost 1977) and has no fundamentally different characteristics (Table 1, Fig. 1).

Apis florea is a very efficient pollinator. In India, up to 75% of the insect visitors in a *Brassica* field may be *A. florea*.

In Oman, successful attempts have been made to domesticate *A. florea* for honey production (Free, 1981). In other countries (Thailand), honey hunters sell the entire combs, with brood and honey, in the markets.

Lindauer found that the communication dances of *A. florea* are similar to those of other species. However, the dances are adapted to the comb structure; the dances are on the platform on top of the comb and point directly to the food source (von Frisch, 1967).

In the composition of *A. florea* melittin (26 amino acids), five residues are different from the melittins of *A. mellifera* and *A. cerana* (which are identical) and five are different from that of *A. dorsata* (Kreil, 1975).

V. THE GEOGRAPHIC RACES OF *APIS MELLIFERA*

A. The Bees of Tropical Africa

Tropical Africa covers about one half of the area occupied by autochthonous *A. mellifera*. In wide zones it offers excellent conditions for honey bees, in spite of the many enemies—including nest-destroying honey hunters, bush fires, and long periods of dearth. "Ethiopia is a heaven for bees" said Gebreyesus (1976) for one part of this zone.

The bee population of tropical Africa is enormous, although figures reported are no more than rough estimates. Gebreyesus (1976) gives the number of colonies of Ethiopia as 3 million, while the quantity of exported wax was 350 metric tons. Using the same relation of exported wax to bee colonies, we can assume the bee colonies in Tanzania to number 2–3 million and those in Angola 5 million. However, the true populations are probably much greater. For the Handeni district (Tanzania) alone, Drescher (1975) calculated the number of exploited colonies as 1 million. In many countries no export of beeswax is organized although the bee populations are very large. It is very likely that the bee population of tropical Africa is far greater than the population of autochthonous *A. mellifera* of Eurasia.

All bees of Africa south of the Sahara are small except for *A. m. monticola*, which lives in higher altitudes and has a medium size (Fig. 3). This is why Buttel-Reepen (1906) proposed a subspecies for the bees of Africa, called "*unicolor* Latreille" as opposed to the subspecies "*mellifica*" comprised of all Mediterranean types. However, this nomination was clearly contrary to the visible phenotype of the multicolored bees of the region. The proposal was never generally accepted. The name "*unicolor*" remained restricted to the uniformly black bees of Madagascar. Instead, the name "*adansonii* Latreille" was generally adopted for the more or less yellow bees of tropical Africa, no matter whether they were from east, west, or south.

However, for one subspecies to occupy such a vast area as tropical Africa with such an extensive variety of altitudes and climates is not very likely with an animal having such a high capacity for specialized adaptation. In the much smaller and more uniform Mediterranean areas, about one dozen well defined geographical races have evolved.

The fictitious uniformity of the bees of tropical Africa is apparently nothing more than a lack of biometric studies. Since morphometric investigations have been started (Smith, 1961; Ruttner, 1975a,b) the ideas concerning uniformity have been fundamentally changed. Clearly, there is a group of morphometrically distinct tropical African races (Fig. 3). However, these bees (as opposed to the races of temperate zones) have many common features which are mainly adaptations to the common tropical environment. Such adaptations include small body size, aggressiveness, excessive swarming, and absconding. These may be called "tropical honey-bee characteristics," as they are found, at least to some extent, in all tropical *Apis* species and races.

Within this African group a clear morpho–geographical differentiation is visible having distances in multivariate analysis similar to those of "established" races (Fig. 3). However, one difference is found in comparisons with bees from, e.g., the Mediterranean area. The groups are not sharply separated from each other, though the samples of different geographic regions are grouped in clearly visible clusters, and there are both transitory types and introgressions into other groups (Fig. 3). Considering the climatic and topographic condition of the African continent between the Sahara and the Cape of Good Hope, it is evident that no efficient barriers exist which could serve as a basis for the evolution of distinct homotypic groups. Instead, an ecological differentiation occurred in the different zones with the areas between them having intermediate types of climate and bees which show permanent hybridization.

A good example is *A. m. monticola*, which lives in mountain regions in altitudes between 2400 and 3200 m in a very disjunct area of Burundi, Tanzania, Kenya, and Ethiopia. Its specific characteristics (see below), clear adaptations to high altitudes as shown by translocation experiments (Gebreyesus, 1976), are maintained by the permanent selection pressures of the environment. However, as long as free hybridization occurs, no completely separated "pure types" can be expected. The situation in tropical Africa is comparable to that in the central USSR or in Iran which shows slow transitions from one type to another.

To underline the importance of ecological factors, it seems justified to use the term "ecogeographic race or subspecies" for this type of intraspecific differentiation.

Some quantitative data of several African races are summarized in Table 2. The enumeration of races as well as their description reflects the present level of knowledge and must by no means be taken as definitive. In 1961, Smith described seven races of honey bees living in Africa. In the present contribution the number is increased to 11. The bees in great parts of the continent still remain unexplored.

1. Apis mellifera litorea *(Smith, 1961: 259)*

This is the bee of the African east coast. First described by Smith from the lowlands of the coast of Tanzania, the same small and very yellow bee was found later on the coast of Kenya (Mombasa, Lamu), in Mozambique (Lorenzo Marques), and elsewhere (Ruttner and Kauhausen, 1985).

The mean values of some characteristics given in Table 2 correspond well with those published by Smith. Several measures show high variability — not a surprising observation if the narrow coast line is considered, since the narrowness may lead to frequent hybridization. The size of worker cells averages 4.62 mm and the spacing of brood comb 28 – 30 mm, which corresponds well to the bees' small body size (Smith, 1961).

The warm and humid climate of the habitat induces an almost continuous flowering and brood rearing throughout the year. Periods of dearth or the abundant occurrence of a wasp *(Palarus latifrons)* induce absconding. When handled with care and appropriate methods, very satisfying honey yields are achieved with this bee in its own habitat.

Morphologically, *A. m. litorea* is very well separated from the other African races (Fig. 3; Table 2) by many characters. However, it is similar to *A. m. adansonii*, a bee from the far distant west coast which is as yellow and nearly as small as *A. m. litorea*. Significantly, *A. m. adansonii* is also a bee of a coastal lowland. On the other hand, *A. m. scutellata*, the direct neighbor of *A. m. litorea* living at higher altitudes, is distinctly different.

Compared to its body size, the proboscis of *A. m. litorea* is long (5.81 mm) and the abdomen is large relative to the length of the hind leg and forewing.

TABLE 2. Some Discriminant Characters of African Races of *Apis mellifera* Arranged According to Size[a]

Character	yemenitica n = 16	litorea n = 11	adansonii n = 23	lamarckii n = 16	scutellata n = 19	monticola n = 9	sahariensis n = 6	intermissa n = 20
Length of hairs (mm)	0.20	0.23	0.24	0.23	0.22	0.26	0.23	0.27
Proboscis (mm)	5.38	5.81	5.68	5.81	5.86	6.06	6.24	6.38
Length of hind leg (mm)	7.10	7.26	7.49	7.47	7.58	7.68	7.70	8.02
Length of fore wing (mm)	8.10	8.40	8.45	8.38	8.66	8.85	8.95	9.19
Length of tergite 3 + 4 (mm)	3.89	3.92	4.02	4.24	4.17	4.17	4.10	4.43
Color of tergite 4 (0.0 = completely dark to 9.0 = completely yellow)	5.06	4.22	4.35	3.37	4.00	2.59	3.63	1.10
Color of scutellum	6.79	6.77	6.13	5.58	5.61	0.97	5.35	1.07
Abdominal slenderness	86.35	85.08	84.47	87.39	85.05	86.05	82.64	81.16
Cubital index	2.25	2.25	2.39	2.37	2.52	2.35	2.62	2.33
Angle of wing venation I 16	92.61	91.31	94.95	96.76	92.40	86.44	97.09	95.77

[a] Numbers are race means; n is the number of samples each with 20 bees.

2. Apis mellifera scutellata *(Lepeletier, 1836: 404)*

The bee of the East African highland is distinctly different from the bee of the west coast (Table 2; Ruttner, 1975b), named *A. m. adansonii* by Latreille in 1804. Thus, it is appropriate to differentiate the East African bee in nomenclature. The correct name is *A. m. scutellata*, given to a bee of *"Afrique Meridionale: de la Caffuerie"* by Lepeletier (Ruttner and Kauhausen, 1985).

Apis m. scutellata is a bee within the African group, with medium body size (Table 2); the values of the body dimensions are situated somewhere between those of the small *A. m. litorea* and the large *A. m. monticola*. The tongue, fore wing, and hind leg are relatively short.

The abdomen is slender, as they are in all the other races of the group. The color on the tergites 1–4 and on the scutellum is yellow, although color shows high variability even among the bees in the same colony (Smith, 1961). *Apis m. scutellata* has the highest cubital index (2.53) of the group.

Considering all measurements collectively, *A. m. scutellata* is a very well defined and relatively uniform race. In the collection at Oberursel it is represented by a great number of samples from Ethiopia, Kenya, Tanzania, Burundi, Zimbabwe, and South Africa. According to Smith (1961), this bee is believed to be the central type of the species *Apis mellifera*, from which all the other races evolved.

Apis m. scutellata is the bee of the *"miombo,"* the open woodland covering large areas of East and South Africa, at altitudes of 500–1500 m. The basic food for the bees comes mainly from *Leguminosae* trees, especially *Brachystegia* and *Julbernardia*. The annual cycle is characterized by a rain period and a long dry period with regular bush fires.

Additionally, enemies, such as ants, birds, honey badgers, and honey-hunting humans, have great importance to the survival of colonies. In spite of all these factors, the bee population density in the *miombo* is high as is shown by the large quantity of exported wax and by the short time which is needed during the swarming period for a bait hive to become occupied. In general it is restricted mainly by lack of nesting sites.

Apis m. scutellata receives more attention than all other African races for several reasons: (1) it is the only tropical race for which scientific data are available; (2) it lives in an area from Transvaal to Ethiopia with very favorable beekeeping conditions and partly with a developed beekeeping technique; (3) it gave the source of the "Africanized bee" in South America, with all its problems and its publicity.

a. Colony Strength and Development. In consequence of the smaller body size, the cell diameter of the worker comb is smaller than in European races [4.7–4.9 mm; Smith (1961)]. Accordingly, the number of cells per square decimeter is about 1000 (Smith, 1961; Tribe and Fletcher, 1976; Chandler,

1976). As the space between combs is reduced to 32 mm and the period of development to 18.5–19 days, the colonies are densely packed with a capacity of rapid development and swarming (Drescher, 1976).

b. Reproductive Swarming. The swarming tendency is high in *A. m. scutellata* as it is with other races of the zone. "Every colony swarms" (Guy, 1972). Frequently swarming is correlated with the flowering of an important honey plant, such as *Eucalyptus* or *Brachystegia*. One important reason for swarming is the narrow dimensions of the nest cavity (Fletcher and Tribe, 1977). Suitable nesting sites are evidently a major limiting factor for the bee population. Thus, during the swarming season, any available place may be occupied by nesting bees, such as electrical poles, metal drums, fire hydrants, holes in the ground and even in the complete open, as well as the more common sites of cavities in trees or houses (Anderson *et al.*, 1973). African bees prefer larger nesting sites than do European bees (Rinderer *et al.*, 1981). Irvine (1957) describes a method of keeping bees in small rooms which can be entered from houses and where colonies of enormous strength develop.

Swarms of tropical Africa are much less "pretentious" than European swarms in the selection of nesting sites. They even accept small holes in the ground. European bees would never do that. Because of abundant swarms, some are forced to occupy small nesting sites; the lack of space in the small nests induces early swarming in consequence of overcrowding—a permanent, self-sustaining chain. In towns, swarms of bees may become a real nuisance. Providing enough space in the hive may reduce the swarming tendency considerably (Chandler, 1976; Fletcher, 1978).

Frequently, swarms are the consequence of the loss of the queen. The number of queen cells is usually not exceedingly great (Tribe and Fletcher, 1976). After repeated swarming, the swarms become very small with only 2000 to 3000 bees, but each swarm and also the mother colony has the chance to survive.

c. Migratory Swarming. Migrating from one place to another in response to lack of honey flow or drought is a typical behavior of honey bees of tropical Africa. In some regions seasonal migrations are said to occur, e.g. between the Rift Valley to the Aberdare Mountains in Kenya (Nightingale, 1976).

The tendency to migrate seems to have a genetic component; in zones with a more humid climate some colonies swarm and some remain stable (Chandler, 1976).

d. Absconding. Absconding is the complete desertion of a colony, leaving behind an empty nest. It is evidently a genus-specific adaptation to conditions, because it is observed with all *Apis* species living in the tropics. Migration is one reason for absconding (seasonal or resource-related absconding). Drastically diminished brood production and waiting until the major part of the sealed brood hatches is the preparative behavior for this type.

The other type of absconding is the immediate response of the colony to different inconveniences [disturbance-induced absconding; Winston *et al.* (1979)]. Also, this type of absconding occurs in all species and races of *Apis* living in the tropics, but rarely in those of the temperate zone.

Swarms of Africanized bees in Venezuela contain an especially high percentage of young bees, thus facilitating the build-up of new colonies (Winston and Otis, 1978). The swarm-to-swarm interval may be as short as 5–7 weeks.

e. Defense Behavior. The "aggressivity" of African bees is their most spectacular characteristic, securing for them general publicity. "The colonies always appear to be alerted, ever ready to defend the hive, and on occasion the whole colony goes berserk and stings every living thing in sight" (Smith, 1958). Though pronounced defensive behavior is found also in races of European and North African origin, there are some specific characters which have been studied experimentally (with *A. m. scutellata* of East Africa and their descendants in South America):

1. Optical stimuli. If presented black disks of different diameter as objects to attack, *A. m. scutellata* prefers small disks (sight angle 4.5°), whereas *A. m. carnica* has its preference at an angle of 9°. Thus, *A. m. scutellata* will attack from a farther distance than *A. m. carnica* (Koeniger, 1978).

2. Olfactory recruitment to attack by alarm pheromones. In *A. m. scutellata* a high level of the secondary alarm pheromone 2-heptanone, produced by the mandibular gland, was found (Kerr *et al.*, 1974) as well as a greater number of pore plates on the antennae (Stort, 1978), indicating a lower recruitment threshold. Isoamyl acetate, the chief alarm pheromone released when a bee stings, is strongly perceived at the entrance of an *A. m. scutellata* colony and releases a general attack (Kigatiira, 1979). This is not the case in a colony of European origin. There is no scientific evidence that *A. m. scutellata* produces more isoamyl acetate.

3. Mass attacks. The number of bees attacking during colony defense is much greater than in other races. It can reach several thousand bees, including recruited bees of neighbor colonies (Kigatiira, 1979).

4. Persistence of attacking the aggressor. The "viciousity" of a colony has been quantified by measuring the distance bees pursue a stung person or a leather ball (Stort, reported in Michener, 1975).

Pronounced defense behavior is by no means a uniform character of *A. m.*

scutellata or, generally, of the bees of tropical Africa. There is a considerable genetic variability (Chandler, 1976), and also a strong influence of nonge-netic intrinsic factors (e.g., colony strength, swarming preparations) and of extrinsic factors (temperature, honey flow, conditioning by frequent distur-bances or, in the opposite direction, by habituation to cautious managing by the beekeeper). Professional beekeepers were able to develop "intelligent" methods to handle a great number of colonies without running into prob-lems with defensive behavior. In a "Beekeeping Handbook" edited for beginners in Botswana, the list of basic equipment includes a simple smoker and a feather. However, it does not include a veil or gloves (Clauss, 1982).

f. The Africanized Bee of South America. In 1956, African honey-bee queens from Transvaal and from Tabora in Tanzania, thus very likely being of the *A. m. scutellata* race, were imported to the south of Brazil (Michener, 1975). The story, which started with the importation of these few queens reports one of the most fascinating unintentional experiments done in biol-ogy and on the overwhelming significance of inherited characters of adap-tation to a certain environment.

At the moment of importation of African queens, South America was inhabited by European bees (mainly *A. m. iberica*). Nevertheless, the African bees multiplied and spread as if the continent was waiting for them without any competition. It was expected that F_1 crosses with the European race would take place and then backcross to the same "Brazilian Europeans" again and again. However, all descriptions of the so-called Africanized bees of Brazil read as if the bees were pure *A. m. scutellata*. Morphologically, they almost are (Daly and Balling, 1978).

This dramatic multiplication and spread of African genes is explained only by the assumption that their selective advantage was so overwhelming that they did not give the European genes the least chance. The Africanized bees proceeded 300 km and more per year, showing that they were adapted to the tropical environment of America as well as to that of Africa. For the Iberian bees, the 300 years after their importation was too short to develop efficient adaptations. They were able to survive in the tropics only as long as there were no other honey bees.

It is worthwhile to note that many attempts have been made to "improve" African apiculture by importing European (primarily Italian) strains. How-ever, no single case achieved success (Ntenga, 1967; Papadopoulo, 1969; Guy, 1972; Nightingale, 1976).

The selective advantage of African bees in the tropics of South America must be due to some of the specific characteristics of *A. m. scutellata* (or of tropical honey bees in general) as described before: (1) acceptance of nesting sites which give a chance of survival in warm climate only (Casas, 1973); (2) rapid build-up of even very small swarms; (3) high numbers of swarms per starting unit and per year; and (4) aggressive behavior toward colonies with less pronounced defense reactions.

If one Iberian colony yields one swarm per year or less (as is perfectly adapted to European conditions) and a predominantly African colony yields 10 swarms and more, the last European gene would soon disappear.

It is absolutely correct to treat the Africanized bee of South America as a separate type. However, in practice all their characteristics nearly equal those of the original *A. m. scutellata* of East Africa. Thus, for the description of the type, we can refer to that given above. Indeed, many of the behavioral characteristics of the African bee have been analyzed and described from Africanized strains.

Morphometrically, Africanized bees are clearly discriminated from European bees (originating from South and North America) by multivariate analysis with 25 characters (Daly and Balling, 1978). It would be of interest to compare them with original *A. m. scutellata* with the same method.

Similar efforts have been made to identify the Africanized bees using enzyme polymorphism. However, only three out of 42 analyzed enzymes have been found to be polymorphic [*Mdh-1*, *Adh*, and an unidentified protein; Sylvester (1982)]. *Mdh-1* allows the identification of individual bees with a probability of 90%. Taking all three loci together, the probability rises to 99%. Where all three loci are considered for samples of colonies of bees, this method appears satisfactory.

One main behavioral characteristic of tropical honey bees is their lack of adaptations to survive long periods of cold weather. Bees from the tropics were observed to start full flight at bright sunshine with the temperature below freezing while neighboring colonies of European races did not move. Therefore it was believed that the distribution of the Africanized bees in South America would be restricted to the tropics and subtropics (Nuñez, 1979). Later, however, Africanized colonies were found in Argentina as far south as 39°S (Dietz *et al.*, 1985). Thus it cannot be excluded that a behavioral pattern differing from that of the parental strain is evolving in the new environment.

Controlled hybrids of Africanized bees with European races and selected strains of the Africanized bees are reported to have a lower tendency to sting and to swarm (Martinez, 1973; Gonçalves, 1978).

3. Apis mellifera monticola *(Smith, 1961: 258)*

This bee was described from the cool forests on the upper slopes of Kilimanjaro, particularly between 2400 and 3100 mm. There, the sun is frequently obscured by cloud and mist and ground frosts occur at night in the grassland above the upper edge of the forest. This mountain bee is, at first sight, very different from the bee of the East African highlands (A. m. scutellata) because of its dark color, large body size (although distinctly smaller than North African bees, Table 2), and gentle temperament (Smith, 1961; Drescher, 1975). The hair cover on the abdomen is longer than in the other tropical races (0.26 mm), but the abdomen remains slender (index 86.05) in spite of its length.

We also were able to find this bee in altitudes higher than 2400 m in Kenya (Mt. Kenya, Meru, Mt. Elgon) and to identify it from Ethiopia and Burundi.

Intermediate types occur and the yellow A. m. scutellata is reported to migrate temporarily to higher altitudes (Smith, 1961). However, the types are kept separate by environmental selection. A really gentle bee for the tropics which can be handled without protecting clothes could solve a lot of problems in African apiculture. However, this may be simplistic. When transported to the rich honey grounds of the highlands, A. m. monticola colonies start to dwindle and become "lazy" (Gebreyesus, 1976). However, this report seems to be the only apicultural experience with this bee so far. In Tanzania the three races (A. m. litorea, scutellata, and monticola) occur within a distance of 250 km from the coast to the altitudes of the Kilimanjaro but in different altitudes.

4. Apis mellifera adansonii (Latreille, 1804: 172)

This bee, first described from Senegal, is clearly distinct from A. m. scutellata of the East African highland in a number of characters of size, color, and wing venation (Table 2). The differences of the bee of the east coast, A. m. litorea, are much less pronounced. Apis m. adansonii is somewhat bigger in all body dimensions (except the relatively short proboscis), with a somewhat darker yellow scutellum and some differences in wing venation. Thus the discrimination of these two clusters in the multivariate analysis creates certain problems. However, the centroids are fairly distant (Fig. 3).

This lowland bee is found in many countries along the West African coast: Senegal, Guinea, Ivory Coast, Upper Volta, Cameron, Togo, Nigeria, People's Republic of the Congo, and Gabon.

Not much is known of the biology of this bee. Absconding and migrating seem to be rather frequent (N'daye, 1975). Schricker (1980) reported interesting experiments to "tame" this bee in Nigeria using the artificial Nasanoff pheromone.

5. Apis mellifera yemenitica (Ruttner, 1975b: 341)

This new type, the smallest of all mellifera bees described so far, was first found in Yemen, and later in Oman (Ruttner, 1975b; Dutton et al., 1981). The further analysis of samples of Sudan and Chad proved to be morphometrically inseparable from the bees of south Arabia. Thus, the *nomen* "*nubica*" [proposed by Ruttner (1975b) for bees of Sudan] has to be withdrawn in favor of "*yemenitica*," which is applied also to the bees of Sudan and Chad. In the meantime, Rashad and El Sarrag (1981) presented data on a much larger number of samples of Sudan than we had at our disposal (18 versus four samples). Their results confirm our classification in every detail (our figures on *A. m. yemenitica* are given in parentheses): proboscis 5.50 (5.38), basitarsis length 2.2 (1.81), tergite 3 + 4 = 3.70 (3.89), abdominal slenderness index 86.0 (86.3), color on tergite three 71.36 (81.69), cubital index 2.37 (2.25).

Evidently, old traditions exist in Yemen apiculture. In Oman the bee is surprisingly gentle. The commercially managed bees there live under extremely hard conditions of drought and heat. Feral populations live exclusively in the mountains, up to 1500 m (Dutton *et al.*, 1981).

Apis m. yemenitica is the African bee of the dry thornbush savanna. The extension of the area of distribution is not yet examined. Its occurrence in south Arabia is not astonishing, as this is part of the Ethiopian zoogeographic region.

Apis m. yemenitica certainly shows the extreme capacity of *A. mellifera* to adapt to a hot dry climate. As observed in Oman (Dutton *et al.*, 1981), wild colonies survive several months of drought and extreme heat in mountain crevices. In dry years with almost no rainfall, nearly no swarms are observed, and colonies without sufficient provisions perish. However, in other years with an average rainfall of about 250–350 mm, the colonies swarm readily. *Apis mellifera* has a very high potential to adapt to xerotherm conditions (another case is *A. m. sahariensis*), thus confirming the statement of Michener (1979) that *Apoidea* have their optimum in general in warm temperate, xeric zones.

6. Apis mellifera lamarckii (*Cockerell, 1906: 166*) [formerly: A. m. fasciata (*Latreille, 1804: 171*)]

This bee belongs clearly to both the group of Africa south of the Sahara and to the North African group (Table 2 and Fig. 4). It is a small bee with a very slender abdomen (index 87.4) and remarkably short legs, wings, and tongue. The contrast of deep black and bright yellow stripes on the abdominal tergites with shining white tomenta is "not surpassed in beauty" (Abushady, 1949) in the beekeeper's eye. Queens of this race were imported to central Europe by about 1860 (Buttel-Reepen, 1906).

Apis m. lamarckii is restricted to the Nile Valley and thus lives under completely different environmental conditions from those found south of the Sahara. The cultivated land is irrigated, allowing a long vegetation period. Here, European honey bees find favorable conditions. Many *A. m. carnica* and *A. m. ligustica* queens were imported and propagated. Bee-keepers claim to get better honey yields from these colonies than from local bees. This is the only case, together with Israel, which has similar agricultural conditions, where European bees were imported with success to a hot country with indigenous, specially adapted bee populations. The reason is a fundamental change of environment by irrigation.

The pure Egyptian type is kept with an elaborate technique developed 4000–5000 years ago (Armbruster, 1921). With this technique bees are housed in piles of clay tubes.

7. Apis mellifera capensis *(Escholtz, 1822: 97)*

In 1912, G. W. Onions, an amateur beekeeper of the Cape District in South Africa, published his observations on the local honey bee, which shows a very unusual reproductive behavior in worker bees. In queenless colonies some workers produced brood resulting in female offspring rather than male. These observations were fully confirmed by Jack (1916), Anderson (1963), and Ruttner (1977). The laying workers are treated as a queen ("pseudo-queen"), as they produce 9-oxodecenoic acid as do normal queens. Moreover, the workers of this race have a higher number of ovarioles (15–20) and a well-developed spermatheca of about half the size of a queen's spermatheca although they are not known to mate. However, *A. m. capensis* workers have been instrumentally inseminated with success (J. Woyke, personal communication).

The thelytokous parthenogenesis in the Cape bee is genetically determined and cytologically the consequence of central fusion of two haploid nuclei during meiosis of the unfertilized egg (Verma and Ruttner, 1983).

Apis m. capensis is about the same size as *A. m. scutellata*, but with a dark abdomen and very different biology and behavior (lower colony strength; gentle disposition, less inclined to abscond). As queens can be reared from thelytokous larvae, this kind of reproduction may be of selective advantage in a climate with high queen losses during the mating flight (Ruttner, 1977).

The fully developed *A. m. capensis* type is found only in a relatively small area, about 50 km around the Cape Peninsula. The whole population is probably not more than an estimated 10,000 colonies.

8. Apis mellifera unicolor *(Latreille, 1804: 168)*

The bee of Madagascar, Mauritius, and Reunion was given a separate taxonomic position very early because of the uniform black body color. Recent biometrical analysis confirmed this concept (Fig. 3); the bees of these islands form a well-separated cluster within the African group. The closest neighbor morphologically is not *A. m. litorea* from the near coast of the continent, but *A. m. scutellata*. The body size of both races is very similar and the cell size [930 cells/dm^2; Douhet (1965)] is identical, although the body measurements of *A. m. unicolor* are generally somewhat smaller, except for the relatively long wings. The tendency of *A. m. scutellata* towards a higher cubital index is still more pronounced in *A. m. unicolor* (2.79, maximum 2.94). Except for size, distinct relations are found also to *A. m. monticola* (color, behavior). *Apis m. unicolor* is completely black, including the scutellum, and the tomenta are so narrow and scarce that they are only indistinctly visible.

Evidently, there are two ecotypes of honey bees in Madagascar. One occurs in the tropical coastal region with a high absconding tendency; the other occurs in the more temperate highlands and behaves similarly to European bees. They do not abscond, are thus easily domesticable, and furthermore are gentle (Douhet, 1965). The biological data available for these bees are rather poor.

B. The Bees of North Africa

1. Apis mellifera sahariensis (Baldensperger, 1922: 59)

This bee was first described and named by the great apiculturalist and experimentator J. Baldensperger, who took the first queen of this race in 1921 from Ain Sefra (Algeria) to Nice. Later this bee was described and imported to Europe by Haccour (1961) and Brother Adam (1983).

Morphometrically, *A. m. sahariensis* is a link between the West African tropical *A. m. adansonii* and the West Mediterranean *A. m. intermissa* (Baldensperger, 1932). The known distribution extends from Ain Sefra in Algeria through the oases of the Sahara south of the Atlas mountains to Figuig in the west. The local populations of the bee seem to be very small, and all indications (empty hives, scarcity of vegetation) show that they have to struggle severely to survive. Most likely, the present *A. m. sahariensis* is a relic of a much larger population of bees that existed when the Sahara was largely covered by savanna several thousand years ago.

Apis m. sahariensis is intermediate in size between the tropical and the West Mediterranean races, and has a tan-yellow body color (Baldensperger, 1932; Adam, 1983). The proportions of the abdomen (82.64) show nearly European values, quite different from those of *A. m. yemenitica* which live in a similar biotop.

The low tendency to defensive behavior is mentioned by all authors.

This race, together with *A. m. yemenitica*, gives another example of the honey bee's capacity for adaptation to extreme conditions of heat and drought. *Apis m. sahariensis* survives not only in human-made hives, but also in cliffs in the walls of *wadis* (Baldensperger, 1922; Ruttner, 1975b), despite an active season of not more than 4 months.

2. Apis mellifera intermissa *(Buttel-Reepen, 1906: 187)*

Known as the "Punic" or "Tellian" bee in the apicultural world, this is the relatively uniform black race of the West Mediterranean African coast *(Maghreb)*, from Tunisia (or from Libya) to the Atlantic. It is not much smaller than European races (some body parts are even bigger than in A. m. ligustica), with a broad abdomen (81.16) and longer hairs than all races of tropical Africa (Table 2).

The coast of northwestern Africa has a typical Mediterranean vegetation, but also has intense climatic extremes. The "Tellian" bee seems to be the only race which can resist the climatic contrasts of North Africa. European races (mostly from Italy and France) were imported to North Africa in large scale. These introductions were complete failures. The colonies did not survive for more than 1 or 2 years, and nature seems to have eliminated all the nonadapted genes within a short time.

The reproductive behavior of A. m. intermissa is well adapted to the environment. In years of drought, 80% of the local population may die even though these bees are resistant to adverse external conditions. However, in the next humid year the losses will be rapidly replaced by an enormous swarming activity. In the oasis of Laghouat, Brother Adam (1983) observed a colony which produced seven swarms during a single season. Finally, the mother colony was reduced to 200–300 bees. Nevertheless, it developed again to a colony of full strength. As in A. m. lamarckii, A. m. intermissa colonies construct up to 100–200 queen cells, and several queens live together until the first queen is successfully mated. *Apis m. intermissa* is the quickest race to develop laying workers after dequeening [only 4–5 days; Ruttner and Hesse (1981)]. Excessive propolis is used. *Apis m. intermissa* shows very strong defensive behavior, but not the mass attacks shown by bees of the tropics. If disturbed, they may follow over 100 m or more. Typically, these bees do not abscond.

The seasonal brood cycle is very typical for the Mediterranean. There is a rapid rise in brood production in spring peaking in April–May, a very deep depression in brood production during the summer drought, and, finally, a second (smaller) peak of production in autumn (Oct.–Nov.) at the start of the humid winter. The failure of imported races was evidently the consequence of their inability to adapt to this rhythm.

3. Apis mellifera major *(Ruttner, 1975b: 331)*

This curiosity is a local type on the slopes of the Rif mountains to the Mediterranean, near El Hoceima east of Tetuan. The dimensions of the body parts exceed those of other races (medium proboscis length, 7.04 mm; maximum value of one colony, 7.12 mm; length of fore wing, 9.52 mm; length of hind leg, 8.42 mm). Thus, it surpasses in size not only the other Mediterranean races which are all of medium size or less, but even the large *A. m. carnica* (Ruttner, 1975a). The data are based on measurements of 80 bees originating from four colonies located in one area of about 40 × 20 km. Nothing is known of the size of the population or of the bees' biology. All colonies observed were kept in traditional hives: tubes of cork bark.

C. The Bees of the West Mediterranean

All parts of this region belong to the temperate zone. Most types of honey bees occurring there are used and studied by beekeepers and scientists in many parts of the world. An extensive bibliography exists, summarized by Buttel-Reepen (1906), Alpatov (1948), Goetze (1964), von Frisch (1967), Adam (1983), and Ruttner (1975a, 1983). Thus, it will be sufficient to mention these races only briefly and to refer to the bibliography.

1. Apis mellifera iberica *(Goetze, 1964: 20)*

This race is sufficiently similar to *A. m. intermissa* in morphology and behavior that it was considered a subvariety of the "Tellian" bee (Adam, 1983). However, biometry reveals substantial differences: it is generally larger in size, with relatively short wings and broad abdomen (80.3). The cubital index is drastically reduced to values below 2.0. These characteristics result in a fairly uniform cluster in the "M" branch of the principle component analysis (Fig. 3).

The biological and behavioral characteristics of the Iberian bee are much less known than those of other European races. Brother Adam (1983) describes it as being rather prolific and having a high swarming tendency, a nervous behavior if disturbed which leads to balling the queen, a high stinging propensity, and an excessive use of propolis.

Imported by Spanish and Portuguese settlers, *A. m. iberica* was distributed throughout South and Central America and it did well there. However, it never really became integrated into the local fauna. Now it is quickly vanishing in the Americas because of the competition from the Africanized bee.

2. Apis mellifera mellifera *(Linn., 1758: 576)*

This bee (the "common black European bee" or "English," "French," or "German" Brown bee) is quite unique owing to the original size of the area of its distribution and its adaptability to different environments. It is found from the Pyrenees to the Ural, from Scotland and South Scandinavia to the Provence. It seems derived from *A. m. iberica*, with some body characteristics modified. Some parts are a little larger, especially the abdomen, although the proboscis is allometrically reduced in length (Table 3). The abdomen is extremely broad, with an abdominal index of 78.6 — by far the lowest of all races examined. Two other measurements also show extreme values: cover hair is more than 0.40 mm and the cubital index is below 2.00. Based on these extremes, it is easy to elaborate simple and efficient methods for the amateur beekeeper to discriminate the three European races used predominantly in apiculture — *A. m. mellifera, A. m. carnica,* and *A. m. ligustica* (Goetze, 1964; Ruttner, 1983).

Apis mellifera mellifera shows some behavioral characters indicating its relation to the North African bees — nervous behavior, irritability, and considerable use of propolis.

The history of *A. m. mellifera* in its present boundaries is the history of the postglacial period in Europe, a time span of not more than 10,000 years. During the last glaciation, no bees were able to exist in the treeless tundra between the Alpine glacier and the northland ice. It was only during the following warm period that bees moved to Central Europe. *Apis mellifera*

TABLE 3. Some Discriminant Characters of European and Eastern Races of *Apis mellifera* Arranged According to Size[a]

Character	iberica $n = 17$	mellifera $n = 9$	syriaca $n = 11$	cypria $n = 8$	sicula $n = 10$	ligustica $n = 35$	carnica $n = 21$
Length of hair (mm)	0.26	0.44	0.23	0.27	0.31	0.28	0.29
Proboscis (mm)	6.44	6.05	6.19	6.39	6.25	6.36	6.40
Length of hind leg (mm)	8.29	8.10	7.83	7.88	7.95	7.97	8.10
Length of fore wing (mm)	9.25	9.33	8.48	8.87	8.98	9.21	9.40
Length of tergite 3 + 4 (mm)	4.56	4.64	4.11	4.24	4.38	4.39	4.51
Color of tergite 4 (0.0 = completely dark to 9.0 = completely yellow)	1.57	2.77	3.75	3.81	1.79	3.78	1.24
Color of scutellum	1.03	1.30	5.95	4.80	0.84	4.18	0.13
Abdominal slenderness	81.03	78.61	84.18	84.58	82.80	83.48	83.46
Cubital index	1.84	1.84	2.28	2.72	2.47	2.55	2.59
Angle of wing venation I 16	96.46	96.36	92.60	89.97	96.51	95.60	96.00

[a] Numbers are race means; *n* is the number of samples each with 20 bees.

mellifera came from the West Mediterranean coast through France and Germany to England, southern Scandinavia, and northern Russia. *Apis mellifera carnica* came from the West Mediterranean coast to the Alps, the Danube valley, and southern Russia.

The Black Bee of the Atlantic coast (the Landés in southwestern France to Norway) is excellently adapted to the typical nectar flow of the area which is primarily from the late summer blooming ling. Louveaux (1969) and Louveaux *et al.* (1966) demonstrated that this coast bee has a very characteristic genetically determined brood cycle, with a peak late in summer. Other genetically determined cycles were found in colonies originating from the south (Provence), central, and west of France. Thus, in spite of the short 10,000 years since recolonization after the last glaciation, distinct ecotypes have developed in France which show the potential for developing new geographic races. Indeed, Cornuet *et al.* (1975) were able to demonstrate that these ecotypes can be distinguished morphologically. As the history of climate and vegetation is exactly dated for this period, these results demonstrate the chronological process of microevolution within the species.

At the other end of the area of its distribution, *A. m. mellifera* survives in the South Ural (Baschkiria) at winter temperatures as low as $-45°C$ and short summers having a rich nectar flow. Certainly, an analysis would reveal special genetically determined adaptations. Thus, a very wide range of adaptations has developed intraracially within the short postglacial period. They extend from those suited to a cool–humid Atlantic climate to those adapted to a continental climate with its severe extremes of temperature.

Most of the experimental work with honey bees in central and western Europe was done with this black bee or its (uncontrolled) hybrids. At present, the *A. m. mellifera* bee is heavily hybridized in many countries, especially those of Central Europe.

D. The Bees of the Near East and Mediterranean

1. Apis mellifera syriaca *(Buttel-Reepen, 1906: 175)*

The bees of former Palestine and Syria are rather uniformly yellow, small (number of cells/dm^2 = 968; wing length = 8.48 mm) and rather aggressive with a high tendency to swarm, constructing 100 queen cells and more (Blum, 1956). In the graphical presentation of multivariate analysis (Fig. 3), the cluster of this race takes a central position, together with *A. m. sahariensis* and the not yet sufficiently analyzed bees of Anatolia.

2. Apis mellifera caucasica *(Gorbatschev, 1916: 39) and* Apis
   mellifera remipes *(Gerstacker, 1862: 61) (Pallas in Litt.)*

Both these Caucasian races are characterized by a long proboscis (Skorikow, 1929), broad tomenta, and short cover hairs. The bee of higher altitudes *(A. m. caucasica)* has a dark body color and, in some locations (Mingre-
lia), a proboscis length of more than 7 mm. Since it is gentle and generally a
good honey producer, *A. m. caucasica* is frequently used worldwide in apiculture. Several authors report it has a high *Nosema apis* susceptibility in
northern parts of Europe (Bilash *et al.*, 1971; Ruttner, 1983). Remarkably,
the colonies use much propolis: during winter they close the entrance with
resin leaving only small openings.

Apis mellifera remipes, the yellow valley bee of the Caucasian region, is
less used in apiculture outside its homeland. Therefore, much less of its
biology is known. For both Caucasian types, a multivariate statistical analysis still remains to be done.

Perhaps several races of bees exist in the northern part of the Near East.
However, the different types, not being isolated by geographical barriers,
are not easily separated and a sophisticated analysis has yet to be done.

In Anatolia, Bodenheimer (1941) and Brother Adam (1983) found types
similar to *A. m. caucasica* and *A. m. remipes,* on the coast of the Black Sea, and
in the southeast a type similar to *A. m. syriaca.* The bee of Central Anatolia is
described as a separate type recommended for apiculture on account of its
hardiness and high honey productivity (Adam, 1983). On the west coast,
the bee of the south (Izmir) is similar to *A. m. adami* and in the north to the
"Greek" bee. Thus a geomorphometrical analysis of the bees of this region
becomes rather difficult. Moreover, a high migratory activity prevails in
Anatolian beekeeping which progressively disturbs any geographic differences.

In Iran a specific geographic subspecies with several local subpopulations
was found: *A. m. meda* (Skorikow, 1928: 261; Ruttner *et al.* 1985). No *A.
mellifera* bees are found east of a line through Mashad – Kerman – Bandar
Abas.

3. Apis mellifera carnica *(Pollmann, 1879: 45)*

The bee of the Balkan peninsula, extending to the Alps and the Black Sea
and passing the Carpath mountains into the Ukraine, is much used in
apiculture and very well known. Therefore, readers are referred to Goetze
(1964) and Ruttner (1975a, 1983). *Apis m. carnica* is a large bee, generally
with a dark pigmented abdomen (occasionally with brownish spots or rings

on tergite 2 and 3), broad dense tomenta, short cover hairs, and a high
cubital index (2.5–2.8). All these are characteristics which place this race at
the extreme end of the C-branch in the graph of multivariate analysis
(Fig. 3).

Since this bee is kept together with A. m. mellifera in several countries of
Europe, it is important to discriminate them. This is easily done using the
characteristics described.

The reasons for keeping this race extensively in apiculture are mainly its
gentleness and its good adaptability to varied climates and honey resources.
Having evolved in the continental climate of southeastern Europe, it shows
quick population increase in spring and a good capacity to collect honey-
dew.

In the Alps the two races A. m. mellifera and A. m. carnica were originally
sharply separated by the mountains: A. m. mellifera north of the Alps, and A.
m. carnica southeast (in the southwest, A. m. ligustica). In Russia no such
barriers exist, and there the races show a slow gradual change from one type
in the north to the other in the south. At first, this was interpreted according
to general zoological rules—a "kline" existed of a short proboscis in the
north to a long one in the south and from a low cubital index to a high one
(Mayr, 1963). However, mapping the geographic races gives a clear interpretation of the phenomenon to be a zone of hybridization between two
very distinct races.

There exist varieties of A. m. carnica, especially the type "carpathica" in
Rumania and the USSR, which are sometimes regarded as a separate race by
some authors (Foti et al., 1965; Avetisyan, 1973).

4. Apis mellifera ligustica (Spinola, 1806: 35)

This race, which is very close to A. m. carnica morphometrically, also
shows good temper and high capacity to adapt to varying environments. It
additionally has a conspicuous yellow color on the abdomen. "A nicer view
is hard to imagine than the high flight activity of the yellow Italians, when
they, illuminated by the bright sun, look like creatures made of pure gold"
(Alpatov, 1948).

Of all continental European races, the Italian bee has the smallest original
area of distribution. It is wedged in by sea and Alps and could not extend
beyond the Apennine peninsula when the glacial period was finished.
Humans made up for this and distributed this race all over the world. "It is
doubtful if modern beekeeping would have made such tremendous forward
strides in the past hundred years had it not been for the Italian bee" (Adam,
1983).

E. The Island Bees of the Mediterranean

The "creativity" of the Mediterranean region in the evolution of the species *Apis mellifera* is well illustrated in three island races. One of them, *A. m. cypria*, has obtained some publicity in the last century. Another, *A. m. adami*, has been discovered only lately to the great surprise of specialists. Thus, these races are mentioned briefly.

1. Apis mellifera cypria *(Pollmann, 1879: 25)*

"For a while this bee has been a favorite of many beekeepers" (Buttel-Reepen, 1906). It was imported to Europe as early as 1866 and to the United States in 1876. In 1880 F. Benton settled temporarily in Cypria to organize the export of queens to Europe.

The main reason for this popularity was the exterior of the bee. It has a reddish-yellow ("carrot") abdominal and scutellar color which contrasts with broad tomenta and a black tip on the slender abdomen. The tongue is long and the cubital index is high (Table 3). With some care from beekeepers it winters north of the Alps. *Apis m. cypria* shows an energetic defensive behavior if disturbed which is another point of its publicity in the apicultural world. However, it is not nervous and remains calm on the combs (Buttel-Reepen, 1906; Adam, 1983).

Apis m. cypria represents a special type showing morphological relations to the bees of the coast of Asia Minor and the Levante. As does *A. m. syriaca*, it has the tendency to construct many queen cells in the swarming season.

2. Apis mellifera adami *(Ruttner, 1975b: 347)*

Although originating in the same region, the bees of Crete have characteristics which are quite different from the bees of Cyprus. The abdominal color is variable with predominantly dark scutellum; body size is larger than medium; the abdomen is rather broad; the proboscis is relatively short; the cubital index is remarkably low (1.9) (Table 3). The bee seems to be intermediate in behavior between oriental and European races (numbers of swarm cells and time to develop ovaries in queenless worker bees). Again, as in the Cyprian bee, a very quiet temper on the comb is combined with a pronounced defensive behavior (Ruttner, 1980).

3. Apis mellifera siciliana *(Grassi, 1881: 277)*

At present, this bee is generally incorrectly referred to as *A. m. sicula* Grassi.

The bees of Sicily are neither black Italians nor European "Tellians," but rather are a distinct race with its own characteristics (Table 3). In cluster analysis they are situated between *A. m. carnica* and *A. m. ligustica* on one side and *A. m. intermissa* of North Africa on the other (Fig. 3) which might reflect the genetic situation.

Unfortunately, only scarce information exists on the biology and behavior of this bee. As with other races of the South and East Mediterranean, the bees of Sicily may construct several hundred (300–400) queen cells during the swarming season. A special supersedure behavior has been observed: despite the presence of many queen cells, the old queen remains in the colony, no swarm issues, and finally a single young laying queen is found. The superfluous virgins and the old mother finally disappear (F. Baumgarten, personal communication). The bee is easy to handle. It uses little propolis and winters with success even in North Germany. Because of high brood-rearing activity until late in autumn (a Mediterranean-type brood rhythm), the colonies winter with strong populations.

The islands of the western Mediterranean (Corsica, Sardegna) seem not to have developed special races, although no final statements can be made since no detailed biometric analyses are available.

REFERENCES

Abushady, A. Z. (1949). Races of bees. *In* "The Hive and the Honey Bee" (R. Grout, ed.), pp. 11–20. Dadant and Sons, Hamilton, Ill.

Adam, Br. (1983). "In Search of the Best Strains of Honey Bees" (2nd ed.). Northern Bee Book, England.

Alpatov, V. V. (1929). Biometrical studies on variation and the races of honeybee. *Q. Rev. Biol.* 4, 1–58.

Alpatov, V. V. (1948). The races of honeybees and their use in agriculture. (In Russian) *Sredi Prirody* 4, 1–65.

Anderson, R. H. (1963). The laying worker in the Cape honeybee, *Apis mellifera capensis. J. Apic. Res.* 2, 85–92.

Anderson, R. H., Buys, B., and Johannsmeier, M. F. (1973), "Beekeeping in South Africa." Bull. 394, S. African Dept. Agric., Pretoria.

Armbruster, L. (1921). Bienenzucht vor 5000 Jahren. *Arch. Bienenkd* 3, 68–80.

Avetisyan, G. A. (1973). Breeds of Soviet Union bees: their selection and protection. *Proc. Inter. Apic. Cong. (Apimondia)* 24, 333–334.

Baldensperger, P. J. (1922). Sur l'apiculture en Orient. *Proc. Inter. Apic. Cong. (Apimondia)* 6, 59–64.

Baldensperger, P. J. (1932). Varietés d'abeilles en Afrique du Nord. *Proc. Inter. Cong. Entomol.* 5, 832–839.

Bilash, G. D., Makarow, I. I., and Strojkow, S. A. (1971). Results of a comparative study of different races of bees in the main zones of USSR. *Proc. Inter. Apic. Cong. (Apimondia)* 21, 397–400.

Blum, R. (1956). Bienen im Nahen Osten. *Proc. Inter. Apic. Cong. (Apimondia)* 16, 28.

Bodenheimer, F. S. (1941). "Studies on the Honeybee and Beekeeping in Turkey." Merkez Ziraat Mucadele Enstitusu, Ankara.

Butler, G. C. (1954). "The World of the Honeybee." Collins, London.

Buttel-Reepen, H. von (1906). Beiträge zur Systematik, Biologie sowie zur geschichtlichen und geographischen Verbreitung der Honigbiene. *Mitt. Zool. Museum Berlin 3*, 117–201.

Casas, C. A. (1973). Against the African bee with Carniolan queen cells. *Proc. Inter. Apic. Cong. (Apimondia) 23*, 256–259.

Chandler, M. T. (1976). The African honeybee—*Apis mellifera adansonii*: The biological basis of its management. *In* "Apiculture in Tropical Climates" Vol. I (E. Crane, ed.) pp. 61–68. Inter. Bee Res. Assoc., Gerrards Cross, England.

Clauss, B. (1982). "Bee Keeping Handbook." Ministry of Agriculture, Gaborone, Botswana.

Cockerell, T. D. A. (1906). New Rocky Mountain bees, and other notes. *Can. Entomol. Ontario 38*, 160–166.

Cornuet, J. M., Fresnaye, J., and Tassencourt, L. (1975). Discrimination et classification de populations d'abeilles à partir de caractères biométriques. *Apidologie 6*, 145–187.

Daly, H. V., and Balling, S. V. (1978). Identification of Africanized honey bees in the Western Hemisphere by discriminant analysis. *J. Kansas Entomol. Soc. 51*, 957–969.

Dietz, A., Krell, R., and Eischen, F. (1985). Preliminary investigation on the distribution of Africanized honeybees in Argentina. *Apidologie 16*, 99–108.

Douhet, M. (1965). Beekeeping in Madagaskar. *Proc. Inter. Apic. Cong. (Apimondia) 20*, 690–710.

Drescher, W. (1975). Bienennutzung in Tansania. *Allg. Dtsch. Imkerztg. 9*, 117–122.

Drescher, W. (1976). The use of movable-frame hives in development programmes (Africa and Latin America). *In* "Apiculture in Tropical Climates" Vol. I (E. Crane, ed.) pp. 23–30. Inter. Bee Res. Assoc., Gerrards Cross, England.

DuPraw, E. (1965). The recognition and handling of honeybee specimens in non-Linnean taxonomy. *J. Apic. Res. 4*, 71–84.

Dutton, R. W., Ruttner, F., Berkeley, A., and Manley, M. J. D. (1981). Observations on the morphology, relationships and ecology of *Apis mellifera* of Oman. *J. Apic. Res. 20*, 201–214.

Escholtz, J. F. (1822). "Entomographien" Vol. I. Reimer, Berlin.

Fabricius, J. C. (1787). "Mantissa Insectorum" Vol. I. Proft., Hafniae.

Fabricius, J. C. (1793). "Entomologia Systematica" Vol. II. Proft., Hafniae.

Fabricius, J. C. (1798). "Supplementum Entomologiae Systematica." Proft. and Storch, Hafniae.

Fahrenhorst, H. (1977). Chromosome number in the tropical honeybee species *Apis dorsata* and *Apis florea*. *J. Apic. Res. 16*, 56–61.

Fletcher, D. J. C. (1978). Management of *Apis mellifera adansonii* for honey production in Southern Africa. *Proc. Symp. Apic. Hot Climates, (Apimondia)*, 86–89.

Fletcher, D. J. C., and Tribe, G. D. (1977). Swarming potential of the African bee, *A. m. adansonii* Latr. *Proc. Symp. African Bees, (Apimondia)*, 25–34.

Foti, N., Lungu, M., Pelimon, P., Barac, I., Copaitici, M., and Mirza, E. (1965). Untersuchungen über die morphologischen Merkmale und die biologischen Eigenschaften der Bienenpopulationen in Rumänien. *Proc. Inter. Apic. Cong. (Apimondia) 20*, 182–188.

Fraser, H. M. (1951). "Beekeeping in Antiquity." Univ. London Press, London.

Free, J. B. (1981). Biology and behaviour of the honeybee *Apis florea*, and possibilities for beekeeping. *Bee World 62*, 46–59.

von Frisch, K. (1967). "The Dance Language and Orientation of Bees." Harvard Univ. Press, Cambridge, Mass.

Gebreyesus, M. (1976). Practical aspects of bee management in Ethiopia. *In* "Apiculture in Tropical Climates" Vol. I (E. Crane, ed.) pp. 69–78. Intern. Bee Res. Assoc., Gerrards Cross, England.

Gerstäcker, C. E. A. (1862). Cited by Buttel-Reepen (1906) loc. cit.

Ghatge, A. I. (1956). Scientific exploitation of the wealth of bees in India. *Proc. Inter. Apic. Cong. (Apimondia) 16*, 45.

Goetze, G. (1930). Variabilitäts- und Züchtungsstudien an der Honigbiene mit besonderer
    Berücksichtigung der Langrüsseligkeit. *Arch. Bienenkd 11*, 135–274.
Goetze, G. (1940). "Die beste Biene." Verl. Liedloff, Loth u. Michaelis, Leipzig.
Goetze, G. (1964). "Die Honigbiene in näturlicher und künstlicher Zuchtauslese. I. Systematik,
    Zeugung and Vererbung." Parey Verlg. Hamburg.
Gonçalves, L. S. (1978). Melhoramento tecnologico e genetico de abelhas *Apis mellifera. Proc.
    Symp. Apic. Hot Climates, (Apimondia)*, 61–69.
Gorbatschev, K. A. (1916). The grey Caucasian bee, *Apis mellifica* var. *caucasica*, and its
    position among other bees. Information of the Station of Silkworm Rearing *Tiflis, 39.*
    (English summary.)
Grassi, B. (1881). Saggio di una monografia delle api d'Italia. *L'Apicoltore 14*, 277–281.
Guy, R. D. (1972). Commercial beekeeping with African bees. *Bee World 53*, 14–22.
Haccour, P. (1961). The bees of the Sahara race. *Proc. Inter. Apic. Cong. (Apimondia) 18*, 75–76.
Hadorn, H. (1948). Betrachtungen über die wilden Bienen in Sumatra. *Schweiz. Bienenzeitg.*,
    308–314.
Irvine, F. R. (1957). Indigenous African methods of beekeeping. *Bee World 38*, 113–128.
Jack, R. W. (1916). Parthenogenesis amongst the workers of the Cape honeybee: Mr. G. W.
    Onion's experiments. *Trans. Ent. Soc. London*, 396–403.
Kerr, W. E., Blum, M. S., Pisani, J. F., and Stort, A. C. (1974). Correlation between amounts of
    2-heptanone and isoamyl-acetate in honeybees and their aggressive behaviour. *J. Apic. Res.
    13*, 173–176.
Kigatiira, K. I. (1979). Behaviour of the East African honeybee. *Proc. Inter. Apic. Cong. (Apimon-
    dia) 27*, 295–299.
Koeniger, N. (1975). Observations on alarm behaviour and colony defence in *Apis dorsata. Proc.
    IUSSI 7th Symp. Dijon*, 153–154.
Koeniger, N. (1976a). The Asiatic honeybee *Apis cerana. In* "Apiculture in Tropical Climates"
    (E. Crane, ed.), Vol. I, pp. 47–49. Intern. Bee Res. Assoc. Gerrards Cross, England.
Koeniger, N. (1976b). Interspecific competition between *Apis florea* and *Apis mellifera* in the
    tropics. *Bee World 57*, 110–112.
Koeniger, N. (1978). Differences in optical releasers of attack flight between *Apis mellifera
    carnica* and *Apis mellifera adansonii. Proc. Symp. Apic. Hot Climates, (Apimondia)*, 58.
Koeniger, N. (1980). Observations and experiments on migration and dance communication of
    Apis dorsata in Sri Lanka. *J. Apic. Res. 19*, 21–34.
Koeniger, N. (1982). Interactions among four species of the genus Apis. *Proc. Congr. IUSSI, 9th
    (Boulder)*, 59–64.
Koeniger, N., and Koeniger, G. (1979). Das Wanderverhalten von *Apis dorsata* in Sri Lanka.
    Proc. Inter. Apic. Cong. (Apimondia), 27, 300–302.
Koeniger, N., and Wijayagunasekera, H. N. P. (1979). Time of drone flight in the three Asiatic
    honeybee species (*Apis cerana, Apis florea, Apis dorsata*). *J. Apic. Res. 15*, 67–71.
Koeniger, N., Weiss, J., and Ritter, W. (1975). Capture, moving and management in
    cages of colonies of giant honeybees *Apis dorsata. Proc. Inter. Apic. Cong. (Apimondia) 25*,
    300–303.
Koeniger, N., Weiss, J., and Maschwitz, U. (1979). Alarm pheromones of the sting in the genus
    Apis. J. Insect Physiol. 25, 467–476.
Koeniger, N., Koeniger, G., and Wijayagunasekara, N. (1981). Observations on the adaptation
    of *Varroa jacobsoni* to its natural host *Apis cerana* in Sri Lanka. *Apidologie 12*, 37–40.
Koltermann, R. (1973). Rassen-und artspezifische Duftbewertung bei der Honigbiene und
    ökologische Adaptation. *J. Comp. Physiol. 85*, 327–360.
Kreil, G. (1975). The structure of *Apis dorsata* melittin: phylogenetic relationships between
    honeybees as deduced from sequence data. *FEBS Lett. 54*, 100–102.

Latreille, P. A. (1804). Des especes d'Abeilles vivant en grande societé, et formant des cellules hexagonales, ou des Abeilles proprement dites. *Ann. Mus. His. Nat. Paris* 5, 161–178.

Lauer, J., and Lindauer, M. (1973). Die Beteiligung von Lernprozessen bei der Orientierung der Hongbiene. *Fortschr. Zool.* 21, 349–370.

Lepeletier, A. L. M. (1836). "Histoire Naturelle Des Insectes. Hyménoptères." Roret, Paris.

Linñe, C. (1758). "System a Naturae." 10th Ed. Vol. I. Lauer. Salvii., Holmiae.

Louveaux, J. (1969). Ecotype in honeybees. *Proc. Inter. Apic. Cong. (Apimondia)* 22, 499–501.

Louveaux, J., Albisetti, M., Delangue, M., and Theurkauff, M. (1966). Les modalités de l'adaptation des abeilles (*Apis mellifera* L.) au milieu naturel. *Ann. Abeille* 9, 323–350.

Maa, T. (1953). An inquiry into the systematics of the tribus Apidini or honeybees (*Hym.*). *Treubia* 21, 525–640.

Martinez, C. R. (1973). Behaviour of hybrids of *lingustica, mellifera, caucasica* and *carnica* queens and *adansonii* drones at Santiago del Estero, Rep. Argentina. *Proc. Inter. Apic. Cong. (Apimondia)* 24, 260–271.

Mayr, E. (1963). "Animal Species and Evolution." Harvard Univ. Press, Cambridge, Mass.

Michener, C. D. (1975). The Brazilian bee problem. *Annu. Rev. Entomol.* 20, 399–416.

Michener, C. D. (1979). Biogeography of the bees. *Ann. Missouri Bot. Garden* 66, 277–347.

Morse, R. A., and Laigo, F. M. (1969). *Apis dorsata* in the Philippines. Monograph. *Philipp. Assoc. Entomol.* 1, Laguna.

Muttoo, R. N. (1956). Facts about beekeeping in India. *Bee World* 37, 125–130.

N'daye, M. (1975). L'Apiculture au Senegal. Thesis, Ecole veter. Dakar.

Nightingale, J. M. (1976). Traditional beekeeping among Kenya tribes, and methods proposed for improvement and modernisation. In "Apiculture in Tropical Climates" (E. Crane, ed.), Vol. I pp. 47–49. Intern. Bee Res. Assoc., Gerrards Cross, England.

Ntenga, N. (1967). The honeybees of Tansania, *Apis mellifera adansonii*. *Proc. Inter. Apic. Cong. (Apimondia)* 21, 253–257.

Nuñez, J. A. (1979). Time spent on various components of foraging activity: comparison between European and Africanized honeybees in Brazil. *J. Apic. Res.* 18, 110–115.

Onions, G. W. (1912). South African "fertile-worker bees." *Agric. J. Union S. Afr.* 3, 720–728.

Papadopoulo, P. (1969). Introduction of foreign queens into *Apis mellifera adansonii* colonies. *Proc. Inter. Apic. Cong. (Apimondia)* 22, 529–536.

Pollmann, A. (1879). "Wert der verschiedenen Bienenrassen und deren Varietäten." Leipzig.

Rashad, S. E., and El Sarrag, M. S. (1981). Morphometrical studies on some of the Sudanese honeybees. In "Apiculture in Tropical Climates" (E. Crane, ed.), Vol. II pp. 302–309. Intern. Bee Res. Assoc., Gerrards Cross, England.

Rinderer, T. E., Collins, A. M., Bolten, A. B., and Harbo, J. B. (1981). Size of nest cavities selected by swarms of Africanized honeybees in Venezuela. *J. Apic. Res.* 20, 160–164.

Roepke, W. (1930). Beobachtungen an indischen Honigbienen, insbesondere an *Apis dorsata* F. *Med. Landbowhoogeschoolte Wageningen* 34, 1–26.

Ruttner, F. (1975a). Races of bees. In "The Hive and the Honeybee" (Dadant and Sons, ed.), pp. 19–38. Dadant and Sons, Hamilton, Ill.

Ruttner, F. (1975b). The African races of honeybees. *Proc. Inter. Apic. Cong. (Apimondia)* 25, 325–344.

Ruttner, F. (1977). The problem of the Cape bee (*A. m. Capensis* Escholtz): parthenogenesis—size of population—evolution. *Apidológie* 8, 281–294.

Ruttner, F. (1980). *Apis mellifera adami* (n.ssp.), die Kretische Biene. *Apidologie* 11, 385–400.

Ruttner, F. (1983). "Zuchttechnik und Zuchtauslese." 5th ed. Ehrenwirth Verlag, München.

Ruttner, F., and Hesse, B. (1981). Rassenspezifische Unterschiede in Ovarentwicklung und Eiablage von weisellosen Arbeiterinnen der Honigbiene *Apis mellifera* L. *Apidologie* 12, 159–183.

Ruttner, F., and Kauhausen, D. (1985). Honeybees of tropical Africa: biological diversification and isolation. *Proc. Int. Conf. Apic. Trop. Climates, 3rd, Nairobi, Kenya* 45–51.

Ruttner, F., and Maul, V. (1969). The cause of the hybridization barrier between *Apis mellifera* L. and *Apis cerana* F. *Proc. Inter. Apic. Cong. (Apimondia)* 22, 510, 562.

Ruttner, F., and Maul, V. (1983). Experimental analysis of the reproductive interspecies isolation of *Apis mellifera* L. and *Apis cerana*. Fabr. *Apidologie* 14, 305–327.

Ruttner, F., Tassencourt, L., and Louveaux, J. (1978). Biometrical-statistical analysis of the geographic variability of *Apis mellifera* L. *Apidologie* 9, 363–381.

Ruttner, F., Pourasghar, D., and Kauhausen, D. (1985). Die Honigbienen des Iran. 2. *Apis mellifera meda* Skorikow, die Persische Biene. *Apidologie* 16, 241–264.

Sakagami, S. F., Matsumura, T., and Ito, K. (1980). *Apis laboriosa* in Himalaya, the little known world largest honeybee (Hym., *Apidae*). *Insecta Matsumurana, New Series* 19, 47–77.

Schricker, B. (1980). Minderung der Aggressivität der zentralafrikanischen Honigbiene (*Apis mellifera adansonii*) durch synthetisches Nasanoff-Pheromon. *Apidologie* 12, 94–96.

Sharma, O. P., Mishra, R. C., and Dogra, G. S. (1980). Management of *Apis mellifera* L. in Himachal Pradesh. *In* "Apiculture in Tropical Climates" (E. Crane, ed.), Vol. II pp. 205–211. Intern. Bee Res. Assoc., Gerrards Cross, England.

Skorikow, A. S. (1929). Eine neue Basis fur eine Revision der Gattung *Apis* L. *Rep. Appl. Entom.* IV, 250–264.

Smith, F. G. (1958). Beekeeping observations in Tanganyika 1949–1957. *Bee World* 39, 29–36.

Smith, F. G. (1961). The races of honeybees in Africa. *Bee World* 42, 255–260.

Sneath, P., and Sokal, R. (1973). "Numerical Taxonomy." Freeman Co., San Francisco.

Spinola, M. M. (1806). "Insectorum Liguriae Species Novae" Vol. I, Genuae.

Stort, A. C. (1978). Aggressive behaviour and sensorial structures in Africanized and Italian honeybees. *Proc. Symp. Apic. Hot Climates, (Apimondia)* 53–55.

Sylvester, H. A. (1982). Electrophoretic identification of Africanized honeybees. *J. Apic. Res.* 21, 93–97.

Tirgari, S. (1971). Biology and behaviour of the Dwarf honeybee (*Apis florea*) of Iran. *Proc. Inter. Apic. Cong. (Apimondia)* 23:344–345.

Tribe, B. C., and Fletcher, D. J. C. (1976). Rate of development of the workers of *Apis mellifera adansonii*. *Proc. Symp. African Bees*. (Apimondia) 116–119.

Verma, S., and Ruttner, F. (1983). Experimental analysis of the reproductive isolation of *Apis mellifera* and *Apis cerana*. *Apidologie* 14, 41–57.

Winston, M. L., and Otis, G. W. (1978). Ages of bees in swarms and afterswarms of the Africanized honeybee. *J. Apic. Res.* 17, 123–129.

Winston, M. L., Otis, G. W., and Taylor, O. R. (1979). Absconding behaviour of the Africanized honeybee in South America. *J. Apic. Res.* 18, 85–94.

CHAPTER 3

Visible Mutants

KENNETH W. TUCKER

I. ANALYSIS OF VISIBLE MUTANTS

A visible mutant is a heritable variant from the wild-type phenotype that is readily visible. Such a mutant can be seen either with unaided human vision or with slight magnification. Most frequently, honey bees have mutants affecting eye color, but also have a few affecting the cuticular color, the loss of body hair, and the morphology of the eyes, wings, and sting.

Almost all the honey bee's visible mutants have occurred spontaneously. These have been found usually by bee geneticists among their own bees or contributed by cooperating beekeepers. If one were to seek spontaneous visible mutants in honey bees, one might expect to look at 10^6 or 10^7 drones for each mutant found (Kerr *et al.*, 1980; Chaud-Netto *et al.*, 1983). One mutant, chartreuse-limão, ch^{li}, was found under conditions where it could have been induced by radiation from ^{60}Co (Soares, 1981a).

Most mutants are discovered in drones. Drones are haploid, so a sample of "brothers" (not true siblings, but fellow gametes) reflects their mother's genotype. Thus, drones show phenotypes of recessive mutations as well as those of dominant mutations that all castes show.

A. Rearing Mutants

To provide drones, workers, and queens for study, one must propagate the mutant. Often a new mutant is encountered as a single drone. This drone must be matured and mated by instrumental insemination to a wild-type queen. The queen daughters of the wild-type queen mated to the mutant drone will all be heterozygotes, and produce mutant drones in large numbers. Sometimes a new mutant is found as abundant drones from a queen already heterozygous in all or part of her germ tract. Further propagation may proceed by alternating generations of wild-type queens instrumentally inseminated to mutant drones and their heterozygous daughters. Queens may also be reared from heterozygotes, but only half of the daughters will include the mutant in their genomes. Also, if daughters are reared from naturally mated queens, sometimes the mutants cordovan or garnet, whose drones can mate, may be found added to the stock unintentionally. However, for drone-sterile mutants only heterozygotes are available for propagation. For easiest stock maintenance, inbreeding and matings producing mutant workers should be avoided.

B. Genic Interrelations

Locating each mutant in the honey bee's genome is a principal reason for conducting breeding experiments with mutants. This reason orders the priority of which objective to pursue first with each new mutant: to determine allelism to known mutants of similar phenotype. If allelism is found, the new mutant shares some features already known for the established allele, such as linkage and similar biochemistry. If allelism is not found, then linkage tests complete the search for the new mutant's genomic location.

In the process of locating a new mutant, some observations can be made on the mode of inheritance, the allelic interaction of dominance or recessiveness, and nonallelic interactions between genes with similar phenotypes. However, it would take additional crosses to complete a comprehensive catalog of all possible interactions.

The following gives a summary of the usual genetic relationships studied and the types of matings conducted in this pursuit.

1. Mode of Inheritance

Nearly all mutants in the honey bee show monogenic inheritance. This is manifest in a heterozygous queen's drone progeny, in which mutant and wild-type drone phenotypes are produced in a 1:1 ratio. This reflects the heterozygous mother's genotype in her gametic ratio, since each drone develops from an unfertilized egg. Monogenic inheritance is evident also from a 1:1 phenotypic segregation in worker progenies of a queen heterozygous for a dominant mutant mated to wild type, or a queen heterozygous for a recessive mutant mated to the same mutant (test cross). Likewise, monogenic inheritance in queen daughters of a heterozygous queen mated to wild type is indicated by a 1:1 genotypic segregation of daughter queens with and without a recessive mutant gene, as determined by their drone progenies. With fully viable mutants, segregations in drone progenies are diagnostic, but with subviability, incomplete penetrance, and drone sterility, evidence from workers and queens sometimes fits expectations better.

The few visible mutants not showing monogenic inheritance have defied analysis and their mode of inheritance is not yet determined.

2. Dominance or Recessiveness

Most honey-bee mutants are recessive to wild type. The wild-type phenotype, not that of the mutant, is visible in heterozygotes. The worker and queen progeny of wild-type queens mated to the mutant are wild type and produce wild-type and mutant drones in a 1 : 1 ratio. One of these heterozygous queens mated to mutant drones will produce mutant and wild-type workers and queens in a 1 : 1 ratio. Then, for all the workers and queens of a progeny to have the mutant phenotype, a homozygous mutant queen must be mated to one or more mutant drones.

Four dominant mutants are known for honey bees, 11% of mutants described so far. With a dominant mutant, not only haploid hemizygotes and diploid homozygotes, but also diploid heterozygotes have mutant phenotypes. Worker and queen daughters of a wild-type queen mated to a dominant mutant all show the mutant phenotype. Moreover, a heterozygote's daughters as well as her drones would be half mutant and half wild-type if she were mated to wild type; but if mated to the same mutant, all daughters and half the drones would have the mutant phenotype.

Interactions between different mutants of an allelic series often include partial dominance, or intermediacy between two alleles of the same series. Several of these interactions are described in Section II,A,6.

Dominance and recessiveness are allelic interactions characteristic only of the diploid bees: workers, queens, and diploid drones. For most purposes, phenotypic segregations of workers suffice. Seldom, as with sterile, subviable drones, the determination of genotypes of queens by progeny test is useful (Laidlaw and Tucker, 1965a). Diploid drones need be used only for questions of gene dosage, linkage to the sex alleles (Woyke, 1973), and for sex-specific phenotypes (Woyke, 1977). In terms of ease and expense of rearing, workers cost less than queens, which in turn cost much less than diploid drones.

3. Allelism

Multiple alleles are common in the honey bee. Of its eye-color mutants, two-thirds are members of three different allelic series.

To test for allelism, a new mutant must be crossed with mutants of a similar phenotypic category. That is, an unknown eye-color mutant would be crossed to known eye-color mutants. With mutants recessive to wild type, an indication of allelism is a mutant phenotype in the worker progeny usually intermediate between the known and unknown mutant. Confirmation of allelism is indicated by the presence of only two phenotypes of drones, the new and the old mutant, from the doubly heterozygous queen daughter of this mating. Indicating nonallelism for recessive mutants would be the appearance of only wild-type workers, and wild-type queens producing four phenotypes of drones (ignoring epitasis) including $\frac{1}{4}$ wild type. For recessive mutants, reciprocal crosses are equally useful and the test queen may be either homozygous or heterozygous. Having the unknown recessive mutant included in the test queen permits conducting tests to several known recessive mutants at one time with one mixed insemination (Tucker and Laidlaw, 1965, 1968).

An allelic series dominant to wild type has not yet been found in honey bees. But whether double heterozygotes in such a series would show complete or partial dominance to one another could be diagnosed by the segregation of drone progenies of a doubly heterozygous queen into two mutant phenotypes (allelism) or into four phenotypes including $\frac{1}{4}$ wild type (nonallelism).

By means of the allelism tests, some mutants could be grouped as multiple alleles. In the eye-color mutants, seven have been assigned to the chartreuse locus, three to the ivory locus, and three to the snow locus. For the body color genes, two alleles have been identified at the black locus. The allelic groupings are reflected in the organization of Table 1. Table 2 gives a list of the numbers of tests for allelism completed or not for each mutant.

4. Linkage

With 16 pairs of chromosomes, the honey bee has a potential 16 linkage groups. So far, three linked pairs of genes have been reported (as listed in Table 2), but there are no indications yet as to the groupings of these pairs.

The indication of linkage is a departure from independent assortment, with an excess of parental types over recombination types. With honey bees, this is usually measured in the drone progeny of a doubly (or triply or quadruply) heterozygous queen. In case of drone subviability or lethality but good viability for workers, the appropriate test cross is made to such a multiply heterozygous queen. This test cross is made to multiply hemizygous recessive drones for recessive genes, to wild-type drones for dominant genes, and the appropriate mix of wild-type and recessive genes in drones for recessive and dominant genes combined (Rothenbuhler et al., 1953a; Mackensen, 1958). With subviability of all castes, close linkage can be dismissed if an excess of wild-type drones is one of the recombination classes of phenotypes (Tucker, 1980).

TABLE 1. Honey Bee Mutants

Name	Symbol	Appearance	Viability[a]	References[b]
Eye colors[c]				
snow group				
snow	s	White	3	8,9,10,11,19,21,28, 34,35,44
tan	s^t	White, darkens to tan	4	8,11,21,22,23,40,42
laranja	s^{la}	Light orange, darkens to reddish brown	3	2,3,4,5,16,40,42,49
bayer	by	White, darkens to reddish orange	3(?)	5,11,24,40,50
ivory group				
ivory	i	White	4	8,9,10,11,19,20,21, 22,23,25,28,33, 34,35,43,44,50
umber	i^u	White to pink, darkens to reddish tan	4	8,9,11,19,49
rose	i^{ro}	Light rose, darkens to deep rose	?	24,50
cream group				
cream	cr	White	4	4,5,8,9,10,11,16,19, 21,22,25,28,34, 35,40,44,50
pearl	pe	White	4	4,5,8,9,10,11,16,19, 21,22,40,44,50
brick	bk	Orangish red, darkens to reddish-brown	2	8,9,10,11,18,19,20, 21,22,23,28,44, 48,49,50
spade	sp	Rose pink, darkens to red	?	5,10,11,16,24,40,50
chartreuse group				
Benson green	ch^B	Greenish chartreuse, darkens to reddish	4	8,9,10,11,21,40
chartreuse-2	ch^2	Chartreuse, darkens to reddish-brown	3	8,9,10,18,20,21,22, 43,44
chartreuse	ch	Chartreuse, darkens to reddish-brown	4	7,28,31,34,35,48,49
chartreuse-1	ch^1	Chartreuse, darkens to reddish-brown	4	8,10,19,20,21,22,23, 44
chartreuse-limão	ch^{li}	Light yellow, darkens to reddish-brown	4	4,5,40
red	ch^r	Purplish red, darkens to reddish-brown	3	2,3,4,5,8,16,20,21, 40,43

TABLE 1. *(continued)*

Name	Symbol	Appearance	Viability[a]	References[b]
cherry	*ch^c*	Variable, red and yellow, darkens to deep red or brownish-red	4	8,10,21,22,23,40
modifier (with *ch¹*)	*m*	Darkens *ch¹*: pink darkens to brown	4	8,10,20,21,22,23,44
garnet group				
garnet	*g*	Deep red-brown, darkens to wild type	2	8,18,19,21,44,47,49, 50
unassigned group				
ocelos claros	*oc*	Compound eyes: rose, darkens to brown. Ocelli: rose changes to glassy white	2	3,4,5,16
pink	*p*	Pink	1	1
white	(none)	White	4	29,30
Nonmutant				
wild type	*mut^{+d}*	Dark brown to black	3	7,8,10,11,25,31,48
Body (cuticle) colors				
yellow	*bl^+*	Partly yellow	4	17,45,51
abdome castanho	*bl^{ac}*	Drones, black; workers, partly yellow	?	2,14,36,51
black	*bl*	Black	3	17,22,23,44,45,51
albino	*a*	Unpigmented, unsclerotized	1	50
cordovan	*cd*	Brown replaces black	3	1,2,3,4,20,22,27,28, 43,44,48,49
yellow face[e,f]	(none)	Yellow clypeus, labrum, genae	?	32
Hair				
haarlos	*H*	Longest hair missing, tomentum present	0	50
hairless	*h*	Hair rubs off	2	2,28,46
schwarzsüchtig	*S*	Longest hair and most tomentum missing	1(?)	6
Eye shape				
cyclops[f]	(none)	Compound eyes merged over vertex	?	12,18,26,50
einäugig[e,f]	(none)	One compound eye missing	?	11
eyeless	*e*	Eye facets vestigial, male sterile	1	11,18,44

TABLE 1. *(continued)*

Name	Symbol	Appearance	Viability[a]	References[b]
facetless	*f*	Eye facets vestigial, male sterile	1(?)	29,30,37
reduced facet number[f]	*rf*	Reduced number of eye facets	?	15,18
Wing				
diminutive	*di*	Wings small, functional on workers, drones, not queens	?	24,47
droopy	*D*	Wings droop	0	28,34
rudimental wing	*Rw*	Wings vestigial	1(?)	13
short	*sh*	Wings small, changed venation	1	15,22,23,44
truncate	*tr*	Wings small, truncated distally	1	22,23,44
wrinkled	*wr*	Wings wrinkled distally, variable penetrance	?	22,23,44
Sting				
split sting[f]	*sps*	Lancets separated from stylet, low penetrance	?	38,39,41

[a] The numbers in this column refer to this five-point rating scale for the incidence of significant subviability in drones: 4 none yet; 3 sometimes; 2 often; 1 always; 0 lethal, but viable in heterozygous workers. The question mark, ?, indicates a lack of data, except for *wr* where all data are confounded by incomplete penetrance.

[b] The numbers in this column refer to the following citations. Numbers in bold face indicate the original description of the mutant. 1. Cale *et al.* (1963); 2. Chaud-Netto (1975); 3. Chaud-Netto (1977); 4. Chaud-Netto (1979); 5. Chaud-Netto *et al.* (1983); 6. Dreher (1940); 7. Dustmann (1966); 8. Dustmann (1969); 9. Dustmann (1973); 10. Dustmann (1975a); 11. Dustmann (1975b); 12. Goebel (1981); 13. Hachinohe and Onishi (1953); 14. Kerr (1969); 15. Kerr and Laidlaw (1956); 16. Kerr *et al.* (1980); 17. Laidlaw and el-Banby (1962); 18. Laidlaw and Tucker (1965a); 19. Laidlaw and Tucker (1965b); 20. Laidlaw *et al.* (1953); 21. Laidlaw *et al.* (1964); 22. Laidlaw *et al.* (1965a); 23. Laidlaw *et al.* (1965b); 24. H. H. Laidlaw, unpublished; 25. Langer *et al.* (1972); 26. Lotmar (1936); 27. Mackensen (1951); 28. Mackensen (1958); 29. Michailoff (1930); 30. Michailoff (1931); 31. Neese (1972); 32. Nolan (1937); 33. Rothenbuhler (1957); 34. Rothenbuhler *et al.* (1953a); 35. Rothenbuhler *et al.* (1953b); 36. Rothenbuhler *et al.* (1968); 37. Schasskolsky (1935); 38. Soares (1977); 39. Soares (1979a); 40. Soares (1981a); 41. Soares (1981b); 42. Soares and Chaud-Netto (1982); 43. Tucker (1958); 44. Tucker (1980); 45. Tucker and Laidlaw (1967); 46. Witherell (1972a); 47. Witherell and Laidlaw (1977); 48. Woyke (1964); 49. Woyke (1973); 50. Woyke (1975); 51. Woyke (1977).

[c] The eye-color mutants are grouped in relation to their biochemistry. See Section II,A,2.

[d] The symbol *mut* stands for any mutant symbol, so *mut*[+] stands for the wild-type allele of any mutant. (However, for *bl*, any allele may be wild type depending on the particular population. See Section II,B.)

[e] Only the phenotype has been described.

[f] The pattern of inheritance is not resolved. Designation as a mutant and designation of symbols should be considered tentative.

TABLE 2. Allelism and Linkage of Honey-Bee Mutants

Mutant	Allelism tests[a]	Linkage tests[b]	Linkage found[c]
s locus	7/10	15,X	None
by	8/10	3,X	None
i locus	8/10	15,X,C	Centromere at 3.6 (43)
cr	9/10	15,X	pe at 0.33 (22)
pe	7/10	4,X	cr at 0.33 (22)
bk	7/10	14,X	None[d]
sp	7/10	0	None
ch locus	9/10	15,X,C	h at 4.1 (28), centromere at 28.8 (43)
m	Specific to ch[1]	4	None
g	8/10	14,X	di at 14.5 (47)
oc	4/10	2	None
p	0/10	1	None
bl locus	1/3	12,X	None
a	0/3	0	None
cd	1/3	16,X	None
Yellow face	0/3	0	None
H	0/2	0	None
h	0/2	8,X	ch at 4.1 (28)
S	0/2	0	None
e	0/1	11	None
f	0/1	0	None
di	3/5	1	g at 14.5 (47)
D	0/5	8,X	None
Rw	0/5	0	A lethal at 31 (13)[e]
sh	3/5	12,X	None
tr	3/5	11,X	None
wr	3/5	13,X	None

[a] Number of tests conducted/number possible to genes with similar phenotypes.
[b] Number of linkage tests of 29 possible, X = test for linkage to sex alleles, C = test for linkage to centromere.
[c] Linkage measurements in map units; the numbers in parentheses cite the pertinent references as coded in Table 1.
[d] The relationship between bk and reduced facet number seems an enhancement of penetrance of reduced facet number by bk rather than linkage between them.
[e] The data for Rw are similar to data from other genes interpreted as subviability rather than linkage.

In terms of bee management, linkage tests usually are an easy sequel to allelism tests. Queens doubly heterozygous in the trans configuration can be reared as all or half the progeny of the allelism cross, depending on whether their mother was homozygous or heterozygous, respectively. Probably the accumulation of linkage data before allelism is established, as sometimes happens, is due to the expediency of using the progeny of allelism tests. However, to test linkage in the cis configuration, an additional sequence (mating queens, then rearing their daughters, then rearing the daughter's drones) is necessary, starting with wild-type queens mated to doubly hemizygous drones. Sometimes, there is an uncertainty in propagating the appropriate progeny for a linkage test because of raising daughters from a heterozygote. This is unavoidable for male sterile and male lethal genes, and may be used otherwise if considered more convenient than obtaining homozygotes. In practice, three matings for each one required usually suffices when the probability of success is 0.5.

The linkage discovered so far and the numbers of different linkage tests conducted are indicated in Table 2. As with tests for allelism, those mutants that were extant only a short time or isolated in time or space from other mutants were little tested.

Linkage to the sex alleles should be evidenced by departure from random segregation in workers of low-survival progenies but not of high-survival progenies, when low survival is due to homozygosity at the sex alleles. The mating usually conducted for this purpose is a series of single-drone inseminations in which each queen and each drone are progeny of the same queen (Mackensen, 1958; Woyke, 1973). Each daughter-by-mother mating (the apparent sister-by-"brother" or "sib" mating) gives either a high- or low-survival progeny. Alternatively, a queen may be inseminated with one of her own drones to give a low-survival progeny (Tucker, 1980). With either sort of mating, the cross to be made is between a heterozygous queen and one mutant drone for a recessive gene, or a heterozygous queen and one wild-type drone for a dominant gene. While worker progenies should suffice, a sample of diploid drones confirms worker counts (Woyke, 1973). However, diploid drones require special rearing techniques, and perhaps should be reared only in those cases where linkage to the sex alleles is suspected. So far, no linkage has been established to the sex alleles; those mutants that have been tested and found not linked are indicated in Table 2.

Linkage between a mutant gene and its chromosome's centromere can be measured by segregation in a heterozygous queen's parthenogenetic workers (Tucker, 1958). The frequency of the appropriate observable homozygote, that is, homozygous recessive for a recessive mutant or wild-type for a dominant mutant, gives an estimate of $\frac{1}{4}$ the proportion crossovers between the gene and its centromere and $\frac{1}{2}$ the proportion of recombination. Thus, the proportion of the observable homozygotes times 2 times 100 gives the distance in map units between the gene and its centromere. So far, the distance between gene and centromere has been estimated for only two mutants (Table 2).

5. Subviability

Fewer adult drones than expected of a specific phenotype are encountered with many honey-bee mutants. Accompanying this is usually some evidence of past mortality in the drone brood from which the adults emerged. If the brood had remained exposed to workers of the colony, all that would show would be a spotty distribution of the remaining drone brood, and perhaps no evidence that the mortality was selective. But if a sealed drone brood completes development away from worker bees in an incubator, selective mortality is readily detected. Some of the deaths happen to completely formed, classifiable adults just before eclosion. Those adults that are dead in their cells are preponderantly those of the subviable phenotype where subviability is severe as with short (*sh*) and truncate (*tr*) wings (Laidlaw *et al.*, 1965b). Moreover, continued mortality of adult drones before sexual maturity occurs with some mutant phenotypes. This can happen to a small proportion of most mutant and even wild-type drones, but occurs with a high proportion of *tr* (Laidlaw *et al.*, 1965b) and pink-eyed (*p*) (Cale *et al.*, 1963) drones.

Viability of workers and queens is usually similar to that of drones with the same mutant. Except for the most severely subviable phenotypes, mutant workers surviving past eclosion seem to function well at least as house bees, and mutant queens seem reasonably productive if instrumentally inseminated. An exception is *p*, in which most mutant workers survived past eclosion only to crawl away from the hives a few days later. Furthermore, all *p/p* queens died before becoming sexually mature (Cale *et al.*, 1963).

Subviability with honey-bee mutants is variable. With most mutants, the proportion dying, or the penetrance towards complete lethality, varies from progeny to progeny (Laidlaw *et al.*, 1965b; Tucker, 1980). This sort of variability is found whether the mutant is always subviable, or significantly subviable in some progenies and not in others. This variability merges, at about 7–12% deviation from expected (or 15–25% of wild type), with similar variation conventionally dismissed as sampling error. In one series of related progenies, however, 2.3% deviation from expected (5% less than wild type) was significant when the progenies were combined (Woyke, 1973).

Also, survival of drones to sexual maturity is variable. An example of this is semen production in *tr*, a poorly viable mutant. Of lots of 60 *tr* drones, each from a separate colony, some would produce no semen, some a few microliters, and occasionally a lot would produce nearly half the 24 μl usually expected of 60 wild-type drones (K. W. Tucker and H. H. Laidlaw, unpublished). Even the usual expectation from wild-type drones is only 40% of every drone being equally productive. The same variability in survival and vigor occurs also in other mutant and wild-type drones except that unfavorable variants are usually less frequent.

There is some evidence that suggests that subviability is a feature of honey-bee development even in the absence of visible mutants. Drones of wild-type and "viable" mutants such as the eye mutants ivory (*i*) and cream (*cr*) also can show a certain level of subviability, as evidenced by drones dead in cells. The amount of mortality has been assumed to be low, but it has not been properly documented, and usually has been ignored since it has not distorted genetic ratios. Studies measuring mortality are few, but in one study, the mortality of drone eggs of wild type varied between 3 and 22% from different queens of one stock, and from 10 to 64% in another stock (Harbo, 1981).

A likely genetic explanation of subviability is an array of polygenes, each with a small effect on viability (Tucker, 1980). In drones, several, and perhaps as many as eight, genes would act jointly to effect each death, with fewer polygenes necessary for death with increased stress. Alternative alleles to each of the polygenes could promote slightly better or slightly worse viability. Possibly, some mutants interact with the polygenes only randomly so that mortality would not be selective when compared to wild type; candidates for this situation are *i*, *cr*, and a third eye color mutant, pearl (*pe*). Some other mutants may be linked to one of the polygenes and thus show selective mortality. With linkage, selective mortality should accompany the wild-type allele of the mutant as well as the mutant itself. Linkage could be inferred for the eye colors brick (*bk*) (Mackensen, 1958) and garnet (*g*) (Tucker, 1980). Still other mutants may have an additional semilethal influence on the polygenic series, somehow amplifying the lethality. The mutants eyeless (*e*), *sh*, *tr*, and *p* are always less viable than the wild type. One environmental influence has been found so far. Mutant drones survived less well in worker cells than in drone cells (Woyke, 1973; Chaud-Netto, 1977).

The comparative viability of the mutants is listed in Table 1. The rating scale is based on the comparative incidence of subviability found so far. The apparently complete lethality of drones for the dominant mutations droopy (*D*) and "haarlos" (*H*) has not been studied.

6. Incomplete Penetrance

Incomplete penetrance has been established for wrinkled wings (*wr*) (Laidlaw *et al.*, 1965b). The typical phenotype for *wr* occurs in only a portion of those bees with the *wr/wr* or drones with the *wr* genotype. The wrinkled phenotype is also variable in expression, and the most extreme expressions of the phenotype coincide with higher penetrance. The penetrance of *wr* is increased when accompanied by *bk*, even when the *bk* phenotype is not expressed, as with *wr; bk, s^t* (*bk* is hypostatic to *s^t*). Several other genes (*s^t*, *ch^1, m, ch^c, i, bl*) seem to provide the same level of penetrance as wild type (Laidlaw *et al.*, 1965a,b). With *g*, however, the penetrance of *wr* was lower than with wild type in one progeny (Tucker, 1980).

Two other mutants, neither known very well, may be subject to variable penetrance. One of these, reduced facet number, is much more frequent with *bk* or with *g* than with wild type. The other mutant, cyclops, may represent a dominant gene with very low penetrance. With these two mutants it is not yet known whether the alternate phenotype is lethality during development or viable wild type (Laidlaw and Tucker, 1965a).

The enhancement of penetrance by *bk* has not been studied. The influence of *bk* applies to the development of two different phenotypes: eye shape, and wing venation. The feature in common between these may be the production of structural proteins; it is known that *bk*'s influence on eye color is a partial lesion in the attachment of screening pigment precursors to the protein of pigment granules, as explained in Section II,A,2.

II. THE VISIBLE MUTANTS

A. Eye-Color Mutants

The listing for eye-color mutants in Table 1 is arranged according to the biosynthetic pathway of the screening pigments of the honey-bee eyes. The allelic groupings of mutants fit within this arrangement, indicating that alleles have some chemistry in common. The diverse colors within an allelic series, as well as some characteristics of eye colors alone and in combinations, can be better discussed after first considering the chemistry of eye color.

1. Chemotype of Wild Type

The brownish-black eye color of wild type is due mainly to ommochrome pigments, and the eye colors of the mutants are due to various departures from the usual pattern of ommochrome biosynthesis (Dustmann, 1966, 1969, 1973, 1975a,b; reviewed by Dustmann, 1981). Moreover, the spectral sensitivities of wild type and several eye-color mutants are similar to the absorption spectra of the ommochrome pigments (Tilson *et al.*, 1972). The eyes of wild type and most mutants also contain colorless pteridines, detectable only after separation and fluorescence under ultraviolet light (Dustmann, 1971). The eye pigments of honey bees do not include the colored pteridines, which are present in flies' eyes (Summers *et al.*, 1982).

In wild type, the ommochrome pigments include five different ommins and xanthommatin. These pigments are found in the compound eyes and ocelli. In the compound eyes, the pigments are bound to protein to form spheroid granules 0.2–0.8 μm in diameter located in the primary pigment cells, the retinula cells, and the secondary pigment cells of each ommatidium, where they function as screening pigments to isolate each photosensitive receptor in the compound eye. The ommins occur mostly in the granules of the primary pigment cells and the retinula cells. These granules [Type I of Dustmann (1975a)] appear black–violet to dark brown with light microscopy, and, when viewed in transmission electron microscope (TEM) micrographs, each granule appears to be formed singly within a vesicle. The xanthommatin is found mainly in the granules of the secondary pigment cells. With light microscopy, these granules appear yellow–brown or carmine–red, due to the oxidized or reduced state, respectively, of xanthommatin. In TEM micrographs, these granules appear arranged in short rows, with each row formed inside a common vesicle [Type II of Dustmann (1975a)]. The formation of pigment extends in time from 4 days before the pupal molt (Dustmann, 1966; Woyke, 1964) to 12 days after the eclosion of adult drones (Dustmann, 1975b). During the pupal stage relatively more ommins than xanthommatin are produced, but more xanthommatin than ommins in young adults.

2. Chemotypes of Eye-Color Mutants

The eye color of most mutants differs from wild type in either a complete absence or decreased amounts of ommochromes. The mutants differ from one another not only in relative amounts of ommochromes, but also in amounts of other related chemical compounds, as well as when and where these differences take place. In brief, each eye-color mutant has a chemotype as part of its phenotype.

The chemotype of an eye-color mutant can be best appreciated from a consideration of its position on the tryptophan-to-ommochrome biosynthetic pathway (Fig. 1). This pathway has been established by studies of mutants of a variety of organisms as well as the honey bee. The fifth step of the pathway is probably subject to further revision, with probably separate pathways for the ommins and for xanthommatin.

The eye-color mutants in honey bees can be grouped according to the step in the biosynthetic pathway at which the block occurs, as follows: the snow group at 1; none at 2, at which mutants in insects are apparently rare

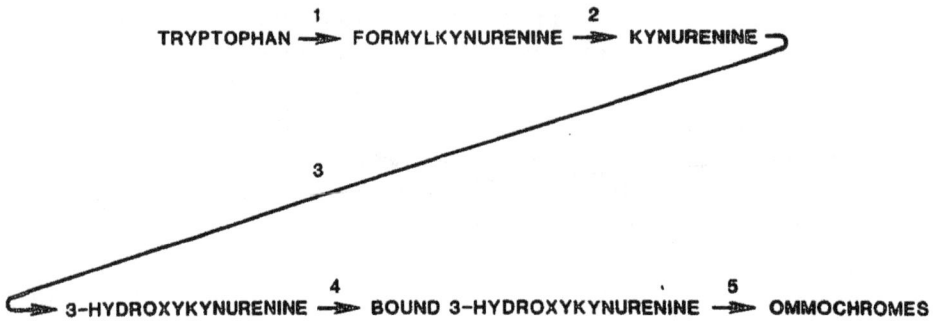

Fig. 1. The tryptophan-to-ommochrome biosynthetic pathway.

(Summers *et al.*, 1982); the ivory group at 3; the cream group at 4; the chartreuse group at 5; and garnet (*g*) beyond 5 (Dustmann, 1969).

The review of the mutant chemotypes in the next five paragraphs is based on Dustmann's work (as cited in Section II,A,1) unless otherwise mentioned.

The snow group includes mutants which have a biochemical block at the step from tryptophan to formylkynurenine (Fig. 1, step 1). Thus, no evidence was found for the activity of the appropriate enzyme, tryptophan oxygenase, for snow (*s*). The substrate immediately preceeding the block, tryptophane, was found present in excess for *s*, tan (*s^t*), and bayer (*by*). Ommochromes are not produced by *s*, but ommochrome granules can be induced in primary and secondary pigment cells following implantation of kynurenine into *s* eyes. The mutant *by* produces some ommochromes on its own but also partially lifts the block for *s*, with which it is neither allelic nor linked. Both *s^t*, and *by* produce their ommochromes after eclosion; those of *s^t* appear in granules of the primary pigment cells only. Laranja (*s^la*), light orange at eclosion, has not been chemotyped, but has been viewed ultrastructurally (Cruz-Landim *et al.*, 1980).

Mutants of the ivory group have a biosynthetic block at the step from kynurenine to 3-hydroxykynurenine (3HOK) (Fig. 1, step 3). The mutant ivory (*i*) was found to have no activity indicating the presence of the appropriate enzyme system, kynurenine 3-hydroxylase. Its allele umber (*i^u*) with an incomplete block at this step showed about 7% of maximum activity for this enzyme at three days after eclosion. The substrate immediately preceeding the block, kynurenine, can be found in excess in newly emerged *i* and *i^u* drones. Ommochromes are not produced with *i*, but can be if 3HOK is implanted into *i* eyes. In *i^u*, ommochrome granules are produced in all three pigment cell types. Rose (*i^ro*) has not been chemotyped.

Mutants of the cream group have a block at the step at which 3HOK binds onto the protein pigment carriers (Fig. 1, step 4). With cream (*cr*), but not with pearl (*pe*), nor with any other genotypes tested so far, the most prevalent pteridines also fail to bind (Dustmann, 1969; Langer *et al.*, 1972). In mutants of the cream group, the 3HOK leaves the eyes and is found concentrated in the rectal fluid of newly emerged bees. With *cr*, this fluid appears yellow, rather than clear, because of the content of 3HOK, which is also high with *pe*, *bk*, and spade (*sp*). The mutants *cr* and *pe*, a tightly linked pair, produce no ommochromes. The red eye-color mutant *sp* produces 25% and *bk* 10% the ommochromes of wild type, with the granules of each mostly in the retinula cells and very sparse in the primary or secondary pigment cells. These mutant eyes have more ommins than xanthommatin, and their xanthommatin is in the red, reduced state, accounting for the bright red eye color of older *bk* and *sp* drones.

Mutants of the chartreuse group have green pigment in their compound eyes due to bound 3HOK (Fig. 1, step 5). If the green color is obscured by red pigment, then green fluorescence under ultraviolet light confirms its presence. The green color is due to 3HOK bound onto protein to form crystalline rodlets (Dustmann, 1975a). Crystal-like structures were also reported for *ch*^{r} by Cruz-Landim *et al.* (1980). As ommochrome granules form [mainly the red, reduced form of xanthommatin (Dustmann, 1969; Neese, 1972)], they form within the green rodlets, mainly in the secondary pigment cells (Dustmann, 1975a). Among the various chartreuse mutants, the amount of bound 3HOK is complementary to the amount of ommochrome produced, with the redder mutants producing ommochromes already as pupae, but with the greenest producing ommochromes only after more than 14 days of adult life. All chartreuse-group mutants become redder with age as some of the bound 3HOK is converted to the reduced form of xanthommatin. Complete conversion of 3HOK can be contrived by enclosing detached heads of *ch*^{B} or *ch*^{2} in pure O_2, whereby the eye color changes from green to deep red in 2 days. The appearance of the eye colors in this allelic series matches decreasing green to decreasing 3HOK, and increasing red to increasing ommochromes, thus from green to red: *ch*^{B}; *ch*^{2}; *ch*; *ch*^{1}, *m*$^+$; *ch*^{1}, *m*; *ch*^{r}, *ch*^{c}. The mutant suppressor *m* partially lifts the block to ommochrome formation for *ch*^{1}, so that the onset of production starts earlier and more ommochrome is produced than with *m*$^+$. The lemon-colored *ch*^{li} (Soares, 1981a) may be expected perhaps to be similar in chemotype to *ch*^{1}, *m*$^+$.

The garnet group includes only garnet (*g*). This mutant is deep red at eclosion, since all of its ommochromes are in the red–violet reduced state. By the third day of adult life, the eye color of *g* has turned black, and both oxidation states of ommochromes are present, just as in wild type.

An unassigned group of eye-color mutants includes those not yet chemotyped or identified as an allele of a chemotyped mutant. Presently unassigned are ocellos claros (*oc*), pink (*p*), and white. These are briefly described and referenced in Table 1.

3. Darkening of Mutant Eye Colors

Mutant eye colors, except for *s*, *i*, *cr*, and *pe*, darken with age. In general, the darker the eye color, the earlier pigmentation begins: for wild type in prepupae, for darker mutants in pupae, and for lighter mutants after eclosion (Dustmann, 1966, 1969; Woyke, 1964). The darkening can continue for as long as three to five weeks or more (Dustmann, 1969; Lee, 1969). This period of time goes byond day 12 of adult life when ommochrome synthesis ceases (Dustmann, 1969). For the various chartreuse mutants, the continued darkening of the eye colors in bees more than 12 days old has been documented as due to oxidation of bound 3HOK to the red, reduced-form xanthommatin (Dustmann, 1969).

The pattern of darkening for the mutant eye color in adults is similar for all mutants. The ventral tips and margins of the eyes darken first and stay darker than the rest of the eye. This pattern is most strikingly noticeable in some variants of *ch^c*. To explain this pattern of darkening, Dustmann (1975a) suggested that the marginal tissues are better oxygenated, promoting increased oxidation of pigments here.

Along with darkening, there is often a qualitative change in the color of the mutant eye. The shift is typically towards increases in red: from chartreuse to yellow to orange to red. Some mutants, such as *ch^B*, darken over nearly the entire range; some over a smaller range, as *s^{la}* over the yellowish-orange to reddish-orange range; some others, as *i^{ro}*, become more intensely red-colored without a qualitative color change. Along with reddening there is often also an addition of black, which changes yellow to olive, orange to brown, reddish orange to maroon, and red to crimson.

The darkening of various chartreuse series mutants parallels increases in ommochromes (Dustmann, 1969). Presumably increased ommochrome and its oxidation to the brown form explain darkening in most mutants. But in addition, Dustmann (1969) found other unidentified red-brown redox pigments for *s^t* and *i^u*.

4. Caste Differences of Eye Colors

The eye color phenotype of different castes may differ or not depending on the mutant (Lee, 1969; Woyke, 1973). They may be nearly the same as with *ch^r*; or they may be similar at eclosion, but with the workers darkening more rapidly than the drones as with *ch^B*, or the workers lighter than drones as with *bk*; or the workers may be brighter at eclosion but darker with age than drones, as with s^t (Lee, 1969). Workers of s^{la} are redder than drones, regardless of whether the drones are haploid or diploid (Woyke, 1973).

5. Border Effects of Eye Colors in Mosaics

The tryptophan-to-ommochrome pathway provides a ready explanation for the border effects seen between eye-color phenotypes in mosaic bees. Also it provides a basis for prediction of the border effects for those combinations not yet seen.

In mosaic-eyed bees, the border between some pairs of phenotypes is sharply distinct, but between certain other pairs of phenotypes it is indistinct and shows a border effect (Rothenbuhler et al., 1953b; Rothenbuhler, 1957). In the border effect of wild type next to i, not only is the border indistinct, but wild-type pigment is stippled into the edge of the i. The same border effect was found with wild type next to s. For ch next to i there was a wild-type border effect into i. Neither i nor wild type produced a border effect into ch, but wild type next to i, ch (combined in the same genome) produced not a wild-type, but a ch border effect. No border effects were observed into cr.

The "diffusable substance" suggested by Rothenbuhler et al. (1953b) as responsible for these border effects can be reasonably expected to be 3HOK and kynurenine. For the wild-type border effect into i and the ch border effect into i, ch, only 3HOK would serve, and for wild type into s either kynurenine or 3HOK should be effective. One should expect strong border effects of cr into i or cr into s, and perhaps a weak border effect of i into s; however, the phenotypes of these mosaics have not yet been reported. The lack of a border effect into ch could reflect a situation in which the infused 3HOK adds to an already adequate supply of 3HOK and induces no increase in pigmentation; however, when the supply of 3HOK is limited to that infusing and resident 3HOK production is blocked, as in i, ch tissue, then a ch border effect could take place. No border effect into cr should be possible, since cr lacks the capacity to bind 3HOK onto pigment granules.

6. Allelic Interaction of Eye Colors

Allelic interactions take place for those loci which have more than one mutant allele. For the eight alleles at the chartreuse locus, nine of 28 possible interactions have been observed, one of three observed for the three alleles at the ivory locus, and two of three observed for the three alleles at the snow locus.

The occurrence of allelic interactions requires a diploid genotype, so these cannot occur in the usual haploid drones. Allelic interactions are usually

observed in workers, but with special preparation can be observed also in
queens and in diploid drones.

For most interacting allelic eye-color mutants seen so far, the heterozygote has a darker phenotype than the lighter homozygote. The exceptions
have been those in which designation of the lighter and the darker homozygote cannot be established by visual criteria alone. In the allelic combinations seen so far, partial dominance, complete dominance, and overdominance have been found.

An eye color intermediate between those of darker and lighter homozygotes has been the most commonly observed phenotype for the heterozygotes. This could be considered a partial dominance of the darker eye color.
Thus, ch^r/ch^1 was reported intermediate between red and chartreuse (Laidlaw et al., 1953); ch^r/ch^{li} as light red and ch^c/ch^{li} as light cherry (Soares,
1981a); ch^r/ch^c as cherry (Laidlaw et al., 1964); and i/i^u as light umber
(Laidlaw and Tucker, 1965b). However, for the dominance relationships
between a greener and less green chartreuse, as the greenish yellow for
ch^B/ch^B to yellowish green for ch^B/ch^{li} to lemon for ch^{li}/ch^{li} (Soares, 1981a),
one must opt for less green as dominant because it probably has more
ommochrome pigment (but less bound 3HOK), than the greener mutant
(see Dustmann, 1969). This designation of lemon as "darker" than greenish
yellow is consistent with the reason for the darker eye color of the redder
mutants.

An eye color for the heterozygote indistinguishable from the darker homozygote occurs in a few allelic interactions. This can be viewed as dominance of the darker allele, even though that allele is recessive to wild type.
Thus, s^{la}/s^t and s^{la}/s^{la} were all classified reddish-eyed and indistinguishable (Soares and Chaud-Netto, 1982). Also, ch^c was reported dominant to
ch^1, but based on the classification of ch^c/ch^1 as cherry, without either
homozygote in the same progeny (Laidlaw et al., 1964).

An eye color for the heterozygote darker than either homozygote is
known for one allelic combination. This represents overdominance for the
darkness of the phenotype. Thus, s/s^t is dark red in distinct contrast to the
white of s/s and the white darkening to tan of s^t/s^t (Laidlaw et al., 1964).
How s and s^t complement one another biochemically in the heterozygote
has not been studied.

7. Nonallelic Interactions of Eye Colors

Nonallelic genes which affect eye color separately interact either by one
precluding the other (epistasis) or by producing a new phenotype. The new
phenotype may represent an enhancement, a blending, or a diminution of
the eye pigmentation.

Data on nonallelic interactions are far from comprehensive. Most such data were observed as adjuncts to tests for linkage, usually with drone progenies. Only about 17% of the possible nonallelic combinations of eye colors have been seen. Even those genes for which the most interactions have been reported, such as *bk, cr*, or *i*, have had no more than 43% of such interactions observed. For seven genes, ch^B, ch^{li}, i^{ro}, *oc, pe, p*, and *sp*, no such interactions have been reported at all.

An enhancement of a mutant eye color to a darker color is known for two combinations. One of these enhancing genes is *m*, and it darkens the eye color of ch^1 from chartreuse to brown in young bees (Laidlaw *et al.*, 1953). This corresponds to increased ommochrome production for ch^1, *m* over ch^1, m^+ in young bees (Dustman, 1969). The gene *m* has little or no effect on other chartreuse alleles, and does not produce a mutant eye color on its own. Another enhancing gene is *by*, which is also an eye-color mutant on its own, white at eclosion and developing a reddish-orange color with age (Laidlaw, cited by Woyke, 1975), and chemotyped in the snow group (Dustmann, 1975b). This gene is neither allelic nor linked to *s*, but behaves as an allele might in the following way. Rather than being hypostatic to *s*, *by* forms an intermediate eye color in combination with *s*. Thus, *by* puts a leak into the *s* block, whereas *m* makes a leaky ch^1 leakier. The molecular explanation of these "mutant suppressors" awaits discovery.

A blending of eye colors to a color intermediate between the two is known for a few nonallelic pairs of genes. An intermediate pink was reported for the interaction of ch^r and *bk* (Laidlaw *et al.*, 1953), and for ch^c and *bk* (Laidlaw *et al.*, 1964). Other interactions that may be intermediates are a buff between *bk* and ch^2 (Laidlaw *et al.*, 1953), and a light buff between s^{la} and *ch* (Woyke, 1973).

There are three nonallelic interactions that seem to be diminutions of eye pigmentation. Each of these is between s^{la} and another eye color: s^{la} with *g* gives white with pink ventral tips, s^{la} with i^u gives a tannish cream (Woyke, 1973), and s^{la} with ch^r gives ivory (Chaud-Netto, 1975). Perhaps these interactions represent complementary exclusions of pigment formation in time or in cell types.

Another nonallelic interaction is the preclusion of one eye color by another, a sort of epistasis. This is the usual interaction for the completely unpigmented mutants in which white precludes any other color. Such has been observed for *cr* precluding *bk*, ch^2, *ch*, ch^c, *g*, and s^t; for *i* precluding *bk*, ch^2, *ch*, ch^r, *g* and s^t; for *s* precluding *bk*, *ch*, ch^1, m^+, and ch^1, *m*; and is a reasonable expectation for *pe* precluding colored mutants (Laidlaw *et al.* 1953, 1964, 1965a; Mackensen, 1958; Rothenbühler *et al.*, 1953a; Tucker, 1980).

In other preclusion interactions, a lighter phenotype precludes a darker one, also called epistasis. Such has been found for s^t precluding bk (Laidlaw et al., 1964), s^{la} precluding bk (Woyke, 1973), and bk or ch^1, m precluding g (Tucker, 1980).

One interaction seemed to indicate a double preclusion, that is a preclusion of each gene by the other. In the interaction of s^t and ch^2, neither typical phenotypes was seen, but rather a white eye color which did not darken to tan with age (Laidlaw et al., 1964). One wonders whether s^t, ch^2 eyes have small amounts of bound 3HOK, and are very faintly chartreuse.

8. Visual Classification of Eye Colors

The classification of eye-color mutants is not always easy. Most such mutants contrast strongly with wild type even when darkened with age, except for g and s/s^t which should be classified as close after eclosion as possible. Light colored mutants, such as by, s^t, and i^u, can be confused with white at eclosion, but not after aging. Similarly colored mutants may be impossible to separate if they occur in the same progeny, but usually such can be avoided by judicious planning. Where greater power of separation is intended, comparison of the bee phenotypes to color classification schemes can be employed. Laidlaw et al. (1953, 1964) and Soares (1981a) used Munsell Color Company Inc. (1976), Woyke (1973) used Ostwald (1933), and another scheme by Maerz and Paul (1930) is available. However, a comprehensive analysis of the classification of bee mutants by these color schemes has not been conducted. In addition to closely similar colors being confused, probably also slight shifts in color values may be expected in contrasting the same eye color in bees with yellow versus black body color.

Two relatively simple observations based on chemical differences enable classification of eye-color mutants into two of the biochemical groups. A green fluorescence under ultraviolet light identifies mutants at the chartreuse locus (Dustmann, 1969). A yellow rather than clear color of the first rectal fluid of newly emerged bees identifies cr (Dustmann, 1975b), and may be useful for the other mutants in the cream group. Other chemical and structural differences are useable, but require expertise in chemistry and ultrastructure.

B. Body-Color Genes

Body color in honey bees is due primarily to the color of the cuticle. This is entirely black in the darkest phenotypes, but in light ones, some of the black is replaced by a brownish or yellowish color in specific areas of the abdomen. Also, in the most yellow phenotypes the scutellum is yellow. These phenotypes are usually described in terms of the color of worker bees because workers are always present. For any degree of yellowness, drone phenotypes are similar in the proportion of yellow to those of workers, except for the sub-Saharan African bee in which the drones are usually darker than the workers. The queens are usually less black than either workers or drones of the same genotype. In a series of worker phenotypes progressing from darker to lighter, the first reduction in the area of black occurs anteriorly on the gaster. In progressively lighter phenotypes, the reduction in the black area spreads posteriorly, to the fifth gastral tergum as an extreme. The sixth gastral tergum is always black. In drones, the first reduction in black is in small lateral spots on each gastral tergum, followed by further reduction of black on the gaster on a generally anterior to posterior gradient. In the progression of intermediate phenotypes in both workers and drones, bands and patches of black are left behind in the yellow areas. Kulinčević (1966) and Woyke (1977) illustrated and categorized this variation, especially for drones; however, they did not include the yellowest possible phenotypes. The yellower phenotypes have yellow also on all gastral sterna but the last. The arrangement of the yellow and black of the cuticle and of the overlying pile give the workers a banded appearance. The three-banded yellow (or "Italian") bees, with yellow bands on gastral terga 1, 2, and 3 of the worker bees, are popular because their queens appear a solid yellow or reddish-yellow color, contrast strongly with the banded worker bees, and hence are easily seen. This is also true of four- or five-banded ("golden") bees. However, bees with some yellow color but less than three-banded have striped queens that are difficult to see among the workers.

There is no single body color that can be considered "wild type," except on a local or regional basis. Even then, these populations are subject to change due to importation of bees from elsewhere. The original distribution of the honey bee was in general, with some exceptions, with the darker body colors common at higher latitudes, higher altitudes, and cooler climates, and with the yellower body colors at lower latitudes, lower altitudes, and warmer climates.

The body color of honey bees seems to be due to the joint action of two independently segregating genic systems: major factor alleles at the black locus modified by a number of polygenes (Woyke, 1977).

The polygenic series provide for a gradually changing series of body color phenotypes within a range specified by whatever black (*bl*) allele or alleles are present. Up to about half the range of possible color types are possible with one allele of *bl*, presumably with maximal heterozygosity of the polygenes (Woyke, 1977). An older view was that the polygenes could be responsible for the entire range of variation from golden to black within a single progeny, without the participation of major factor genes (Roberts and Mackensen, 1951; Kulinčević, 1966). However, the frequency distribution of segregation in drone progenies of F_1 queens from crosses between extreme body colors were distinctly bimodal with a paucity of drones in phenotypes where the old hypothesis should dictate a high frequency. On the other hand, the bimodality could be readily explained as being due to the modification by the polygenic series of two different alleles of major influence (Woyke, 1977). The number of polygenes was estimated at seven by Roberts and Mackensen (1951). Woyke (1977) estimated six or seven polygenes, and postulated that the segregation of three pairs of polygenes is adequate to explain the observed variation. Each polygenic locus would have a yellow or black alternative, with the increments of yellowness or blackness cumulative between polygenes. For haploid drones, the phenotype would depend on the additive expression of the three polygenes. For diploids (workers, queens, and diploid drones), Woyke (1977) suggested a dominance of black over yellow at each of the three polygenic loci, then a sum of the expressions of the three pairs of polygenes as they modify whatever *bl* gene is present.

Of the genes with major influence on body color, two have been identified so far. These are black, *bl*, and *abdome castanho* (brown abdomen), bl^{ac}. That these are allelic to one another and to bl^+ (nonblack and nonbrown) was indicated by Woyke (1977). However, he used these different symbols: y^{bl} for *bl*, y^{ac} for bl^{ac}, and Y for bl^+.

The gene *bl* is most easily identified if it is maintained in "golden" stock (four- to five-banded yellow workers). Then, this gene can be seen to behave as a Mendelian recessive with black phenotypes always distinct from the yellow (Laidlaw and el-Banby, 1962). Presumably here the polygenes approach homozygosity of the yellow alternatives. But with considerable heterozygosity of the polygenes, a bimodal distribution is formed, with the darker yellow and lighter black phenotypes grading into one another in the trough between the two modes. The range of variation for *bl* given by Woyke (1977) included only the darker phenotypes, thus modifiable by the polygenes over a limited range. However, that reported by Laidlaw and el-Banby (1962) included a greater replacement by yellow. Two possible differences could explain this: that either there were different numbers of polygenes for yellow, or that the *bl* genes were not strictly isogenic in each case. In golden stock, *bl* can be studied as a marker and employed in genetic tests along with mutant genes (Laidlaw *et al.*, 1965a; Tucker, 1980).

Another major factor gene for body color is bl^{ac} (Kerr, 1969). This gene is typical of the Africanized bee, and is expressed differently in drones and in workers: the drones dark, the workers yellow. According to Woyke (1977), in drones the haploid bl^{ac} and diploid bl^{ac}/bl^{ac} are black, but usually not as dark as with *bl*, and vary over a wider range by polygenic influence than with *bl*. The bl^{ac}/bl heterozygote diploid drone is black and variable, similar to bl^{ac}/bl^{ac}; hence bl^{ac} is partially dominant to bl. The heterozygous bl^{ac}/bl^+ diploid drones are yellow enough that bl^{ac} seems recessive to bl^+. In workers, the phenotypes of all the foregoing diploid genotypes are yellow. However, Stort (1977) found Africanized workers distinctly darker than "Italian" workers (and much lighter than European brown workers) when the dark and light cuticular areas were quantified. This suggests that even in worker bees bl^{ac} is intermediate between bl^+ and bl in the color range of its phenotypes.

There might be more than one other allele at the bl locus. Those pheno-types with considerable yellow showing have been designated bl^+ (or Y), but there is no assurance of isogenicity, and there might be several bl^+'s. The distribution of pheotypes for the Saharan bee (Kulinčević, 1966) suggests an allele of bl between bl^+ and bl^{ac}. Also, on the dark side of bl there might be other bl alleles for even darker phenotypes. Such a darker allele of bl might provide a better explanation for jet black than that proposed by Tucker and Laidlaw (1967). The finding that Kulinčević (1966) required several differ-ent scales of phenotypic categories to describe different geographic popula-tions also suggests a variety of alleles at the bl locus.

It seems clear that further progress in understanding the genetics of body color in honey bees depends on studying stocks with pedigrees for their bl alleles and for their polygenes.

A tentative explanation of body color in honey bees in terms of biochemi-cal genetics can be attempted, despite the fact that this subject has not been studied in honey bees and, indeed, is poorly understood for any insect (Andersen, 1979). Pivotal to this explanation are the following assumptions about the chemistry of the honey bee's body color and the specific functions of the bl alleles and the polygenes.

The first assumption is that the chemical likely to account for all of the honey bee's body colors is melanin. The colors for melanin solutions of increasing concentrations are yellow, brown, and black (Needham, 1978), precisely those colors of the honey bee's integument. Thus considered, the difference between yellow and black is quantitative, rather than qualitative, as has been supposed previously (e.g., Laidlaw and el-Banby, 1962). Thus, in the descriptions above, black could be considered as most melanin, brown as less melanin, and yellow as the least melanin. This ought to simplify as well as clarify a genetic analysis of body color.

The second assumption assigns functions to the bl alleles and to the polygenes. Inasmuch as bl alleles are dominant in the direction of reduced production of melanin, they should have a controlling function, and limit production of melanin in some way, such as the rate or the duration of the reactions, or the adjustment of reactant concentrations. The polygenes, however, show dominance in the direction of increased production, thus they might be genes for enzyme structures. Melanin is derived from tyrosine metabolism, a pathway necessary also for sclerotization of the cuticle [e.g., Chapman (1982), p. 523]. This pathway provides at least six reaction steps, and thus is adequate to accommodate the polygenes.

Based on the foregoing conjecture, all body colors could be due to melanin produced in the tyrosine-to-melanin biochemical pathway. The different body colors might result from different restraints imposed on this biochemical pathway, so that the least restrained would yield the darkest phenotypes and the most restrained the lightest. Whatever *bl* alleles are present would provide the restraint, and the polygenes would produce melanin at more or less that level, depending on the alternative alleles at the polygene loci. The ability of only certain abdominal and scutellar cuticular areas to show the yellow and brown of melanin's lower concentrations could be due to differential concentrations of substrates or enzymes within the epidermal cells of these areas, and perhaps be related to the patterns of the hemolymph's circulation and to the arrangement of underlying tissues.

C. Body-Color Mutants

Three body-color mutants have been reported in honey bees. Two of these, albino (*a*) and cordovan (*cd*), affect the whole body's cuticle, but the third, yellow face, is restricted to part of the head. Apparently, *cd* and yellow face affect only the cuticular color, but *a* also seems to interfere with the cuticle's sclerotization. However, none of these affect the color of the compound eyes. All three mutants are recessive, but only *cd* has been studied extensively.

Drones that appear white but have apparently normal compound eyes and wings were called albino, *a*, by F. Ruttner [cited by Woyke (1975)]. The cuticle of *a* drones is neither pigmented nor sclerotized, which might be due to a biochemical lesion in tyrosine metabolism. The *a* drones were fewer than expected, indicating semilethality for *a*. Moreover, no mature sperm was obtained from these drones. (This condition is other than the so-called albino drones of older bee literature, which were drones with mutant eye colors.)

The mutant cordovan was discovered and named by W. J. Nolan (unpublished), shown to be a recessive mutant by Mackensen (1951), and symbolized *cd* by Laidlaw *et al.* (1953). In the *cd* phenotype, the black cuticular areas of wild type are replaced by various shades of brown. This difference is distinguishable 1 day before eclosion, but not earlier (Woyke, 1964). Cordovan makes yellow bees seem yellower, and in golden stock a possible confusion of *cd* and *cd⁺* is best resolved by viewing the bees' antennae where there is no overlying pile to obscure the cordovan or black cuticular color. The *cd* phenotype in dark bees in bronze, and sometimes reflects incident light with a purple iridescence. Whether the difference at the molecular level between *cd* and *cd*⁺ is qualitative or quantitative remains to be discovered. The cordovan gene is probably widely distributed; for many years it was included in the genome of every commercially available Starline Hybrid queen.

A partial replacement of the black cuticular color with yellow on the bee's head was called yellow face by W. J. Nolan (unpublished). Nolan (1937) reported that the yellow cuticular area included the bees' clypeus, genae, labrum, and sometimes the antennal bases.

D. Body-Hair Mutants

Three different mutants for loss of body hair have been found in honey bees. Two conditions have been discovered: that hair is not produced, or that hair is produced but easily lost. Relationships between these mutants have not been studied.

The dominant mutant schwarzsüchtig, *S*, was described by Dreher (1940). The phenotype differs from wild type in being distinctly less hairy. Few of the very long, plumose hairs (overhair) and substantially less of the very shortest hairs (tomentum, fuzz, or underhair) develop. Other types of hairs, bristles, and spines are similar to wild type. The missing hairs are not formed at all (in contrast to being formed and then lost), but the sockets of the missing hairs appear normal. Otherwise, these mutant workers take slightly longer to develop than wild-type, the mutant workers and drones are subviable, and the mutant drones often have crumpled wings.

The recessive mutant, hairless, *h*, was described by Mackensen (1958). With *h*, the hair is formed, but appears matted on newly emerged bees, and soon rubs off. Some hair is left in protected places, but exposed hair is lost all over the body. Witherell (1972a) reported that on drones the tuft of hair on the last abdominal tergite remains unaffected. The long hairs fringing the corbiculae on worker bees are lost, but the spines of the metatarsal combs and tibial takes remain. Witherell (1972a) reported also that the spindle hair is missing from the centers of the corbiculae. Laidlaw [cited by Witherell (1972a)] found that *h/h* queens tend to lose tarsal segments easily.

The dominant mutant *haarlos*, *H*, was reported by F. Ruttner [cited by Woyke (1975)]. On workers, the long plumose overhair is missing from all parts of the body, but the tomentum, or underhair, is present. (Whether the overhair is not formed or it lost posteclosion was not stated.) The bristles of mutant workers' pollen combs are brittle. Drones of this mutant are not viable.

E. Eye-Shape Mutants

The shape of the compound eyes can differ from wild type. Of the five such phenotypes described so far, two have been a loss of facets, one other a reduction in the number of facets of the compound eyes, and a fourth the complete loss of one of the two eyes. The fifth mutant is a developmental error in the placement of the compound eyes. How such morphological anomalies relate to genes at the molecular level is not well understood (Fristrom, 1970).

The absence of some of the eye facets is characteristics of a phenotype called reduced facet number (Laidlaw and Tucker, 1965a). The number of facets missing is variable, from a few to almost all. A frequent variant of this phenotype is the one in which the facets are missing in the midlateral part of the eye, giving the eye a dumbbell shape. This phenotype occurs almost exclusively in drones and rarely in queens. Its inheritance is not clear, but is possibly polygenic. Its occurrence is increased by the presence of brick or garnet eye colors.

The apparent absence of all facets of the compound eye have been reported for two recessive mutants, eyeless, e (Laidlaw and Tucker, 1965a), and facetless, f (Michailoff, 1930; Schasskolsky, 1935). In both of these mutants, the compound eye facets appear to be missing; however, the interfacetal hairs are present. Crumpled remnants of ommatidia were observed for e by Dustmann (1975b). The missing facets make the head look small for both e and f, less rounded and more triangular. This also accentuates the appearance of the hair around the head to make the drone's head appear more hairy, even though the same amount of hair is present around wild-type heads. There have been two pleiotropic conditions reported for e. One of these is lack of sperm and an apparent lack of testes (Laidlaw and Tucker, 1965a). Similarly, f drones produced no sperm and were found to have very small testes (Michailoff, 1930). The mutant e was found to be semilethal in hemizygous drones (Laidlaw and Tucker, 1965a; Tucker, 1980), and the mutant f could also have been semilethal to account for the significantly fewer f drones than expected in the progenies of laying workers (Schasskolsky, 1935).

The entire absence of one of the two compound eyes of drones was reported by Dustmann (1975b). This was clearly not cyclops, but "drones in which one compound eye is normally developed, but the other completely absent." The genetic nature of "einäugig" is not known.

An error in the placement of the compound eyes is what seems to be wrong with cyclops (Lotmar, 1936). In most extreme phenotypes, the compound eyes are joined into one eye smoothly rounded over the top of the head, and may also project a bit backward medially. This is the most common phenotype. Less frequently, there are variations between this extreme and wild type. The ocelli are still to be found externally in the less drastically changed phenotypes, but in the common, extreme phenotype, the ocelli are displaced inside the head capsule, producing the appearance of a lack of ocelli. The brain is also changed in these phenotypes by variable fusion of the optic lobes and various displacements of the mushroom bodies. However, the changes in the brain relate to the external phenotype somewhat inexactly (Lotmar, 1936). The genetic basis for cyclops is not clear. That it occurs not randomly, but at low frequency within certain progenies, and that it was reported to occur in succeeding generations [Dittrich, cited by Lotmar (1936); Ruttner, cited by Woyke (1975)] suggest some means of inheritance. If it is genic, it would seem to be dominant because it appears in workers and queens as well as drones (Lotmar, 1936; Laidlaw and Tucker, 1965a; Ruttner, cited by Woyke, 1975; Goebel, 1981). Several attempts to propagate cyclops have failed (Kerr and Laidlaw, 1956; Laidlaw and Tucker, 1965a), but Ruttner [cited by Woyke (1975)] reported a low level of cyclops for five successive generations.

F. Wing Mutants

Of the six wing mutants of the honey bee, five affect wing morphology, and one changes wing comportment. The morphological mutants are wrinkled (*wr*), diminutive (*di*), short (*sh*), truncate (*tr*), and rudimental wing (*Rw*). The mutant changing the way the wings are held is droopy (*D*). *Rw* and *D* are dominant, and the others recessive; *wr*, *sh*, *di*, and *tr* are nonallelic, and *Rw* and *D* were not tested to the other wing mutants.

The phenotype of *wr* varies from wild type to one in which the distal part of the bee's fore wing appears wrinkled (Laidlaw *et al.*, 1965b). The wrinkling is in the region of the wing cells first R_s, second R_s, and second M, where there is also anomalous venation with several extra crossveins and spurs. The extent of wrinkling and anomalous venation is variable, from none at all, through very minor, to distortion so bad that flight is crippled. Thus, *wr* has variable and incomplete penetrance as discussed in Section I,B,6.

The wing mutant *di* has small wings (Witherell and Laidlaw, 1977). At a glance, *di* bees resemble *sh*, but *di* wings are closer to wild type in shape, have wild-type venation, but only about 63% the area of wild type. The wings of *di*/*di* queens curve up over the body convexly on their long axis. This arrangement of the wings may be the reason the *di*/*di* queens cannot fly. Workers and drones can fly, and the sound of their wings in flight is pitched distinctly higher than that of wild type.

In the mutant *sh*, the wings are short and do not reach the tip of the abdomen (Laidlaw *et al.*, 1965b). The wings of drones and workers are also narrower than wild type. The venation of the forewing of *sh* is abnormal in that the wing cells second R_s and second M are open distally, because of a partial or complete loss of crossveins. The wings of *sh/sh* queens are tiny stumps (K. W. Tucker and H. H. Laidlaw, unpublished). Short-winged bees cannot fly. Short-winged drones take a day longer to develop to eclosion than wild type. Short phenotypes are subviable (see Section I,B,5).

Drones and workers of *tr* have much reduced wings (Laidlaw *et al.*, 1965b). The typical phenotype shows wings that are squared off at their tips and cover only the anterior part of the abdomen. The wing veins are all present, but their pattern is compacted. In some progenies the typical phenotype is seen infrequently, and many drones have only scaly stumps for wings. So far only workers and drones of this phenotype have been produced. The drones are poorly viable (see Section I,B,5). Only young *tr/tr* workers were seen, and they stayed in the warm part of the broodnest; a few attempts to raise *tr/tr* queens failed (K. W. Tucker and H. H. Laidlaw, unpublished).

Bees with *Rw* have straplike vestiges of their wings (Hachinohe and Onishi, 1953). The wing rudiments extend no further than the anterior part of the abdomen and are sometimes curved upward distally. The phenotype is similar for workers, queens, and drones. The lethality associated with *Rw* may be a linked lethal, as Hachinohe and Onishi (1953) conclude, or may be subviability measured at 45 and 85% of wild type for two small progenies. Some *Rw* drones produced functional sperm.

The mutant *D* affects the positioning of the wings (Rothenbuhler *et al.*, 1953a). The worker bees carry their wings spread from the longitudinal body axis and drooping from their points of attachment. The wings appear otherwise normal, but D/D^+ bees cannot fly or right themselves when turned over on their backs on a flat surface. The D/D^+ bees are often distended with feces, since they cannot manage cleansing flights. No *D* drones were seen, so presumably *D* is lethal in the hemizygous condition. Only workers and queens were known to be droopy, and only heterozygotes have been possible.

G. A Sting Mutant

One mutant has been described for the sting. This phenotype has been designated ferrão aberto, open stings, and split sting by Soares (1977,

1979a, 1980, 1981b) and symbolized *sps* by Soares (1981b) and presumably not as *s/s* (Soares, 1979b).

The phenotype of split sting is the detachment of one or both of the lancets from the stylet. When the detachment is complete, this disables the sting and the bee's use of it as a defensive weapon. Otherwise workers with split sting performed normally both as house bees and as foragers.

The frequency of split sting is variable. The incidence when first discovered was 3.5%, and has been increased to as high as 62% by selection. Split sting was always found more frequently in queens than in workers of the same progeny, and queens reared from 3-day-old larvae had fewer split stings than those reared from 2-day-old larvae (Soares, 1979a). Temperature changes during pupal and prepupal development gave fewer split stings when lower, and more split stings when higher than usual (Soares, 1979a).

The mode of inheritance of split sting is not clear. The options reported so far are (1) genes with incomplete penetrance (Soares, 1977), (2) dominant genes with incomplete penetrance (Soares, 1979a), and (3) nonnuclear because it is less easily transmitted by drones than queens (Soares, 1980). Split sting could have been induced by radiation of an ancestor (Soares, 1981b); however, the same phenotype also has occurred spontaneously (Simpson, cited by Soares, 1981b).

III. FUNCTIONAL CAPABILITIES OF MUTANTS

The worker bees of most mutant phenotypes seem to be competent house bees. The only evidence for asserting so is that they are present in the colony at least until they attempt orientation flight. That they actually conduct duties of a house bee or that they are physiologically equipped to do so is in need of study. However, workers of a few mutants may not be competent house bees. Only quite young workers of *p/p* (Cale *et al.*, 1963) and of *tr/tr* (K. W. Tucker and H. H. Laidlaw, unpublished) have been seen, and the *p/p* workers crawled away from their colonies and became lost a few days after eclosion (Cale *et al.*, 1963).

As field bees, those with mutant phenotypes range from fully competent to incapable, depending on the mutant. Cordovan, *cd/cd*, workers seem to behave as well as wild type, and often have been employed in experiments when a contrasting body color to wild type is required. However, a detailed comparison of the capabilities of cordovan and wild type has not been conducted. One would expect the capabilities of eye-color mutants to diminish with decreasing darkness of eye colors, but the comparisons have not been made. For chartreuse, the one eye color that has been studied,

ch/ch, worker bees were at a disadvantage (Neese, 1968, 1969). The eye color of ch/ch darkens from chartreuse to red by usual flight age, but even so, most ch/ch workers became lost. A small proportion were trained to feeding stations, and these responded less well to contrasts in darkness, flew more slowly, danced more slowly, and quit foraging earlier in the evening than wild type, ch/ch^+, sisters. Even trained ch/ch workers were severely disoriented by the brightest sunlight. Workers of diminutive, di/di, with 63% the area of wild-type wings, gathered smaller pollen loads and performed a higher proportion of round dances than wild type (Witherell and Laidlaw, 1977). Also at a disadvantage in collecting pollen were hairless, h/h, workers, probably because their corbiculae are defective (Witherell, 1972a). Similarly, haarlos, H/H^+, workers gather only small pollen loads, probably because the metatarsal combs break off [Ruttner, cited by Woyke (1975)]. Workers homozygous for the white-eyed mutants and for sh cannot fly, but crawl and flutter away from the colony to become lost at the time when their wild-type sisters perform orientation flights (K. W. Tucker, unpublished).

As queens, mutant phenotypes have a full range of competency, depending on the mutant. Cordovan queens are fully capable of all behavior including natural mating (Taber and Wendel, 1958). Of 20 red-eyed queens, ch^r/ch^r and ch^r/ch^B, 15 mated naturally (Chaud-Netto and Stort, 1980). It seems likely that other dark-eyed mutants could mate naturally as well, but they have not been tested. Even though di/di workers can fly, di/di queens cannot, presumably because of the convex curvature of their wings (Witherell and Laidlaw, 1977). Almost all mutants, including those incapable of natural mating, function satisfactorily as instrumentally inseminated queens. However, h/h queens are prone to supersedure (Witherell, 1972a). Pink, p/p, queens died before reaching sexual maturity (Cale et al., 1963), and tr/tr larvae have not yet been reared into adult queens (Tucker and Laidlaw, unpublished).

In drones, the full range of competency also exists, depending on the mutant. Cordovan presumably manages to mate with queens (Taber and Wendel, 1958). Probably garnet drones also mate, because garnet has been found accidentally added to wild type by natural mating (K. W. Tucker, unpublished). Possibly also ch^r, ch^c, h, and di drones may be capable of mating, but evidence is lacking. Among eye-color mutants, Witherell (1972b) found that the ability of drones to avoid becoming lost is much better for the darker eye colors than the lighter: that h and di drones avoided loss better than wr drones. Mature drones of most mutants go to the entrance and seem to try to fly, even if incapable, at the time wild-type drones are flying (K. W. Tucker, unpublished). Only the eyeless type has difficulty reaching the entrance (Witherell, 1972b). Almost all mutants mature semen, most in quantities comparable to wild type. However, usually proportionately few *p* and *tr* drones produce semen, and *e* is sterile.

REFERENCES

Andersen, S. O. (1979). Biochemistry of insect cuticle. *Annu. Rev. Entomol.* 24, 29–61.

Cale, G. H., Jr., Gowen, J. W., and Carlile, W. R. (1963). Pink—an eye-color and viability gene in honey bees. *J. Hered.* 54, 163–166.

Chapman, R. F. (1982). "The Insects." 3rd. ed., Harvard Univ. Press, Cambridge, Mass.

Chaud-Netto, J. (1975). Estudos de segregação com o gene laranja em *Apis mellifera* (Hymenoptera, Apidae). *Ciên. Cult. (São Paulo)* 27, 1227–1230.

Chaud-Netto, J. (1977). Ocelos claros (*oc*): a new mutation in *Apis mellifera adansonii*. *Ciên. Cult. (São Paulo)* 29, 316–318.

Chaud-Netto, J. (1979). Estudos biologicos com o mutante "ocelos claros" em *Apis mellifera*. In "Apicultura em Clima Quente" (Apimondia, ed.), pp. 147–148. Apimondia, Bucharest.

Chaud-Netto, J., and Stort, A. C. (1980). Successful matings of chartreuse-eyed queens of *Apis mellifera* (Hymenoptera, Apidae). *Ciên. Cult. (São Paulo)* 32, 1542–1543.

Chaud-Netto, J., Kerr, W. E., and Bezerra, M. A. F. (1983). Mutation in honeybees 2. Average rate of mutation based on seven genes for eye colour. *J. Apic. Res.* 22, 17–21.

Cruz-Landim, C., Chaud-Netto, J., and Gonçalves, L. S (1980). Comparative studies on pigment granules and pigment distribution in the compound eyes of wild type and mutant worker honeybees: an ultrastructural analysis. *Rev. Bras. Genét.* 3, 115–122.

Dreher, K. (1940). Eine neue, dominant wirkende Mutation "schwarzsüchtig" (S) bei der Honigbiene (*Apis mellifera* L.) *Zool. Anz.* 129, 65–80.

Dustmann, J. H. (1966). Über Pigmentuntersuchungen an den Augen der Honigbiene *Apis mellifica*. *Naturwissenschaften* 53, 208.

Dustmann, J. H. (1969). Eine chemische Analyse der Augenfarbmutanten von *Apis mellifica*. *J. Insect Physiol.* 15, 2225–2238.

Dustmann, J. H. (1971). Pteridine bei der Honigbiene *Apis mellifica* Isolierung neuer Lumazine. *Hoppe-Seyler's Z. Physiol. Chem.* 352, 1599–1600.

Dustmann, J. H. (1973). Kynurenin-3-hydroxylase in den Augen der Honigbiene *Apis mellifica*. *Hoppe-Seyler's Z. Physiol. Chem.* 354, 1068–1072.

Dustmann, J. H. (1975a). Die Pigmentgranula im Komplexauge der Honigbiene *Apis mellifica* bei Wildtyp und verschiedenen Augenfarbmutanten. *Cytobiologie* 11, 133–152.

Dustmann, J. H. (1975b). Quantitative Untersuchungen zur Tryptophan→Ommochrom-Reactionskette bei Wildtyp und Mutanten der Honigbiene *Apis mellifera*. *Insect Biochem.* 5, 429–445.

Dustmann, J. H. (1981). Farbmutationen der Bienenaugen. *Imkerfreund* 36, 152–153.

Fristrom, J. W. (1970). The developmental biology of *Drosophila*. *Annu. Rev. Gent.* 4, 325–346.

Goebel, R. L. (1981). Cyclops mutation in the worker honey bee. *Am. Bee J.* 121, 590.

Hachinohe, Y., and Onishi, N. (1953). On the new mutation "rudimental wing" in the honeybee (*Apis mellifica* L.). *Bulletin of the National Institute of Agricultural Sciences, Series G (Animal Husbandry) Number 7, October, 1953 (Chiba, Japan)*, pp. 139–145. (Japanese, English summary and tables).

Harbo, J. R. (1981). Viability of honey bee eggs from progeny of frozen spermatozoa. *Ann. Entomol. Soc. Am.* 74, 482–486.

Kerr, W. E. (1969). Genética e melhoramento de Abelhas. In "Melhoramento e Genética." (Univ. São Paulo, ed.), pp. 263–297. Edit. Univ. São Paulo e Edicões Melhoramentos, São Paulo.

Kerr, W. E., and Laidlaw, H. H. (1956). General genetics of bees. Adv. Genet. 8, 109–153.

Kerr, W. E., Chaud-Netto, J., and Silva, Â. T. (1980). Mutação em abelhas. I. Taxa de mutação reversa para genes que afetam a cor dos olhos nas abelhas. Rev. Bras. Genét. 3, 275–284.

Kulinčević, J. (1966). Die Phänoanalyse der Abdominal-tergite von Drohnen verschiedener geographischer Rassen der Honigbiene (Apis mellifica L.) und ihrer Kreuzungen. Ann. Abeille 9, 115–152.

Laidlaw, H. H., and el-Banby, M. A. (1962). Inhibition of yellow body color in the honey bee Apis mellifera L. J. Hered. 53, 171–173.

Laidlaw, H. H., and Tucker, K. W. (1965a). Three mutant eye shapes in honey bees. J. Hered. 56, 190–192.

Laidlaw, H. H., and Tucker, K. W. (1965b). Umber eye-color — a new mutant in honey bees. J. Hered. 56, 271–272.

Laidlaw, H. H., Green, M. M., and Kerr, W. E. (1953). Genetics of several eye color mutants in the honey bee. J. Hered. 44, 246–250.

Laidlaw, H. H., el-Banby, M. A., and Tucker, K. W. (1964). Five new eye-color mutants in the honey bee. J. Hered. 55, 207–210.

Laidlaw, H. H., el-Banby, M. A., and Tucker, K. W. (1965a). Further linkage studies in the honey bee. J. Hered. 56, 39–41.

Laidlaw, H. H., el-Banby, M. A., and Tucker, K. W. (1965b). Three wing mutants of the honey bee. J. Hered. 56, 84–88.

Langer, H., Schneider, L., und Täuber, U. (1972). Vergleichende Untersuchungen am Augen-pigmentsystem von Wildform und weissäugigen Mutanten der Honigbiene (Apis mellifica). Cytobiologie 6, 427–438.

Lee, G. L. (1969). The effect of gene dosage on variability in the honeybee. 1. The eye colour mutants. J. Apic. Res. 8, 75–78.

Lotmar, R. (1936). Anatomische Untersuchungen an Cyklopen-Bienen. Rev. Suisse Zool. 43, 51–72.

Mackensen, O. (1951). Viability and sex determination in the honey bee (Apis mellifera L.). Genetics 36, 500–509.

Mackensen, O. (1958). Linkage studies in the honey bee. J. Hered. 49, 99–102.

Maerz, A., and Paul, M. R. (1930). "A Dictionary of Color." McGraw-Hill, New York.

Michailoff, A. S. (1930). On two mutations in the honey bee. Zhurnal Opytnaya Paseka 5, 291–293. (Russian, English summary)

Michailoff, A. S. (1931). Über die Vererbung der Weissäugigkeit bei der Honigbiene (Apis mellifera). Z. Indukt. Abstammungs Vererbungsl. 59, 190–202.

Munsell Color Company Inc. (1976). "Munsell Book of Color; Glossy Finished Collection." Munsell Color, Macbeth Division of KollMorgen Corp., Baltimore, Md.

Needham, A. E. (1978). Insect biochromes: their chemistry and role. In "Biochemistry of Insects." (M. Rockstein, ed.), pp. 233–305. Academic Press, New York.

Neese, V. (1968). Zur optischen Orientierung der Augenmutante "Chartreuse" von Apis melli-fica L. I Teil. Z. Vergl. Physiol. 60, 41–62.

Neese, V. (1969). Zum Verhalten der Augenmutante chartreuse von Apis mellifica L. Proc. Congr. 6th IUSSI (Bern), pp. 195–200.

Neese, V. (1972). Die Altersabhängigkeit des Ommochromgehalts in Komplexauge der Bienen. Ein quantitativer Vergleich zwischen Wildtyp und der Augenmutante Chartreuse. J. Insect Physiol. 18, 229–236.

Nolan, W. J. (1937). Bee breeding. United States Department of Agriculture Yearbook, pp. 1396–1418.

Ostwald, W. (1933). "Color Science, Part 2." (Translated by J. Scott Taylor), Winsor and Newton, London.

Roberts, W. C., and Mackensen, O. (1951). Breeding improved honey bees. II. Heredity and variation. *Am. Bee J. 91*, 328–330.

Rothenbuhler, W. C. (1957). Diploid male tissue as new evidence on sex determination in honey bees. *J. Hered. 48*, 160–168.

Rothenbuhler, W. C., Gowen, J. W., and Park, O. W. (1953a). Allelic and linkage relationships of five mutant genes in honey bees (*Apis mellifera* L.). *J. Hered. 44*, 251–253.

Rothenbuhler, W. C., Gowen, J. W., and Park, O. W. (1953b). Action of eye-color mutations as revealed in mosaic honey bees (*Apis mellifera* L.). *Genetics 38*, 686.

Rothenbuhler, W. C., Kulinčević, J. M., and Kerr, W. E. (1968). Bee genetics. *Annu. Rev. Genet. 2*, 413–438.

Schasskolsky, D. W. (1935). Genetische Analyse der Biene nach der Nachkommenschaft der Arbeitsbienen. *Arch. Bienenkd. 16*, 1–8.

Soares, A. E. E. (1977). Ferrão aberto: "uma nova mutação em *Apis mellifera* L." *Anais do 4° Congresso Brasileiro de Apicultura*, pp. 127–132.

Soares, A. E. E. (1979a). Fatôres que alteram a expressividade da mutação "ferrão aberto" em *Apis mellifera* L. *In* "Apicultura em Clima Quente" (Apimondia, ed.), pp. 144–145. Apimondia, Bucharest.

Soares, A. E. E. (1979b). Banco de mutantes da Faculdade de Medicina de Ribeirão Preto, USP.: Marcadores bioquimicos e as novas mutações obtidas em *Apis mellifera* no Brasil. *In* "Apicultura em Clima Quente" (Apimondia, ed.), pp. 149–151. Apimondia, Bucharest.

Soares, A. E. E. (1980). A mutation preventing bees from stinging. *Am. Bee J. 120*, 834–835.

Soares, A. E. E. (1981a). Chartreuse-limão: first eye mutation induced by gamma radiation with ⁶⁰Co in the honeybee. *J. Apic. Res. 20*, 137–139.

Soares, A. E. E. (1981b). Split-sting: a new honeybee character. *J. Apic. Res. 20*, 140–142.

Soares, A. E. E., and Chaud-Netto, J. (1982). Laranja—an additional eye color gene in the snow series of *Apis mellifera* L. *J. Hered. 73*, 80.

Stort, A. C. (1977). Análise da coloração abdominal em três linhagens puras e em híbridos de *Apis mellifera*. *Anais do 4° Congresso Brasileiro de Apicultura*, pp. 155–165.

Summers, K. M., Howells, A. J., and Pyliotis, N. A. (1982). Biology of eye pigmentation in insects. *Adv. Insect Physiol. 16*, 119–166.

Taber, S., III., and Wendel, J. (1958). Concerning the number of times queen bees mate. *J. Econ. Entomol. 51*, 786–789.

Tilson, R. L., Judson, C. L., and Strong, F. E. (1972). Electrophysiological responses in mutant-eyed drones of *Apis mellifera* to selected wavelengths of light. *J. Insect Physiol. 18*, 2441–2447.

Tucker, K. W. (1958). Automictic parthenogenesis in the honey bee. *Genetics 43*, 299–316.

Tucker, K. W. (1980). Tests for linkage and other interactions in the honey bee. *J. Hered. 71*, 452–454.

Tucker, K. W., and Laidlaw, H. H. (1965). Compound inseminations to abbreviate tests for allelism in honey bee queens. *J. Hered. 56*, 127–130.

Tucker, K. W., and Laidlaw, H. H. (1967). Honey bee drones with jet black bodies. *J. Hered. 58*, 184–185.

Tucker, K. W., and Laidlaw, H. H. (1968). More allelism tests in honey bees by compound inseminations. *J. Hered. 59*, 145–146.

Witherell, P. C. (1972a). Can hairless honey-bees collect pollen? *Am. Bee J. 112*, 129, 131.

Witherell, P. C. (1972b). Flight activity and natural mortality of normal and mutant drone honeybees. *J. Apic. Res.* 11, 65–75.

Witherell, P. C., and Laidlaw, H. H. (1977). Behavior of the honey bee (*Apis mellifera* L.) mutant, diminutive-wing. *Hilgardia* 45, 1–30.

Woyke, J. (1964). Genetic characters in immature stages of wild and mutant honeybees. *J. Apic. Res.* 3, 91–98.

Woyke, J. (1973). Laranja: A new honey bee mutation. *J Hered.* 64, 227–230.

Woyke, J. (1975). Genetische Aspecte der künstlichen Besamung. *In* "Die instrumentelle Besamung der Bienenkönigin" (F. Ruttner, ed.), pp. 93–106. Apimondia, Bucharest.

Woyke, J. (1977). The heredity of color patterns in the honey bee. *In* "International Symposium on Genetics, Selection, and Reproduction of the Honey Bee" (Apimondia, ed.), pp. 49–55. Apimondia, Bucharest.

CHAPTER 4

Sex Determination

JERZY WOYKE

I. HISTORY

A. Discovery of Parthenogenesis

Dzierzon (1845) published a brief note stating that worker bees and queens develop from fertilized eggs while drones develop from unfertilized ones. Several decades later (1898, 1899) he described how he made this discovery. In 1835, 10 years before the first publication, two colonies swarmed in his apiary: one with an old queen and one with a virgin. The swarms united and the queens fought; the old queen was killed and one wing of the virgin was damaged. As a result, the virgin could not make a mating flight and she remained uninseminated. She deposited eggs from which only drones developed. After Dzierzon killed her, he found waterlike fluid in her spermatheca rather than the white semen found in normal inseminated queens.

Dzierzon, who usually kept black bees, imported a yellow *Apis mellifera ligustica* colony from Italy in 1853. One year later (Dzierzon, 1854), he reported that yellow queens mated to black drones produced exclusively yellow drones while black queens mated to yellow drones produced exclusively black drones. In both cases, the workers were hybrids which showed biparental influences on their coloration. These findings supported his hypothesis.

Siebold (1852) confirmed that Dzierzon's hypothesis was consistent with the anatomy of reproductive organs of queen bees. Leuckart (1855) found micropyles in eggs collected from both worker and drone cells. Thus, the eggs were not completely different as was thought by some opponents of the hypothesis. But Leuckart only found spermatozoa on two eggs collected from worker cells and none on those from drone cells. These findings were insufficient to fully support Dzierzon's hypothesis. Siebold (1856a,b) found spermatozoa in 30 of 52 eggs collected from worker cells and none in 27 collected from drone cells. These findings confirmed Dzierzon's hypothesis.

B. Classical Investigations

Since the development of drones from unfertilized eggs had been proven, further discussion concerned sex determination and whether or not drones can also arise from fertilized eggs. Many fantastic hypotheses were launched. The best known is that of Dickel (1897), who questioned the value of Siebold's findings. He kept bee colonies on drone combs exclusively and found drone brood as well as worker brood in drone cells (Dickel, 1898a). After he reciprocally transferred eggs between drone and worker cells, he reared brood and adults which corresponded to the kind of cells in which they were reared (Dickel, 1898b). He concluded that unfertilized eggs are only laid by uniseminated queens and laying workers. Drones develop from those eggs. However, all eggs laid by inseminated queens are fertilized regardless of the type of cell receiving them. According to Dickel, the sex of developing larvae depends on the kind of saliva which is deposited by the nurse bees over the eggs. He also affirmed that the sex of larvae can be changed depending upon the food which the larvae receive.

Thus, further studies were necessary. Cytological investigations (Blochman, 1889; Paulcke, 1899; Petrunkewitsch, 1901) showed that drones develop from unfertilized eggs and also when they are laid by inseminated queens. Nachtsheim (1913) found 16 chromosomes in unfertilized eggs and 32 in fertilized ones. Meves (1907) found 16 chromosomes in spermatogonia. No reduction of the number of chromosomes occurs during spermatogenesis. As a result, the spermatids, which later develop into spermatozoa, also contain 16 chromosomes. Adam (1912) and Bresslau (1906) described the anatomy and function of the spermatheca which allow the queen to store semen and have control over which eggs will be fertilized. Zander et al. (1916) found that the sex of just hatched larvae is determined and can not be changed by different foods. Thus, it seemed that the questions of parthenogenesis and sex determination in the honey bee were fully answered. But the future showed that this was not true.

II. UNUSUAL BEES

A. Females from Unfertilized Eggs

Honey-bee females (queens or workers) regularly develop from fertilized eggs. However, sometimes they also develop from unfertilized eggs. Jack (1916) and Onions (1912) reported that workers developed from unfertilized eggs deposited by *Apis mellifera capensis* egg-laying workers. Several years later, this phenomenon was investigated in *A. m. capensis* by Anderson (1963), Ruttner (1977), and Woyke (1979b).

Prior to these studies, Mackensen (1943) found that 23% of virgin *A. m. caucasica* queens, 9% of virgin *A. m. ligustica,* and 57% of golden virgin queens from an *A. m. ligustica* stock produced unfertilized eggs, of which less than 1% developed into workers. Woyke (1962a) reported that 2% of the unfertilized eggs laid by *A. m. mellifera* egg-laying workers developed into workers. Triasko (1965) found 6.4% of unfertilized eggs laid by queens developing into workers after selection for this characteristic. According to Tucker (1958), most such workers occurred among the queen's initial brood where they constituted from less than 1% to as much as 7% of the total progeny. Hachinohe *et al.* (1958) induced impaternal females by an injection of 0.02 cm² of 0.01% colchicine in the abdomen of a honey-bee queen.

The origin of females from unfertilized eggs is explained by the occurrence of the fusion of two haploid egg nuclei. Ordinarily, four haploid nuclei result from the meiotic egg division. According to Tucker (1958), two internal nuclei unite and form a diploid cleavage nucleus. Ruttner (1977) and Verma and Ruttner (1981), studying *A. m. capensis,* confirmed that the two central nuclei unite and form the diploid cleavage nucleus.

B. Mosaics and Gynandromorphs

Artificial insemination, together with mutant genes used as markers, has made it possible to determine the origin of unusual bees depicted in Fig. 1.

Several unusual bees originating from unfertilized eggs have been described. Unfertilized binucleate eggs result in mosaic males. The union of two pronuclei in unfertilized eggs produces parthenogenetic females. Two haploid egg pronuclei can divide at least once before union. Next two haploid nuclei unite and form a diploid cleavage nucleus, which develops into female tissues; the other haploid nuclei develop into mosaic male tissues. Thus, a gynandromorph is formed.

Also, several unusual bees originating from inseminated eggs have been described. Diploid males develop when the sex alleles are homozygous. If

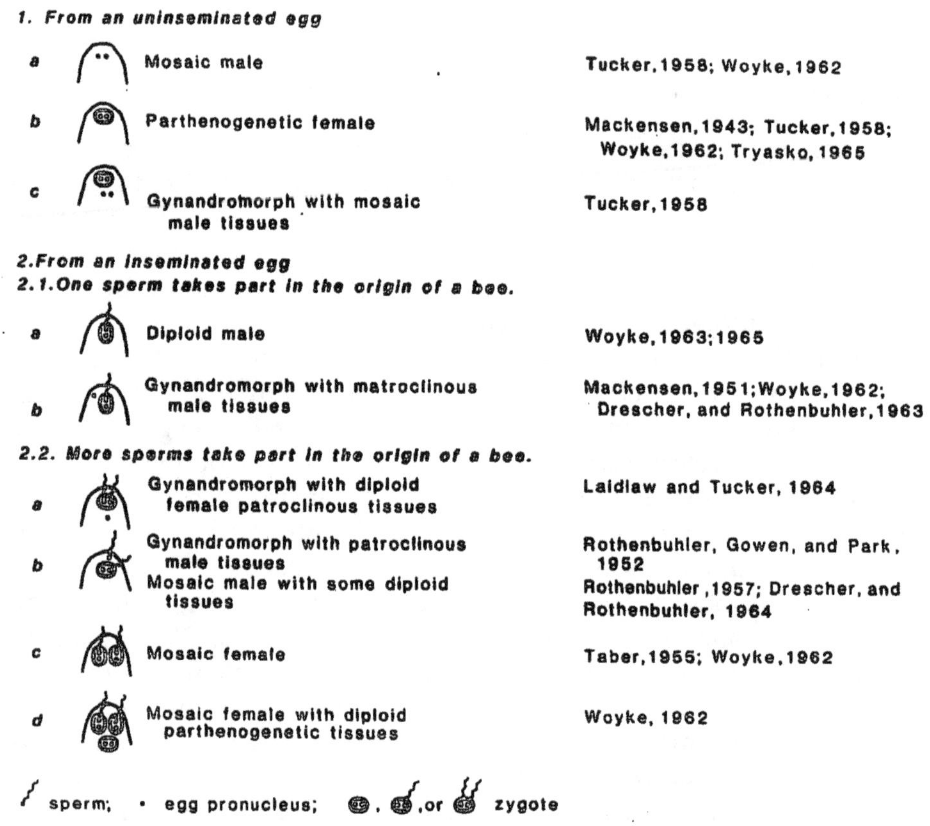

Fig. 1. Origins of unusual honey bees. [Redrawn by Sandra Kleinpeter from J. Woyke (1969c), with permission. Copyright 1969 by Apimondia.]

only one of the two pronuclei in a binucleate egg is fertilized, the developing gynandromorph possesses male tissue of matroclinous origin and female tissue of biparental origin.

Polyspermy also may be the cause of unusual bees. Two spermatozoa in an egg can unite while the egg pronucleus remains unfertilized. This results in a gynandromorph in which diploid female tissues originate from two fathers without the participation of the mother. Polyspermy of an egg with one nucleus results in a fertilization of the nucleus. However, sometimes one or more accessory spermatozoa do not degenerate but develop into male tissues. Thus, a different type of gynandromorph develops. Female diploid tissue originates from both parents, and haploid male tissue develops from a spermatozoon or spermatozoa. A mosaic female can result from poly-spermy in a binucleate egg if spermatozoa of different fathers unite with the two pronuclei. An even more complex bee results when two haploid egg pronuclei first divide, then two of them are fertilized by different spermatozoa, and the remaining two egg pronuclei unite. This results in a female that is partly parthenogenic and partly has two fathers.

C. Diploid Drones

1. Discovery and Characteristics

In the 1950s some queens were found producing scattered brood after they were inseminated with their siblings' semen (Hachinoe and Jimbu, 1958; Mackensen, 1951). Daily counts of brood from these queens showed that the greatest loss of brood occurred at hatching (between the third and fourth day after egg laying). It was concluded that the eggs were homozygous at the X locus, did not hatch, and were removed by the workers. Eleven (Mackensen, 1955) and twelve (Laidlaw et al., 1956) lethal alleles responsible for this phenomenon were reported.

Mosaic drones, in which some patches of eye tissue were diploid, were found by Rothenbuhler (1957) and Drescher and Rothenbuhler (1964). They suggested that this diploid tissue was able to survive only by virtue of its association with normally viable haploid tissue in the drones. Throughout the 1960s Woyke conducted a series of investigations which showed that the eggs that were homozygous at the X locus were viable and that the larvae were removed by worker bees within 6 hours after they hatched (Woyke, 1962b). Histological studies showed that these larvae were drones (Woyke, 1963b,c,d), and cytological studies showed them to be diploid and developed from inseminated and fertilized eggs (Woyke and Knytel, 1966; Woyke et al., 1966). Thus, diploid males could develop from fertilized eggs.

Diploid drone larvae do not survive in nature, and investigations revealed that the workers eat them within a few hours of their hatching (Woyke, 1963e). It was thought that diploid-drone larvae might be less viable than the haploid larvae or the normal worker larvae but this proved to be incorrect. In fact, their viability may be even higher than that of normal haploid drones (Woyke, 1963f, 1965b). Another possibility was that these larvae were destroyed because they were in worker cells. However, when larvae from sibling-mated queens (which produced up to 50% diploid drones) were transferred to drone cells, the diploid drones were still eaten while female larvae were reared normally (Woyke, 1965d).

In other studies, diploid-drone larvae were placed into cells already containing normal worker larvae (Woyke, 1967). Both larvae were eaten whether or not the diploid-drone larvae were alive or dead when they were added to the cells. However, if the diploid-drone larvae were washed in a lipid solvent before being placed into cells containing normal worker larvae, the worker larvae were not destroyed. This indicated that diploid-drone larvae might be producing a pheromone which induced worker bees to eat them. This pheromone, which Woyke called "cannibalism substance," was produced mainly during the first day of larval life and to a lesser extent during the second day (Woyke, 1967).

A technique for rearing diploid drones in the colony was then developed (Woyke, 1969a,b). Eggs were hatched in the incubator, the larvae were placed on royal jelly and held in the incubator for 2–3 days after hatching, and the larvae were then transferred to a colony where they were reared normally. Several thousand diploid drones were reared in this way. The biparental origin of these drones was demonstrated (Woyke, 1965a; Woyke and Adamska, 1972). Mutant queens were mated to wild-type drones or to drones carrying a mutant gene at a different locus from that carried by the queen. Normal haploid drones produced from eggs laid in drone cells by these queens differed phenotypically from those reared from eggs laid in worker cells. The participation of the father in the origin of these drones was shown quite clearly. Androgenesis was also excluded, and the biparental origin of the diploid drones was proved.

Diploid drones (Fig. 2) are heavier and larger than haploids (Woyke, 1977, 1978a,b). They have smaller testes, which maybe only one-tenth of the testicular volume of haploids, and contain less and shorter testicular tubules than haploids (Woyke, 1973a).

Sex determination in *A. cerana* was investigated by Woyke (1979a) in India. He found that the mechanism of sex determination was very similar to that of *A. mellifera*. However, the diploid drone larvae of *A. cerana* are eaten by worker bees within 2–3 days after hatching (Woyke, 1980a), which is later than in *A. mellifera*. Hoshiba *et al.* (1981) confirmed the production of diploid drone larvae in *A. cerana*. They found 16 chromosomes in the haploid and 32 in diploid *A. cerana* drones.

2. Spermatogenesis and Spermatozoa

Spermatogenesis in diploid drones is very similar to that in the haploid ones (Milne, Chapter 8). No pairing or reduction of chromosomes occurs in diploid drones during spermatogenesis. In haploid drones, 16 chromosomes were found in all stages of spermatogenesis; in diploid drones, 32 chromosomes occurred in all stages, including anaphase II (Hoshiba, 1979; Woyke and Skowronek, 1974). The DNA content of the heads of spermatozoa from diploid drones was twice as high as that in those from the haploids (Woyke, 1975).

2n♂

Fig. 2. A diploid drone. [Reproduced from J. Woyke (1977), by permission. Copyright 1977 by The International Bee Research Association.]

Diploid spermatozoa are longer (312 μm) than haploid spermatozoa (242 μm). The heads of diploid spermatozoa are also longer (7.4 μm) than those of the haploids (4.8 μm). Thus, the entire diploid spermatozoa and their heads alone are, respectively, 130 and 155% longer than these lengths for haploid spermatozoa (Woyke, 1983).

The ultrastructure of diploid spermatozoa is very similar to that of haploids (Woyke, 1984a). The diploid tail also contains one axoneme and two mitochondrial derivatives of unequal length (Fig. 3). The fibrils in the axoneme are arranged according to the $9 + 9 + 2$ pattern—single periferals, double inners, and two centrals. Many multiple spermatozoa are found among single diploid spermatozoa (Fig. 3). Double spermatozoa contain two axonemes and four mitochondrial derivatives; triple spermatozoa contain three axonemes and six mitochondrial derivatives. The diameters of the tails of single, double, and triple diploid spermatozoa are, respectively, 115, 190, and 220% the diameter of single haploid spermatozoa. However, the diameter of each axoneme in multiple spermatozoa is equal to that of the one axoneme in single diploid spermatozoa. The arrangement of utrastructures inside the tail supports the conclusion that the spermiogenesis (sperm formation) of multiple spermatozoa occurs in unpartitioned spermatocytes and spermatids.

Thus, diploid drones produce diploid spermatozoa. It might be possible to produce triploid bees if a queen were inseminated with the semen of diploid drones.

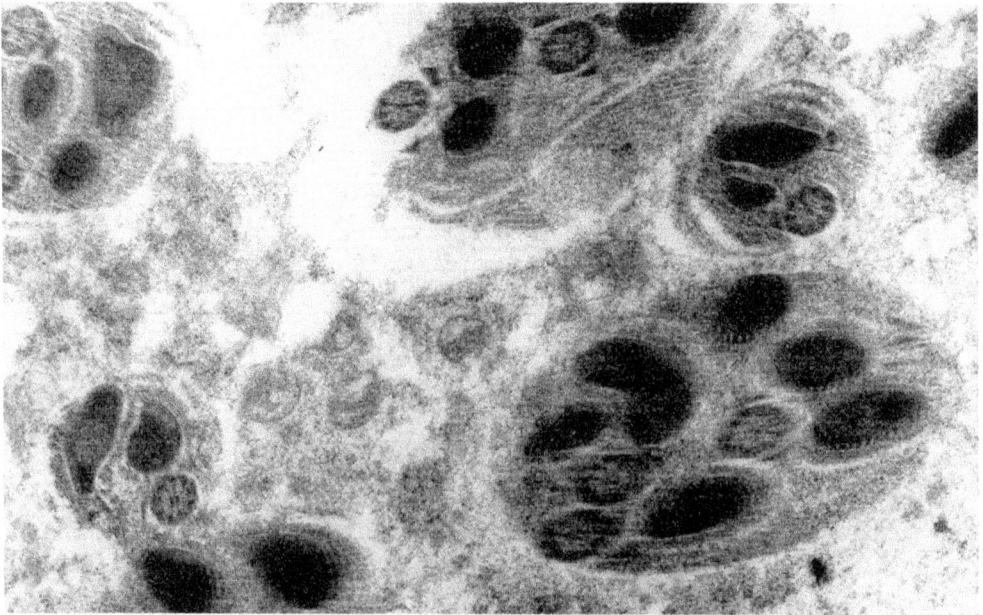

Fig. 3. Cross section through the tails of single, double, and triple diploid spermatozoa.

III. SEX DETERMINATION HYPOTHESES

A. Sex Chromosome

Sanderson and Hall (1948) noticed some individual characters of honey-bee chromosomes. They found one hooked chromosome in haploid drone cells. Manning (1949) considered this to be the sex male chromosome and the remaining 15 as autosomes (1 X + 15 A). According to Manning, during spermatogenesis all the autosomes duplicate but the sex chromosome does not and is eliminated from the spermatocyte with the second polar body. As a result, only 15 autosomes remain in the spermatide and consequently in the spermatozoon. Further on, Manning found 31 chromosomes in oogonia before meiosis (1 X + 30 A) and 16 in mature unfertilized eggs. Of those, one was the hooked sex chromosome and the remaining were the 15 autosomes.

According to Manning's hypothesis, the action of the one sex chromosome is stronger than that of the 15 autosomes (1 X > 15 A). Thus, a male develops from an unfertilized egg. During fertilization the 15 autosomes of the spermatozoon unite with the 16 chromosomes of the egg: a zygote is formed with one sex chromosome and 30 autosomes. Here, the action of 30 autosomes is stronger than that of the one sex chromosome (1 X < 30 A). As a result, a female develops from a fertilized egg.

Sanderson and Hall (1951a,b) criticized those results and confirmed their earlier findings that 16 and 32 chromosomes are present in reproductive bee cells. Ris and Kerr (1952), using the Feulgen reaction which stains only chromatin, could not find the elimination of the sex chromosome during spermatogenesis. Rothenbuhler *et al.* (1952), studying gynandromorphs, also could not confirm Manning's hypothesis. Development of additional spermatozoa in an egg resulted in male tissues rather than the female ones predicted by the hypothesis. Hoshiba (1984), using the C-banding technique, could not find the sex chromosome in the honey bee.

B. Multiple Sex Alleles

After Whiting (1940, 1943) published the hypothesis of multiple alleles determining sex in *Habrobracon*, Mackensen (1951) tested it in the honey bee. The ratio of females reared agreed with the hypothesis, but lethals were reported instead of males. Rothenbuhler (1957) supported the hypothesis after finding patches of diploid male tissues in the eyes of mosaic drones. However, the hypothesis was not proven until Woyke (1962b, 1963b,c,d) showed that the larvae which hatched from fertilized eggs, homozygous at the sex locus, were diploid drones.

According to this hypothesis there are multiple alleles (X^a, X^b, X^c, etc.) at

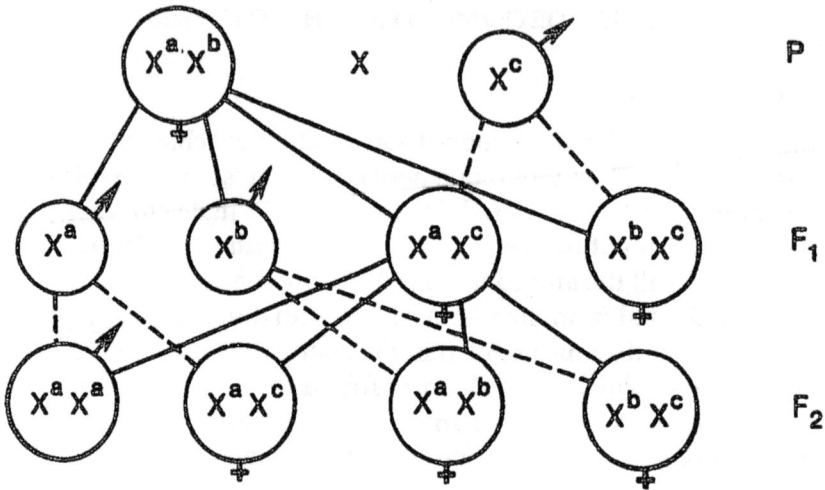

Fig. 4. Sex determination in the honey bee. Females (♀) are diploids which are heterozygous at the sex-locus X. Males are hemizygous (♂, small) or are homozygous (♂, large). An F_1 from a three-allelic parental mating is shown. In the F_2, a two-allelic mating results in half of the progeny being homozygous diploid drones and the other half heterozygous females. Mating with both drones results in $\frac{1}{4}$ diploid drones and $\frac{3}{4}$ females.

the sex locus (X). Males develop from unfertilized hemizygotes or from fertilized X homozygotes. Females develop from X heterozygotes. Fig. 4 presents examples: A queen inseminated by one drone ($X^aX^b \times X^c$) produces haploid males X^a and X^b from unfertilized eggs and heterozygote females X^aX^c and X^bX^c from fertilized eggs. No diploid males are produced from this mating. If the virgin queen X^aX^c is inseminated by her brother X^a, she produces two types of fertilized eggs: heterozygotes (X^aX^c) from which females develop, and homozygotes (X^aX^a) from which diploid males develop. Half of the diploid progeny from this mating is female and the other half is male.

No X locus allele has known pleotropic effects or is linked to other genes which make it easily identifiable. Identification must be infered from the diploid sex ratios of selected matings. Other possible F_1 matings, for example $X^aX^c \times X^b$ (which cannot *a priori* be distinguished from $X^aX^c \times X^a$), would produce female diploids exclusively.

Thus, when sister virgin queens are each mated to one of their brothers, half of the inseminated queens produce normal brood with females developing from all fertilized eggs. The other half of the queens produce fertilized eggs, half of which develop into females and half of which develop into diploid males. The diploid males are eaten by the workers, which results in scattered brood.

The number of sex alleles in various honey-bee populations was estimated by several researchers using different methods. Mackensen (1955) found 11 sex alleles; Laidlaw et al. (1956) found 12 ± 3.5, and Adams et al. (1977) found 18.9. Woyke (1976), in a study of the bee sanctuary on Kangaroo Island, only found six alleles. He presented a simple formula to calculate the number of sex alleles (N) in a population whose mean production rate of homozygous eggs $(a = 0.0 - 0.5)$ is determined by

$$N = 1/a \tag{1}$$

Thus, the number of sex alleles in a population is the reciprocal of the average rate of homozygotes produced.

C. Genic Balance

1. Two Versions

Da Cunha and Kerr (1957) presented an initial version of a genic-balance hypothesis concerning sex determination. The sex determination by arrhenotokus parthenogenesis may be explained by assuming a series of genes (m) for maleness having no additive effects and a series of genes (f) for femaleness having additive effects. The effect of m would be the same in the hemizygotes m as in the homozygotes mm and may be represented as M in both. The effects of f, since they are additive, would be F in the hemizygotes f and $2F$ in the homozygotes ff. Sex would then be determined by the relation $2F > M > F$. In haploid individuals, $M > F$ and males result; in diploids, $2F > M$ and females result. This hypothesis is not contrary to the multiple sex alleles hypothesis. It is only more general. The sex alleles X can be considered major f genes, which lose their additivity in homozygous conditions. Kerr (1974a,b) developed this hypothesis further in some recent papers.

Woyke (1965c, 1973a) showed that diploid drones have much smaller testes than do haploid drones. Also, Kerr and Nielsen (1967) reported the external rudiments of reproductive organs to be less developed in diploid-drone larvae than in haploid larvae. Thus, the diploid males possess femalelike chracteristics. This led Kerr (1974a,b) to the conclusion that some f genes are also additive when X alleles are homozygous. However, Woyke (1973a) found the external parts of reproductive organs of adult diploid drones to be better developed than those of haploids. Also, the femalelike characteristics of diploid drones were not confirmed by studies on bees with the sex-limited ac gene, in which females are yellow and haploid drones are black. Diploid drones reared by Woyke (1971a,b) were also black. Thus, they showed male characteristics.

Detailed morphological investigations on diploid drones showed that many of their body parts were larger than those of the haploids (Woyke, 1971a,b). This was confirmed by Chaud-Netto (1975) and Chaud-Netto and Moura-Duarte (1975), who concluded that diploid drones are meta-males from estimates of Mahalanobis generalized distances. This indicated that the m genes are slightly additive. Chaud-Netto (1975) presented an explanation of the genic-balance hypothesis on which Kerr has been working for several years:

$$\text{haploid drones are } m_1 + m_2 > X_1 + f$$
$$\text{diploid drones are } 2m_1 + 2m_2 > X_1X_1 + ff$$
$$\text{diploid workers are } X_1X_2 + ff > 2m_1 + 2m_2$$

Thus, the genic-balance hypothesis for sex determination in Hymenoptera originally proposed by Cunha and Kerr (1957) was modified to a second version that involves additivity of m genes. According to Kerr (1974a,b), sex determination is seen as the result of a balance between nonadditive (or slightly additive) male-determining genes and totally additive (or almost totally additive) female-determining genes. The X alleles are interpreted as major female-determining genes that have lost the additive property, unless they are heterozygous.

2. Sexuality of Diploid Drones as a Test

Diploid drones differ in many characteristics from the haploid ones. Therefore, questions arose concerning their sexuality. Are diploid drones normal males, supermales, or intersexes? Is the sexuality of diploid drones the result of genic balance between feminine ($X + f$) and masculine ($m_1 + m_2$) genes, or of another mechanism? According to the gene-balance hypothesis, the sexuality of different body parts of diploid drones should be nearly constant. To be consistent with the first version of the hypothesis (additivity of feminine genes), diploid drones should show intersex characteristics; to be consistent with the last version (additivity of masculine genes), they should show supermale characteristics.

Using almost 38,000 measurements or counts, Woyke (1977, 1978a,b) compared body parts of diploid drones (D) with those of haploid drones (H) and workers and queens (F). Except for testes and bristles on wings, body parts were larger in diploid males than in haploids males (D > H). Head and thorax parts, normally larger in haploid males than in females, showed supermale characteristics in diploid drones (D > H > F). Mouth parts and some of the abdomen parts normally larger in females than in haploid males showed intersex characteristics in diploid drones (F > D > H). Characteristics ranging from supermale through male, intersex, intercaste, and female to superfemale were found among body parts of diploid drones.

It was impossible to conclude that the dimensions of different body parts of diploid drones result from a new genic balance between masculine and feminine genes, because no consistent sexuality was found among the characters. Neither version of the genic-balance hypothesis proved acceptable.

A new hypothesis, consistent with the data (Woyke, 1980c), suggested that the sizes of diploid drone body parts result from higher ploidy numbers and higher gene dosage. The size relation of different body parts in diploid and haploid drones should be reasonably constant (D > H). However, the sexuality (relation to both haploid drones and females) should shift. This hypothesis enables a forecast to be made of the sexuality of body parts not yet investigated. Higher poliploidy will cause larger body parts in diploid than in haploid drones. Diploid drone body parts will show supermale characteristics if the parts are larger in the haploid drones than they are in females (D > H > F), and will show an inclination toward female characteristics if the parts are smaller in haploid drones than they are in females (F > D > H).

High endopolyploidization of the somatic tissues occurs in all four castes of honey bees. The rate of polyploidization is different in these castes. The average degree of polyploidization in adults has been measured as the volume of cell nuclei and their DNA content (Woyke and Król-Paluch, 1985). These values, as proportions to the values of haploid drones, are 0.95 for workers, 1.20 for diploid drones, and 1.45 for queens. The polyploidization found in body tissues of all four castes of bees agrees quite well with the sizes of body parts measured.

D. Multiple Heterozygous Loci

Crozier (1971) suggested that sex in haplo–diploid species is determined by a number of loci with several alleles. Females are heterozygous at one or more loci, while males are homozygous at all loci. In the honey bee, one sex locus perhaps predominates and heterozygosity at the other loci has little effect. The remaining effect of heterozygosity at other loci might, however, explain the feminization of diploid drones (Kerr and Nielsen, 1967).

Woyke (1974) tested whether heterozygosity of sex-limited or multiple-sex loci has any influence on the degree of feminization of diploid drones as expressed by the size of their testes. He reared diploid drones having 8, 25, and 27% homozygosity. According to the hypothesis, the drones with less homozygosity should be more femalelike (small testes) and the drones with more homozygosity should be more malelike (larger testes). However, in a failure to support the hypothesis, the decrease in homozygosity did not decrease the size of diploid testes.

IV. UNEQUAL ADDITIVITY AMONG SEX ALLELES

Differences were found between individuals in the specific nature of their diploid-male organs [ommatidia of mosaic drones: Dresher and Rothenbuhler (1964), Rothenbuhler (1957); rudiments of the copulatory apparatus of diploid-drone larvae: Kerr and Nielsen (1967)]. This led (Kerr, 1967) to the conclusion that different X alleles have lost their additive ability to different degrees. Thus, for example, X^aX^a individuals could be more feminine and X^dX^d individuals more masculine. Significant differences in the size of testes found in diploid drones having different origins (Woyke, 1973a) were consistent with this hypothesis, since different additive abilities of X alleles could have caused the observed differences.

Woyke (1974) tested the hypothesis in crosses and backcrosses of *A. m. ligustica* and *A. m. scutellata*. *Apis m. scutellata* diploid drones had relatively large testes and *A. m. ligustica* diploid drones had relatively small testes. The nature of the crosses enabled Woyke to determine the X allele composition of the offspring. Testes size did not follow the allele composition of progeny. Thus, the unequal additivity hypothesis was not confirmed.

V. GENE DOSAGE

Van Pelt (1966), working with *Habrobracon*, found two alleles at the garnet locus (eye-color mutation) which exhibited darker pigmentation in the haploid male than in either the diploid male or female. The darker haploid male coloration was much closer to that of the normal black-eyed wild strain. This phenomenon is explained by gene dosage (the effect expressed by the two alleles in the diploid is stronger than that of the single allele in the haploid). Thus far, in 10 other cases of gene dosage reported in *Habrobracon*, diploids always have a more extreme mutant phenotype.

Lee (1969) investigated gene dosage effects in the honey bee. The expressions of five eye color mutants of the honey bee were found to differ in an irregular fashion between workers and drones.

Woyke (1973b) found *laranja* (an orange eye mutation) in Africanized bees in Brazil. The eyes of drones were light orange and those of the workers were dark orange-red. Diploid drones had light orange eyes similar to those of the haploids. Thus, gene compensation was not detected. The differences found between haploid males and diploid females did not result from different gene dosages but rather from differences in anatomical structure.

The ability to study haploid and diploid honey bees of the same sex is an excellent tool for the study of gene dosage. Certain differences between haploid and diploid drones may be found to be the result of gene dosage. However, care must be taken in such studies since there is an unequal poliploid development of body cells during the larval life of haploid and diploid drones. The final number of chromosomes in the adults must govern gene-dosage interpretations rather than the germinal number.

VI. SEX ALLELES, MATING, AND BROOD SURVIVAL

A. Participation of Multiple Drones in Offspring Production

Since a queen is mated by several drones, consequences of this phenomenon must be considered in sex allele genetics. One drone produces about 10–11 million spermatozoa. A queen returning from a mating flight has on average 80 million spermatozoa in her oviducts, of which only 5 million enter the spermatheca. The question arises, do all drones contribute the same proportion of spermatozoa to the queen's spermatheca? Woyke (1963g) inseminated queens having mutant alleles for several loci, with semen from several different mutant and normal drones. He concluded that each of the drones fathered similar proportions of offspring. However, investigations on the number of spermatozoa entering the spermatheca of once or twice inseminated queens (Mackensen, 1964; Woyke, 1960) revealed that more spermatozoa enter a queen's spermatheca from drones participating in the first mating episode. This may cause differences in the percentage of offspring of different fathers produced by individual queens, but has no significance when an average of several queens is considered.

B. Controlled Mating

In order to obtain highly productive bee colonies, attention should be given to the consequences of the sex alleles. Maximum productivity is obtained from colonies with a high survival rate of brood. The eggs of a queen mated with an unrelated drone ($X^aX^b \times X^c$) are all heterozygous and have 100% survival. The queen with X^aX^b sex alleles mated with an X^a drone will lay two kinds of fertilized eggs: X^aX^a and X^aX^b. Worker bees will emerge

TABLE 1. Combinations of Sex Alleles When the Same Queen Is Used in the Drone-Producing Colony for 2 Consecutive Years

Year	Queen of the mother colony	Queen of the drone-producing colony	Daughter queens	Survival rate of the brood from each daughter queen (%)
1	$X^aX^b \times X^c, X^d$	$X^eX^f \times \cdots$	$X^aX^c \times X^e, X^f; X^aX^d \times X^e, X^f; X^bX^c \times X^e, X^f; X^bX^d \times X^e, X^f$	100
2	$X^aX^c \times X^e, X^f$	$X^eX^f \times \cdots$	$X^aX^e \times X^e, X^f; X^aX^f \times X^e, X^f; X^cX^e \times X^e, X^f; X^cX^f \times X^e, X^f$	75

from the heterozygous eggs, and diploid males, which are eaten by worker bees, from the homozygous eggs. Consequently, only 50% of the hatched eggs will develop into worker bees. The brood of such a queen will be scattered and the colony will be too weak to yield a good honey crop.

Naturally mated or commercially acceptable instrumentally inseminated queens receive semen from several drones. When a virgin is mated to both types of her brothers ($X^aX^c \times X^a, X^b$), she will lay four kinds of fertilized eggs: X^aX^a, X^aX^c, X^aX^b, and X^bX^c (Fig. 4). Worker bees will develop from 75% of the heterozygous eggs if both types of brothers are equally represented in the progeny. Such matings are frequent in a mating station where only one drone-producing colony exists and the virgin queens are related to the queen producing the males. Resulting colonies will never have strong populations.

A number of problems attend controlled mating (Woyke, 1963a, 1972). The first arises when all males are produced by one queen alone and two generations of virgins mate with them (Table 1). The mother queen is inseminated by unrelated drones. The queen of the drone-producing colony has sex alleles which differ from those of the mother queen. (The sex alleles of drones which inseminated the queen of the drone-producing colony are of no concern to us, since drones develop by parthenogenesis.) In the first generation, all young queens mate with drones having different X alleles and will produce brood with 100% survival. All of them have the chance to become high productive queens. However, when the second generation of queens is reared and mated with drones from the original drone-producing colony, only 75% of the eggs will be heterozygous and the survival of brood will be 75%. Queen breeders should not keep the same queen in the drone-producing colony for two seasons.

Problems also arise when the drone-producing colony is requeened every year with a daughter queen from the previous season (Table 2). In the first generation the brood produced by all the daughter queens will have 100% survival. In the following generation results will depend on the chance

TABLE 2. Combinations of Sex Alleles When the Queen of the Drone-Producing Colony Is Replaced with a Daughter

Year combination	Queen of the mother colony	Queen of the drone-producing colony	Daughter queens	Survival rate of brood from specific colonies[a]
1, 1	$X^aX^b \times X^c, X^d$	$X^eX^f \times X \cdots$	$X^aX^c \times X^e, X^f; X^aX^d \times X^e, X^f; X^bX^c \times X^e, X^f; X^bX^d \times X^e, X^f$	All, 100%
2, 1	$X^aX^c \times X^e, X^f$	$X^aX^c \times X \cdots$	$X^aX^e \times X^a, X^c; X^aX^f \times X^a, X^c; X^cX^e \times X^a, X^c; X^cX^f \times X^a, X^c$	All, 75%
2, 2	$X^aX^d \times X^e, X^f$	$X^cX^c \times X \cdots$	$X^aX^e \times X^a, X^c; X^aX^f \times X^a, X^c; X^dX^e \times X^a, X^c; X^dX^f \times X^a, X^c$	One half, 75%; one half, 100%
2, 3	$X^bX^c \times X^e, X^f$	$X^eX^c \times X \cdots$	$X^bX^e \times X^a, X^c; X^bX^f \times X^a, X^c; X^cX^e \times X^a, X^c; X^cX^f \times X^a, X^c$	One half, 75%; one half, 100%
2, 4	$X^bX^d \times X^e, X^f$	$X^cX^e \times X \cdots$	$X^bX^e \times X^a, X^c; X^bX^f \times X^a, X^c; X^dX^e \times X^a, X^c; X^dX^f \times X^a, X^c$	All, 100%
3, 1	$X^bX^e \times X^a, X^c$	$X^bX^e \times X \cdots$	$X^bX^e \times X^b, X^c; X^bX^c \times X^b, X^c; X^eX^e \times X^b, X^c; X^eX^c \times X^b, X^c$	All, 75%
3, 2	$X^bX^f \times X^a, X^c$	$X^bX^f \times X \cdots$	$X^bX^a \times X^b, X^c; X^bX^c \times X^b, X^c; X^fX^a \times X^b, X^c; X^fX^c \times X^b, X^c$	One half, 75%; one half, 100%
3, 3	$X^dX^e \times X^a, X^c$	$X^bX^f \times X \cdots$	$X^dX^a \times X^b, X^c; X^dX^c \times X^b, X^c; X^eX^a \times X^b, X^c; X^eX^c \times X^b, X^c$	One half, 75%; one half, 100%
3, 4	$X^dX^f \times X^a, X^c$	$X^bX^b \times X \cdots$	$X^dX^a \times X^b, X^c; X^dX^c \times X^b, X^c; X^fX^a \times X^b, X^c; X^fX^c \times X^b, X^c$	All, 100%

[a] In the second and third years, if all combinations have equal representation, half the daughters produce brood with 75% survival and half produce brood with 100% survival.

selection of parents from among the inseminated queens. The beekeeper has one chance in four of selecting queens having the same sex alleles (Table 2, second year, combination 1). Here, all young queens will produce brood with 75% survival. In two cases of four, the beekeeper is likely to select parents having one similar allele (second year, combinations 2 and 3). Half of the young queens will produce brood with 75% survival and the other half will produce brood with 100% survival. In the last of the four cases, parents are selected with no similar alleles and a mated queen will produce brood with 100% survival (second year, combination 4). Hence, there is a slight possibility of parental selection that will not result in problems. However, the most frequent parental selections will produce some queens having brood with 75% survival. When the virgin queens descend from several queens, all combinations are likely; half of them will produce normal brood while the other half will produce scattered brood.

The third year a beekeeper would certainly rear a third generation of virgin queens only from queens producing normal brood. Nonetheless, the results again depend on chance selection of the parents. Even when the brood in both colonies has 100% survival (if some sister queens produce a brood of 75% survival), only one-fourth of the new queens will produce brood with 100% survival (third year, combinations 1 and 2 together).

When queens producing virgins and drones are selected from two different groups, in which some queens produce scattered brood, then three-fourths of the new queens may produce brood with 100% survival (third year, combinations 3 and 4 together).

A virgin queen may be mated with drones from several queens. Woyke (1963a, 1972) presented various examples of mating with drones from two, three, or four colonies. For example, a mating of a queen with four drones, only one of which carries a sex allele of the queen, will result in $\frac{1}{8}$ of the eggs being homozygous and an 87.5% survival of brood. Thus, an increase in the number of sex alleles in a population increases the average brood's survival. However, while this is true, the more often a queen mates in a population with several sex alleles the more likely she is to mate at least one drone having a sex allele similar to one of hers. This decreases the percentage of queens producing only diploids heterozygous at the X locus, with 100% brood survival.

C. Results of Multiple Matings

With multiple mating, the survival rate of brood produced by specific queens is the result of two factors: the number of X alleles in the population and the number of drones mating the queen. The frequency distribution of queens mated to drones with different sex alleles is given by the binomial

$$Q = (p + q)^k \qquad (2)$$

where p is the proportion of drones having one allele identical to one of the queen's alleles, q is the proportion of drones with other alleles, and k is the number of matings (Shaskolsky, 1968).

Any queen has two of N alleles present in a given population. The probability (p) of randomly mating a drone having one of the two alleles of the queen is $2/N$. (This is a biallelic mating, b.) The complementary probability (q) of mating a drone having one of the other alleles present in the population ($N - 2$) is equal to $(N - 2)/N$ (triallelic mating, t). The frequency of all the possible combinations of random mating with k drones is given by the expression

$$Q = [(2/N) + (N - 2)/N]^k \qquad (3)$$

In order to recognize the type of mating, the letter b (biallelic) or t (triallelic) may be added to the equation. Below an example is given for $N = 12$ alleles and $k = 3$ matings of each queen:

$$Q = \{(2/12)b - [(12 - 2)/12]t\}^k = [(1/6)b - (5/6)t]^3$$
$$= 0.001b^3 + 0.069b^2t + 0.347bt^2 + 0.578t^3 = 1 \qquad (4)$$

This produces four classes of queens. The survival rate of brood produced by the queens in these classes depends upon the type of mating (b^3, b^2t, bt^2, or t^3) and is 50, 67, 83, or 100%, respectively. With queens mating three times in a population having 12 sex alleles, most of them (0.578) produce brood having 100% survival.

The frequency distribution of queens mated to drones with various sex alleles may also be calculated in another way (Page and Laidlaw, 1982). The proportion of the queen's alleles to all sex alleles in a population (N) is $p = 2/N$. The probability that a queen mating k times will mate y males with a sex allele identical to one of hers is

$$P_{(p,k,y)} = [k!/(k - y)!y!](2/N)^y(1 - 2/N)^{k-y} \qquad (5)$$

Assuming a constant number of matings (e.g., $k = 8$) and a constant number of alleles in a population (e.g., $N = 12$), it suffices to replace the y by 0, 1, 2, 3, . . . to k to get the frequency distribution of queens mating with y drones. The frequencies calculated for the example are 0.232, 0.372, 0.260,

0.103, 0.026, 0.004, 0.0004, 0.00002, and 0.0000006 for the proportion of queens mated to zero through eight drones with similar alleles. Of course, for other populations the number of sex alleles as well as the number of matings may be different.

To convert the different classes of matings into brood survival (S), Page and Metcalf (1982) used $S = 1 - y/2k$. The survival of brood for the examples' distribution of queens would be 100.00, 93.57, 85.50, 81.25, 75.00, 68.75, 62.50, 56.25, and 50.00%.

Page and Metcalf (1982) also discussed the influence of the number of matings on the distribution of queens producing brood with different survival rates. The increase of matings increases the number of classes of queens producing brood with different survival rates. With 1 mating there are 2 classes; with 2, 3, 4, and k matings, there are 3, 4, 5, and $k + 1$ classes, respectively. The increase of matings decreases the frequency of extreme classes, especially those of queens producing brood with 50 and 100% survival. As a result, the frequencies of some classes are so low that they practically disappear. At the same time there is an increase in frequencies of classes with queens producing brood surviving at rates close to the average for the whole population. Thus, the number of matings by a single queen has an influence on the number of classes of queens but has no effect on the mean survival of brood in the whole population. An increase of the number of sex alleles in a population increases the frequencies of classes with queens producing brood of higher survival rate.

D. Average Brood Survival

For many purposes, the average brood survival of the entire population is an important consideration. A population with three sex alleles (X^a, X^b, X^c) includes different diploid queens which may be mated by a three-allele cross (e.g., $X^a X^c \times X^b$) as well as by a two-allele cross (e.g., $X^b X^c \times X^b$).

Woyke (1976) considered a population with several sex alleles (X^a, X^b, X^c) having frequencies distributed as p, q, and r, respectively, where $p + q + r = 1$. After a random cross $(p + q + r)^2$, the following frequencies of genotypes occur in the offspring, whose sum is unity:

$$p^2 X^a X^a + q^2 X^b X^b + r^2 X^c X^c + 2pq X^a X^b + 2pr X^a X^c + 2qr X^b X^c \quad (6)$$

All the homozygotes are eliminated by the worker bees. Thus, the proportion of remaining survival heterozygotes (S) in the population is

$$S = 1 - p^2 - q^2 - r^2 \quad (7)$$

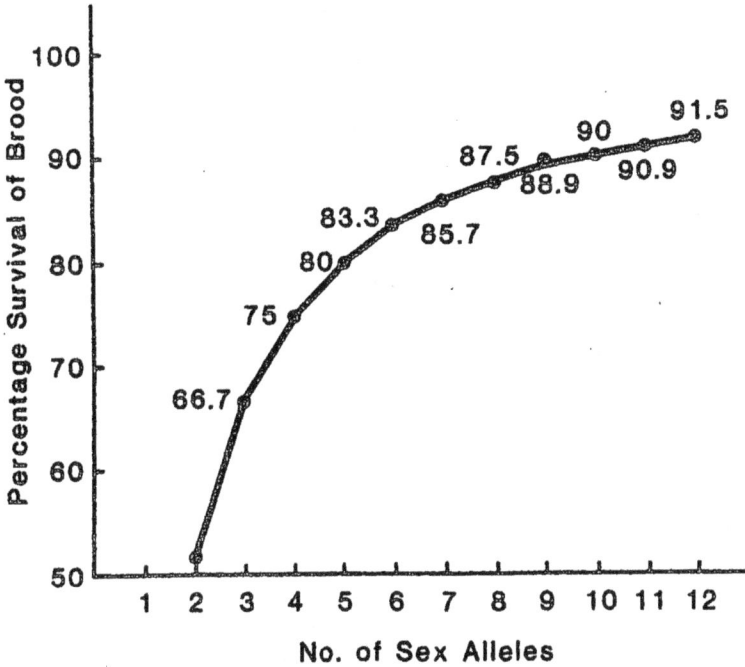

Fig. 5. Relationship between the number of sex alleles in a honey bee population and the survival rate of brood. [Redrawn by Sandra Kleinpeter from Woyke (1976), with permission. Copyright 1976 by the International Bee Research Association.]

The percentage survival ($S\%$) is $100S$.

When there are N alleles equally distributed throughout the population, the frequency of each is $1/N$. Hence Eq. 7 may be transferred to

$$S = 1 - N(1/N^2) = 1 - 1/N = (N - 1)/N \tag{8}$$

The survival percentage is

$$(S\%) = 100\,[(N - 1)/N] \tag{9}$$

This equation was used to calculate the estimates presented in Fig. 5, which shows that an increase in the number of sex alleles to about five results in a rapid increase in the percentage of surviving brood in the population (5–17% increase in survival per added allele). A further increase from five to eight alleles will improve the survival of brood much less — under 3.5% per added allele. Thereafter, the survival of brood is increased by less than 1.5% per added allele. Thus, the increase of the number of sex alleles above 10 is very ineffective in increasing the average brood survival of a population.

The survival for 20, 50, and 100 alleles would be 95, 98, and 99%, respectively.

According to Page and Laidlaw (1982), the variance of the average brood survival (σ_s^2) is

$$\sigma_s^2 = (1/2k)(1/N)(1 - 2/N) \qquad (10)$$

where k is the number of matings and N is the number of sex alleles in that population.

The importance of sex alleles to honey bee biology and practical beekeeping is considerable. Their population changes under various conditions have been extensively studied (Woyke, 1976). This subject is reviewed by Cornuet (Chapter 9).

VII. APICULTURAL CONSEQUENCES OF SEX ALLELES

A. Colony Populations and Honey Production

The question arises, how far does the diminished survival of brood affect colony population and honey production? The simplest expectation is that both the number of offspring and honey production will be proportional to brood survival. Woyke (1980b, 1981) investigated this question with three groups of colonies having queens producing brood with 100, 75, and 50% survival. The results were rather unexpected (Fig. 6; Table 3). Seasons of the year interacted with the amount of brood. Similar brood areas were found in all three groups of colonies during spring and autumn. In summer, colonies having brood with 75 and 50% survival produced about 82 and 68%, respectively, of the brood area of normal colonies. When queens in normal colonies laid fewer than 1000 eggs/day, those that produced brood of lower survival rates were able to replace nonsurviving larvae with new eggs. This was not the case when the normal number of eggs was higher.

Colonies with 75 and 50% survival rates had 88 and 79%, respectively, of the normal spring worker-bee population. In summer they had 93 and only 35%, and in autumn had 92 and 65% of the normal worker population.

Colonies having brood with 75 and 50% survival produced 103 and 50%, respectively, of the surplus honey harvested from normal colonies in 1978. Most of the honey harvested originated from spring-time nectar of rape and acacia. If the weather conditions in the second half of the season had been better, the honey production relationships of the autumn harvest—100 to 50 to 8%—would have had a greater influence on the total amount of

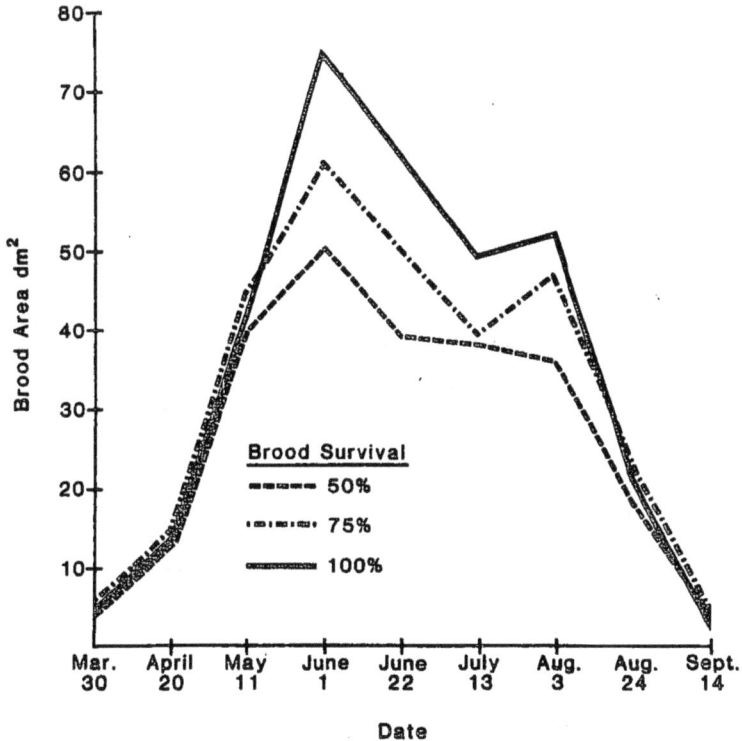

Fig. 6. Average brood area in groups of colonies with 50, 75, and 100% survival through 1978. [Redrawn by Sandra Kleinpeter from Woyke (1980b), with permission. Copyright 1980 by the International Bee Research Association.]

honey harvested. In the second year the three groups of colonies produced honey in the relationship of 100 to 87 to 75%.

These results show that in the spring, when queens are able to replace eaten larvae, the differences in honey production between colonies with low or high brood survival are low. Later in the season, when differences in the amount of brood produced are higher, the differences in honey harvested are also higher.

B. Brood-Comb Use

The common belief is that in colonies with queens producing brood with 100, 75, and 50% survival, comb cells among those used for brood rearing are 0, 25, and 50% empty. This is true only for combs containing exclusively 1-day-old larvae, or when the brood combs are separated from the queen after she has laid eggs. Queens do not lay new eggs in empty cells immedi-

TABLE 3. Influence of Brood Survival on the Numbers of Progeny and Honey Produced through the Active Seasons of 2 Years

Parameter and season	Favorable year, 1978, survival of brood			Unfavorable year, 1979, survival of brood		
	Actual measurement, 100% group	As a percentage of the 100% group		Actual measurement, 100% group	As a percentage of the 100% group	
		75% Group	50% Group		75% Group	50% Group
Brood area (dm²)						
spring	14.7	103	97	6.4	94	108
summer	78.4	82	68	58.3	81	67
autumn	22.6	111	93	20.2	87	94
Worker population (thousand)						
spring	13.8	88	79	9.1	77	73
summer	30.3	93	35	18.9	87	65
autumn	11.8	92	65	7.9	92	85
Honey (kg)						
spring	2.5	144	80	—	—	—
summer	4.6	111	26	10.6	86	70
autumn	2.4	50	8	2.3	87	77
left in hive	2.4	96	108	0.4	125	150
Total honey	11.9	103	50	13.3	87	75

ately after adults have emerged, and therefore, some empty cells may be found among brood even in colonies producing brood having total survival. On the other hand, in colonies with brood having 50% survival, the queen does not wait for 3 weeks until all adults emerge. She lays new eggs into cells from which 1-day-old diploid drone larvae were eaten.

A 2-year investigation (Woyke, 1984b) showed great variations in replacement egg-laying between colonies. Distinct differences resulted from X locus homozygosity, although these differences were influenced by the season of the year. In spring, colonies producing brood with 100, 75, and 50% survival had about 7, 10, and 20% of the brood cells empty, respectively. In summer about 10, 14, and 24% of the brood cells were empty, and in autumn, about 10, 13 and 16% were empty. Thus, the overall percentage of empty cells in the brood area was higher in colonies producing low-survival brood than in colonies with high brood survival. Nonetheless, the percentage of empty cells in colonies with brood of low survival is lower than the percentage of diploid larvae produced by queens and eaten by workers. Hence, queens replace these larvae with new eggs before the adults emerge from other cells of that comb. The percentage of empty cells in all three groups of colonies is high in the summer when a large comb area is used in brood rearing. When the brood area is smaller, the queen is more effective at replacement egg laying. Thus the differences between the three groups of colonies were higher in summer than in autumn.

REFERENCES

Adam, A. (1912). Bau und Mechanismus des Receptaculum seminis bei den Bienen, Wespen und Ameisen. Zool. Jahrb., Abt. II. 35, 1–74 + 3p1.

Adams, J., Rothman, E. D., Kerr, W. E., and Paulino, Z. L. (1977). Estimation of the number of sex alleles and queen matings from diploid male frequencies in a population of Apis mellifera. Genetics 86, 583–596.

Anderson, R. H. (1963). The laying worker in the Cape honey-bee Apis mellifera capensis. J. Apic. Res. 2, 85–92.

Blochman, R. (1889). Über die Zahl der Richtungskörper bei befruchteten und unbefruchteten Bieneneiern. Morp. Jahrb. 15, 85–96.

Bresslau, E. (1906). Der Samenblasengang der Bienenkönigin. Zool. Anz. 29, 299–323.

Chaud-Netto, J. (1975). Sex determination in bees. II. Additivity of maleness genes in Apis mellifera. Genetics 79, 213–217.

Chaud-Netto, J., and Moura-Duarte, F. A. (1975). Sex determination in bees. V. The action of sexual genes in Apis mellifera. Ciên. Cult. (São Paulo) 27, 125–129.

Crozier, R. H. (1971). Heterozygosity and sex determination in haplodiploidy. Am. Nat. 105, 399–412.

Cunha, A. B. Da, and Kerr, W. E. (1957). A genetical theory to explain sex-determination by arrhenotokus parthenogenesis. Forma Functio 1, 33–36.

Dickel, F. (1897). Über die Beweiskraft der V. Siebold'schen Untersuchungsergebnisse von Bieneneiern im Jahre 1855. Bienenzeitung 53, 249–252.

Dickel, F. (1898a). Unter normalen Verhältnissen werden alle Bieneneier befruchtet; ihr Schicksal wird entschieden durch die Einflüsse der Arbeitsbienen. Bienenzeitung 54, 50–53, 68–71, 83–85.

Dickel, F. (1898b). Der geschlechtsauslösende Einfluss der Arbeitsbienen ist gebunden an die Wirkung verschiedenartiger Drüsensekrete; er beginnt, nachdem die Königin das Ei in die Zelle abgesetzt hat und schliesst ab, sobald die Larve die normale Grösse erreicht hat. Bienenzeitung 54, 99–101, 114–115, 134–136.

Drescher, W., and Rothenbuhler, W. C. (1964). Sex determination in the honey bee. J. Hered. 55, 90–96.

Dzierzon, J. (1845). Gutachten über die von Herrn Direktor Stöhr in ersten und zweiten Kapitel des General-Gutachtens aufgestellten Fragen. Bienenzeitung 1, 109–113, 119–121.

Dzierzon, J. (1854). Fernere Beobachtungen über Vermährung und Reinerhaltung der italienischen Bienen. Bienenzeitung 10, 251–254.

Dzierzon, J. (1898). Widerlegung der jüngst gegen meine Theorie der Fortpflanzung der Bienen erhobene Einwande. 43 Wanderversammlung, Salzburg. Bienenzeitung 54, 299–302.

Dzierzon, J. (1899). Grundlosigkeit und Haltlosigkeit der neuen Lehre. Erste gemeinsame Wanderversammlung . . . in Köln a. Rh. Bienenzeitung 55, 318–320.

Hachinohe, Y., and Jimbu, M. (1958). Occurrence of lethal eggs in the honey bee. Bull. Natl. Inst. Agric. Sci., Ser. G. Chiba, Japan 14, 123–130.

Hachinohe, Y., Jimbu, M., and Ohnishi, N. (1958). Inducement of impaternate females in the honeybee. Bull. Natl. Inst. Agric. Sci., Ser. G. Chiba, Japan 14, 117–122.

Hoshiba, H. (1979). Chromosome of the diploid and haploid drone honeybee. *Proc. Inter. Apicultural Cong. (Apimondia)* 27, 253–256.

Hoshiba, H. (1984). The C-banding analysis of the diploid male and female honeybee (Apis mellifera). *Proc. Japan. Acad.* 60 (Ser. B) 238–240.

Hoshiba, H., Okada, J., and Kusanagi, A. (1981). The diploid drone of *Apis cerana japonica* and its chromosomes. *J. Apic. Res.* 20, 143–147.

Jack, R. W. (1916). Parthenogenesis amongst the workers of the Cape honey-bee: Mr. G. W. Onions' experiments. *Trans. R. Ent. Soc. London* 64, 396–403.

Kerr, W. E. (1967). Genetic structure of the populations of hymenoptera. *Ciênc. Cult. (São Paulo)* 19, 39–44.

Kerr, W. E. (1974a). Advances in cytology and genetics of bees. *Annu. Rev. Ent.* 19, 253–268.

Kerr, W. E. (1947b). Genética da determinação do sexo em abelhas X. Programação da atividade dos genes determinadores de sexo e casta. *Proc. Congr. Bras. Apicultura, 3rd*, Piracicaba, 179–187.

Kerr, W. E., and Nielsen R. A. (1967). Sex determination in bees (Apidae). *J. Apic. Res.* 6, 3–9.

Laidlaw, H. H., Gomes, F. P., and Kerr, W. E. (1956). Estimation of the number of lethal alleles in a panmictic population of *Apis mellifera* L. *Genetics* 41, 179–188.

Lee, G. L. (1969). The effect of gene dosage on variability in the honeybee. I. The eye colour mutants. *J. Apic. Res.* 8, 75–78.

Leuckart, R. (1855). Seebacher Studien. *Bienenzeitung* 11, 199–211.

Mackensen, O. (1943). The occurrence of parthenogenetic females in some strains of honeybees. *J. Econ. Entomol.* 36, 465–467.

Mackensen, O. (1951). Viability and sex determination in the honey bee (*Apis mellifera* L.). *Genetics*, 36, 500–509.

Mackensen, O. (1955). Further studies on a lethal series in the honeybee. *J. Hered.* 46, 72–74.

Mackensen, O. (1964). Relation of semen volume to success in artificial insemination of queen honey bees. *J. Econ. Entomol.* 57, 581–583.

Manning, F. J. (1949). Sex determination in the honeybee. *Microscope* 7, 175–180.

Meves, F. (1907). Die Spermatocytenteilung bei der Honigbiene (*Apis mellifica* L.) nebst Bemerkungen über Chromatinreduction. *Arch. Mikrosk. Anat. Entwg.* 70, 414–491 + 5 tables.

Nachtsheim, H. (1913). Cytologische Studien über die Geschlechtsbestimmung bei der Honigbiene (*Apis mellifica* L.). *Arch. Zellforsch.* 11, 169–241.

Onions, G. W. (1912). South African "Fertile-Worker Bees". *Agric. J. S. Afr.* 3, 720–728.

Page, R. E., and Laidlaw, H. H. (1982). Closed population honeybee breeding. 1. Population genetics of sex determination. *J. Apic. Res.* 21, 30–37.

Page, R. E., and Metcalf, R. A. (1982). Multiple mating, sperm utilization and social evolution. *Am. Nat.* 119, 263–281.

Paulcke, E. (1899). Zur Frage der partenogenetischen Entstehung der Drohnen (*Apis mellifica*). *Anat. Anz.* 16, 185–187.

Pelt, G. S. Van (1966). The problem of gene dosage in *Habrobracon*. *Genetics* 54, 367–368.

Petrunkewitsch, A. (1901). Die Richtungskörper und ihr Schicksal im befruchteten und unbefruchteten Bienenei. *Zool. Jahrb., Abt. Anat. Ontog. Thiere* 40, Separ. 1–36 + 4 pls.

Ris, H., and Kerr, W. E. (1952). Sex determination in the honey bee. *Evolution* 6, 444–445.

Rothenbuhler, W. C. (1957). Diploid male tissue as new evidence on sex determination in honey bees. *J. Hered.* 48, 160–168.

Rothenbuhler, W. C., Gowen, J. W., and Park, O. W. (1952). Androgenesis with zygogenesis in gynandromorphic honeybees (*Apis mellifera* L.). *Science* 115, 637–638.

Ruttner, F. (1977). The problem of Cape bee (*Apis mellifera capensis Escholtz*): parthenogenesis—size of populations—evolution. *Apidologie* 8, 281–294.

Sanderson, A. R., and Hall, D. W. (1948). The cytology of the honey bee (*Apis mellifica* L.). *Nature (Lond.)* 162, 34–35.

Sanderson, A. R., and Hall, D. W. (1951a). Sex determination in the honeybee. *Evolution 5,* 414–415.

Sanderson, A. R., and Hall, D. W. (1951b). Sex in the honey-bee. *Endeavour 10,* 33–39.

Shaskolsky, D. V. (1968). The distribution of a series of multiple alleles in theoretical populations, as related to the biology of reproduction in the honey bee *(Apis mellifera). Genetica 4,* 41–55.

Siebold, C. Th. (1852). III Wanderversammlung in Brieg und Karlsmarkt. *Bienenzeitung 8,* 56–57.

Siebold, C. Th. (1856a). "Wahre Parthenogenesis bei Schmetterlingen und Bienen." Engelmann, Leipzig.

Siebold, C. Th. (1856b). Die Drohneneier sind nicht befruchtet. *Bienenzeitung 12,* 181–184.

Triasko, V. V. (1965). Natural parthenogenesis of honeybee. *XX Int. Jub. Beekeep. Congr. Kolos, Moskva* 356–362.

Tucker, K. W. (1958). Automictic parthenogenesis in the honey bee. *Genetics 43,* 299–316.

Verma, S., and Ruttner, F. (1981). Cytologische Analyse der Thelytoken parthenogenesis bei der Kapbiene *(Apis mellifer capensis* Escholtz). *Apidologie 12,* 88–89.

Whiting, P. W. (1940). Multiple alleles in sex determination of *Habrobracon. J. Morphol. 66,* 323–355.

Whiting, P. W. (1943). Multiple alleles in complementary sex determination of *Habrobracon. Genetics 28,* 365–382.

Woyke, J. (1960). Naturalne i sztuczne unasienianie matek pszczelich [Natural and artificial insemination of queen honeybees]. *Pszczel. Zesz. Nauk, 4,* 183–275. Summarized in *Bee World* (1962), *43,* 21–25.

Woyke, J. (1962a). Geneza powstawania niezwykłych pszczół. [The origin of unusual bees.] *Pszczel. Zesz. Nauk. 6,* 49–63.

Woyke, J. (1962b). The hatchbility of "lethal" eggs in a two sex allele fraternity of honeybee. *J. Apic. Res. 1,* 6–13.

Woyke, J. (1963a). Determinacja płci a kontrolowany dobór pszczół. [Sex determination and controlled mating of honey bees]. *In* "Hodowla Pszczół [Honeybee Breeding]" pp. 670–678. Państwowe Wydawnictwo Rolnicze i Leśne, Warszawa.

Woyke, J. (1963b). Drones from fertilized eggs and the biology of sex determination in the honeybee. *Bull. Acad. Pol. Sci. Cl. V. Serie des Sciences Biologiques 9,* 251–254.

Woyke, J. (1963c). Drone larvae from fertilized eggs of the honeybee. *J. Apic. Res. 2,* 19–24.

Woyke, J. (1963d). Drones from fertilized eggs and the biology of sex determination in honey bee. *Proc. Inter. Apic. Cong. (Apimondia) 19,* 704–714.

Woyke, J. (1963e). What happens to diploid drone larvae in a honeybee colony. *J. Apic. Res. 2,* 73–76.

Woyke, J. (1963f). Rearing and viability of diploid drone larvae. *J. Apic. Res. 2,* 77–84.

Woyke, J. (1963g). Contribution of successive drones to the insemination of a queen. *Proc. Inter. Apic. Cong. (Apimondia) 19,* 715–718.

Woyke, J. (1965a). Genetic proof of the origin of drones from fertilized eggs of the honeybee. *J. Apic. Res. 4,* 7–11.

Woyke, J. (1965b). Study on the comparative viability of diploid and haploid larval drone honeybees. *J. Apic. Res. 4,* 12–16.

Woyke, J. (1965c). The diploid drones. *Proc. Inter. Apic. Cong. (Apimondia) 20,* 152–154.

Woyke, J. (1965d). Do honeybees eat diploid drone larvae because they are in worker cells? *J. Apic. Res. 4,* 64–70.

Woyke, J. (1967). Diploid drone substance—cannibalism substance. *Proc. Inter. Apic. Cong. (Apimondia) 21,* 471–472.

Woyke, J. (1969a). A method of rearing diploid drones in a honeybee colony. *J. Apic. Res. 8,* 65–71.

Woyke, J. (1969b). Rearing diploid drones on royal jelly or bee milk. *J. Apic. Res. 8*, 169–173.

Woyke, J. (1969c). Genetic aspects of artificial insemination. *In* "The Instrumental Insemination of the Queen Bee" (F. Ruttner, ed.), pp. 51–58. Apimondia Publ., Bucharest.

Woyke, J. (1971a). New experimental data in the honey bee genetics. *Proc. Inter. Apic. Cong. (Apimondia) 23*, 110–111.

Woyke, J. (1971b). "Biology of Reproduction as a Basis for Production of New Varieties of Honeybees: Final Technical Report for USDA," 154 pp. Bee Culture Laboratory, Warsaw.

Woyke, J. (1972). Sex alleles and controlled mating. *Proc. Inter. Symp. on Controlled Mating and Selection of the Honeybee.* Lunz. am. See 69–74.

Woyke, J. (1973a). Reproductive organs of haploid and diploid drone honeybees. *J. Apic. Res. 12*, 35–51.

Woyke, J. (1973b). Laranja: a new honey bee mutation. Gene dosage and maleness of diploid drones. *J. Hered. 64*, 227–230.

Woyke, J. (1974). Genic balance, heterozygosity and inheritance of size of testes in diploid drone honeybees. *J. Apic. Res. 13*, 77–85.

Woyke, J. (1975). DNA content of spermatids and spermatozoa of haploid and diploid drone honeybees. *J. Apic. Res. 14*, 3–8.

Woyke, J. (1976). Population genetic studies on sex alleles in the honeybee using the example of the Kangaroo Island bee sanctuary. *J. Apic. Res. 15*, 105–123.

Woyke, J. (1977). Comparative biometrical investigation on diploid drones of the honeybee. I. The head. *J. Apic. Res. 16*, 131–142.

Woyke, J. (1978a). Comparative biometrical investigation on diploid drones of the honeybee. II. The thorax. *J. Apic. Res. 17*, 195–205.

Woyke, J. (1978b). Comparative biometrical investigation on diploid drones of the honeybee. III. The abdomen, and weight. *J. Apic. Res. 17*, 206–217.

Woyke, J. (1979a). Sex determination in *Apis cerana indica. J. Apic. Res. 18*, 122–127.

Woyke, J. (1979b). New investigations on *Apis mellifera capensis. Proc. Inter. Apic. Cong. (Apimondia) 27*, 319–321.

Woyke, J. (1980a). Evidence and action of cannibalism substance in *Apis cerana indica. J. Apic. Res. 19*, 6–16.

Woyke, J. (1980b). Effect of sex allele homo-heterozygosity on honeybee colony population and their honey production. I. Favourable development conditions and unrestricted queen. *J. Apic. Res. 19*, 51–63.

Woyke, J. (1980c). Genic background of sexuality in diploid drones. *J. Apic. Res. 19*, 89–95.

Woyke, J. (1981). Effect of sex allele homo-heterozygosity on honeybee colony populations and their honey production. II. Unfavourable development conditions and restricted queens. *J. Apic. Res. 20*, 148–155.

Woyke, J. (1983). Length of haploid and diploid spermatozoa of the honeybee and the question of production of triploid workers. *J. Apic. Res. 22*, 146–149.

Woyke, J. (1984a). Ultrastructure of single and multiple diploid honeybee spermatozoa. *J. Apic. Res. 23*, 123–135.

Woyke, J. (1984b). Exploitation of comb cells for brood rearing in honey bee colonies with larvae of different survival rates. *Apidologie 15*, 123–136.

Woyke, J., and Adamska, Z. (1972). The biparental origin of adult honeybee drones proved by mutant genes. *J. Apic. Res. 11*, 41–44.

Woyke, J., and Knytel, A. (1966). The chromosome number as proof that drones can arise from fertilized eggs of the honeybee. *J. Apic. Res. 5*, 149–154.

Woyke, J., and Król-Paluch, W. (1985). Changes in tissue polyploidization during development of worker, queen, haploid and diploid drone honeybees. *J. Apic. Res. 24*, 14–224.

Woyke, J., and Skowronek, W. (1974). Spermatogenesis in diploid drones of the honeybee. *J. Apic. Res. 13,* 183–190.

Woyke, J., Knytel, A., and Bergandy, K. (1966). The presence of spermatozoa in the eggs as proof that drones can develop from inseminated eggs of the honeybee. *J. Apic. Res. 5,* 71–78.

Zander, E., Löschel, E., and Meier, K. (1916). Die Ausbildung des Geschlechtes bei der Honigbiene (*Apis mellifica* L.) *Z. Angew. Entomol. 3,* 1–20.

CHAPTER 5

Genetics of Bees Other than Apis mellifera

ROBIN F. A. MORITZ

I. INTRODUCTION

Detailed genetic work is done on only very few of the 20,000 species of bees (Apoidea). On one hand, this might be due to their slow generation cycle and the relatively complex breeding methods. On the other hand, those bee species having direct commercial value are genetically well investigated. Thus, in spite of their important role in pollination, the genetics of most bees has rarely been studied. This chapter will mainly deal with the eusocial bees of the subfamily Apinae: bumble bees (Bombini), stingless bees (Meliponini), and honey bees (Apini) will be discussed.

II. THE BUMBLE BEES (BOMBINI)

Most genetical study of the tribe Bombini is done on the genus *Bombus.* Mostly, bumble bees live in cool or temperate regions. They are only rarely found in the American tropics and there are no reports of them in southern Africa and Australia. Most bumble-bee species build annual colonies in underground nests. They often protect their cocoons and food cells by a wax roof in the nest which is supported by various other nesting materials. At the end of a season, the colony dies and only young fertilized queens hibernate

BEE GENETICS
AND BREEDING

after burrowing in the soil (Alford, 1975; Free, 1982). In the tropics, colonies may live for several years (Sakagami and Zucchi, 1965; Saraceni, 1972). In such colonies polygyny seems to be common; however, aggressiveness between the laying gynes rises during the season, queens disappear one by one, and monogyny is usually established.

Bumble bees are some of the most important pollinators of commercial crop plants. In spite of this, they are not systematically used for commercial pollination. One of the main problems is the collection and overwintering of young queens. Another problem is the development of strong colonies early in the season (Holm, 1966).

A. Mating and Reproduction

Bumble-bee queens and males are produced late in the season when colonies are populous. Males leave their nest at the age of 2–4 days (Free, 1982) and never return. Once gone they forage for their own food. However, matings usually occur in the nest or in its close vicinity. In some species the males gather in front of nests where they mate queens; in other species they establish definite mating routes (Frank, 1941).

Such circuits were analyzed by several authors (Haas, 1946, 1949a,b, 1952; Stein, 1963a; Svensson, 1979). Males mark their routes at intervals with a scent produced in their mandibular glands. These substances have been identified for several species (Bergström et al., 1967, 1981; Kullenberg et al., 1970; Stein, 1963a,b; Svensson and Bergström, 1977, 1979). The compounds secreted in the mandibular glands range from monoterpenes and short-chain fatty acid derivatives to diterpenes and long-chained fatty acid derivatives (Bergström, 1982). Bergström et al. (1973) used differences in the volatile marking secretions to separate Bombus lucorum into two separate taxonomic forms. The cephalic marking secretions function as territory recognition marks, territorial flight stimulators, and excitants (Bringer, 1973; Free, 1971; Kullenberg, 1956; Kullenberg et al., 1973; Kullenberg and Bergström, 1975). Honk et al. (1978) suggested that a sex pheromone also is produced in the mandibular glands of queens.

The phenomenon of distinct flight routes of bumble-bee males (Bombus lucorum) was observed by Darwin (1886). The pattern and height of the flight paths are different in different species. Bombus hortorum and B. hypnorum fly almost at ground level while other species such as B. pomorum, B. terrestris, B. lucorum, and B. mycorum fly at the level of foliage (Haas, 1949a; Krüger, 1951; Svensson, 1979). Awram (1970) found males of B. lucorum and B. terrestris flying at 20 m as does B. ruderatus (Sladen, 1912). The biological consequences of these routes are not clear. Wynne-Edwards (1962) proposed that the flight-path phenomenon might regulate population density.

When a male mates a queen he usually knocks her to the ground, climbs on her thorax, and copulates. After copulation, the queen protrudes her sting and the male falls back while remaining attached. Attachment may last from several minutes to longer than an hour (Alford, 1975). In some species [*B. hypnorum*: Pouvreau (1963); *B. huntii*: Hobbs (1967)] multiple matings are reported, but in most species single mating seems to be the rule.

B. Population Genetics

Most genetic work on bumble bees is done at the level of the population. A genetically determined color polymorphism was recently found by Owen and Plowright (1980) in *Bombus melanopygus* Nylander. *Bombus melanopygus* usually has a reddish-orange pile on the second and third abdominal tergites in both sexes (Stephen, 1957). Nevertheless, *B. melanopygus* sometimes has the black coloration of the closely related and sympatric species *B. edwardsii* Cresson. In the progeny of 25 red and 13 black *B. melanopygus* queens, clear-cut segregations in the pile color were found.

An F_1 crossing experiment testing the one-gene model confirmed that the color dimorphism has a simple one-locus, two-allele, Mendelian basis. In some colonies the male segregation differed significantly from the 1 : 1 ratio that would be expected if the original queen had produced all the male offspring. Laying workers, common in bumble bees, probably are the origin of these different ratios in males (Owen and Plowright, 1980). Worker-produced males impose nontrivial consequences on the population genetics of bumble bees. The speed of approach to Hardy–Weinberg equilibrium is increased; the equilibrium gene frequency between mutation and selection is altered; the gene frequencies in a polymorphism are shifted (Owen, 1980).

Another color polymorphism is known in *B. rufocinctus* (Plowright and Owen, 1980). In this case at least two loci are involved. Since there are many intermediates, this polymorphism might be a pleiotropic character affected by additional modifier genes. Pekkarinen (1979) found an extensive color variation in *B. lucorum*. He, as did several others (Friese and Wagner, 1912; Krüger, 1932; Pittioni, 1940, 1941; Vogt, 1909), showed that the coloration of bumble bees correlates to their environment. He found a tendency to albinism in steppe areas and to melanism in arctic or mountain regions. The thermal effect of dark colors seems to be advantageous for insects during isolation in cold climates (Digby, 1955). However, much color variation does not follow climatic gradients (Pekkarinen, 1979). In many cases, a selective value in avoiding predation seems to be a more likely explanation of this variation (Plowright and Owen, 1980; Stiles, 1979).

Isoenzyme studies give additional information about the genetic composition of bumble-bee populations. Pamilo *et al.* (1978) studied six species of

Bombus and found the mean proportion of polymorphic loci was 0.109 and the mean heterozygosity per locus was 0.0124. These values are fivefold smaller than those for diploid insects (Selander, 1976). Pekkarinen (1979) presented similar data for eight different bumble-bee species in Fennoscandia. He obtained a mean proportion of polymorphic loci of 0.17 and a mean heterozygosity per locus of 0.03. Data from *Apis mellifera* are similar. Snyder (1974) did not find *any* allozymic variation in *B. americanorum* when studying 25 enzymes.

Electrophoretic methods were used to find genetic differences between species of *Bombus* and *Psithyrus* in allomorphs of α-glycerophosphate dehydrogenase (α-GPDH) (Fink *et al.*, 1970). Pamilo *et al.* (1981) and Pekkarinen *et al.* (1979) constructed a phenogram using differences in nine different enzyme systems. Two enzyme systems, α-GPDH and isocitric dehydrogenase (ICDH), were used to separate *Bombus* and *Psithyrus* (also see Stephen and Cheledin, 1973). The taxonomic result obtained by cluster analysis agrees with similar models based on morphological studies (Sakagami, 1976). Using different enzymes, Obrecht and Scholl (1981) showed similar results in the social Bombini. Scholl and Obrecht (1983) succeeded in electrophoretically identifying two species in the *B. lucorum* complex. They found differing motilities for phosphoglucomutase (PGM) and esterase-1 (EST-1) in Switzerland, Germany, and France. This indicates a strong reproductive isolation, since no heterozygotes were found. The authors suggested adding the species *B. magnus* Vogt to *B. lucorum*. These findings show the importance and possibilities of allozyme electrophoresis in bumble-bee systematics and population genetics.

C. Sex Determination

Bumble bees have a sex-determination mechanism differing from that of *Apis mellifera*. Garofalo (1972, 1973) showed that after inbreeding, diploid drones ($2n = 40$; Kerr and Silveira, 1972) appear in *B. atratus*. A 3:1 ratio (diploid drones to workers) was found in the progeny of mother–son matings. Garofalo suggests a genetic model where two independent sex loci determine sex. Males are homozygous at least at one of the two loci. In further work, Garofalo and Kerr (1975) mated queens of *B. atratus* with diploid males and successfully reared nine workers which produced parthenogenetic offspring. From the sex and chromosome numbers of these offspring, they argued that the two-loci model explained sex determination in *Bombus*. On the other hand, Crozier (1971, 1977) presented data fitting the single locus model.

Colonies with a high ratio of diploid males show a significantly slower development than non-inbred colonies (Plowright and Pallett, 1979). The numerical sex ratio in natural colonies is highly male-biased [6 to 1: Owen *et al.* (1980); 4 to 1: Pommeroy (1979)] and might be due to laying workers. There is a positive correlation between sex ratio and colony size in *B. tericola*, which implies that the highest proportion of worker-laid males were produced in smaller colonies. Nevertheless, in some cases equal investment in both sexes might be possible (Webb, 1961).

III. THE STINGLESS BEES (MELIPONINI)

Meliponini are social bees living in colonies ranging from 500 to 4000 individuals in *Melipona* and from 300 to 80,000 in *Trigona*. These are the two main genera in the tribe. Most stingless bees exist in the tropics of the American continent (ca. 185 species); another 35 species exist in Africa, 42 exist in Asia, and 20 in Australia (Kerr and Maule, 1964). The tribe of Meliponini is generally divided into five different genera: (1) *Melipona* lives only in tropical America; (2) *Meliponula*, of which only a single species is found in Africa; (3) *Trigona* is spread in all the tropics and occupies 16 subgenera having a large variation in morphology and biology; (4) *Dactylurina* has only a single species in Africa; (5) *Lestrimelitta* is found in Africa and tropical America. Michener (1974) proposes two separate genera for this last group: *Lestrimelitta* for the American and *Cleptotrigona* for the African species.

The Meliponini got their name "stingless bees" because of their vestigial sting which they cannot use for defensive purposes (Radovic, 1981). This does not imply that they are not able to defend their colonies effectively. Workers of stingless-bee species use other means of defense very impressively. In some species, workers eject burning liquids while biting the predator. Many *Trigona* species show a very extraordinary defensive behavior which is at least as effective as that of *Apis mellifera*.

Most stingless bees build combs in a horizontal arrangement. Some species build simple cell clusters while others build regular horizontal combs. Vertical, double-layered combs are built by *Dactylurina* which are similar to those of *Apis*. Honey is stored in separate storage pots. Workers generally use "cerumen" (a mixture of wax and propolis) in cell construction. Stingless bees mostly nest in hollow trees or comparable cavities. The brood nest is often covered by the involucrum and both brood and storage areas are protected by several layers of "batumen" (plates consisting of propolis, hard cerumen, and mud or earth) (Fig. 1).

Before *Apis* was imported to South and Central America by humans, Meliponini were the only honey producers in these areas. Schwarz (1948) describes how native Indians used honey and wax from stingless bees. Even

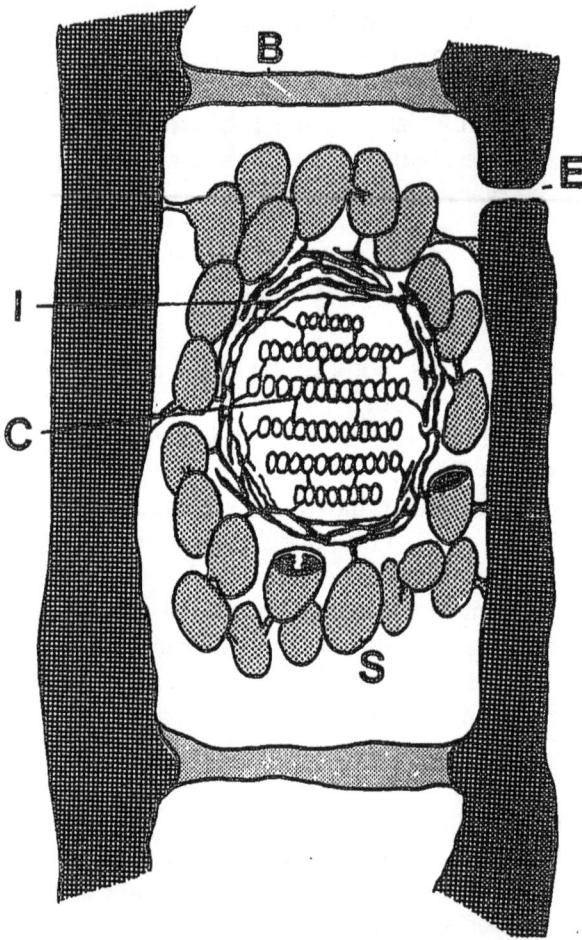

Fig. 1. Semischematic of a nest of *Melipona* species. The brood nest is covered by the layered involucrum. Storage pots are built in the next cavity which is protected by the batumen plates. B, batumen; E, nest entrance; I, involucrum; C, brood cells; S, storage pots.

today apiaries of Meliponini exist where nests may contain up to 2 kg of honey (Ly, 1981). Nogueira-Neto (1970) gives a detailed view of the management and domestication of several stingless-bee species. Nevertheless, the quality of honey produced by stingless bees varies widely and mostly is rather poor when compared to that of *Apis*. Thus, *Apis mellifera* has all but replaced Meliponini for honey production. In spite of this, the stingless bees are of great importance for the pollination of many wild flowers (Augspurger, 1980; Nogueira-Neto, 1970). The importation of *A. mellifera*, especially *A. m. scutellata*, seems to have had negative effects on the population density of stingless bees (Roubik, 1978).

A. Mating and Reproduction

When swarming, the young queens of stingless bees leave the nest. Before the flight of the virgin queens, workers from the mother colony will have prepared the new nesting site. Except for brood cells, the new nest is complete when the young queen arrives. Building the new nest may last several weeks. Alternatively, there are species (*Trigona julianii, T. varia,* etc.) which do only a little construction in the new nest by the time the young queen arrives.

After an orientation flight the virgin queens leave their nest on a mating flight which may last up to 102 min. Males often swarm in front of the nest entrance. In *Melipona plebeia* two types of males (normal and giant) can be distinguished (Cortopassi-Laurino, 1979) showing the same swarming behavior. Kerr (1975b) supposed a pheromone produced by queenless colonies attracts the males.

Direct observations by Kerr and Krause (1950) and spermatozoa counts by Kerr *et al.* (1962) suggest that queens of stingless bees only mate once. The number of spermatozoa of one drone (about 1,100,000) is quite similar to the number of spermatozoa in the oviducts of freshly mated queens. Also, the phenotypic segregations found by Kerr and Nielsen (1966, 1967) support the hypothesis of single mating. The hypothesis was further supported by da Silva *et al.* (1972), who found that a queen cannot remove her mating sign (male genitalia and seminal vesicles) during her nuptial flight.

In the genus *Lestrimelitta*, mating behavior seems to be completely different. The laying queen is mated by several males in the nest (Sakagami and Laroca, 1963). Multiple matings must not be restricted to *Lestrimelitta*, since gravid queens of *Trigona* sp. offered to swarming males outside the nest were readily mated (Michener, 1974). Stingless bees do not necessarily need to fly during mating. Camargo (1972a) obtained controlled matings of *Melipona quadrifasciata* in closed boxes (10 × 10 × 7 cm). Virgin queens were put together with 8- to 12-day-old males. In copulation lasting about 2 sec, the male mounts the queen from the back, separates, and leaves a mating sign.

B. Population Genetics

The genetic load (proportion of genes which are lethal recessives in a population) in haplo–diploid populations should be low, since only female-limited recessive lethal factors can exist. The genetic load of Meliponini populations is described by Almeida (1974) and Kerr (1975a). Their method of estimating the genetic load depends on a model proposed by

TABLE 1. Genetic Load in Stingless Bees[a]

Species	Genetic load, B	
	Kerr and Almeida (1981)	Kerr (1975a)
Melipona scutellaria	1.131	0.1557
Melipona rufiventris	1.330	0.1582
Melipona seminigra	1.080	0.1054
Melipona nigra	1.134	0.1593
Melipona quinquefasciata	1.240	0.1324
Melipona quadrifasctiata	1.107	—
Scaptotrigona xanthotricha	1.079	0.1042
Lestrimelitta limao	1.081	0.1056

[a] The X gene is treated in the left column as lethal and in the right column as vital.

Morton et al. (1956). The number of surviving brood in colonies having different degrees of inbreeding are determined. In species having a large genetic load, inbred colonies should have a low survival rate compared to non-inbred colonies. The probability for homozygosity at one of the recessive lethal alleles increases due to inbreeding. The genetic load B [lethal equivalents per gamete, according to Morton (1975)] is calculated by $B = -\log_e S - A$, where A is lethal effects of environment and S is rate of survival.

The results for several species of Melipona, Trigona, and Lestrimelitta are summarized in Table 1. The genetic load might be overestimated, since the method is very sensitive to effects of general inbreeding depression due to loss of variability (Fischer, 1978). Kerr (1975a) did not regard the sex-determining X gene in Melipona as a lethal gene, which seems to be correct according to Camargo (1977a). However, diploid drones of Melipona, though they do develop to adults, show a significantly lower viability than normal drones (Camargo, 1982). Kerr and Almeida (1981) treated the X alleles as lethal factors and added this to the genetic loads calculated by Kerr (1975a) (Table 1). The new values of B in stingless bees are almost the same as the value for Apis mellifera [$B = 1.3$; Kerr (1969a, 1974b)].

Several authors predict a low genetic variability in haplo–diploid populations (Hartl, 1971; Suomalainen, 1950, 1962; White, 1954). A source of variability in stingless-bee populations (as in all Hymenoptera) might be female-sex limited genes (Crozier, 1977; White, 1954). Kerr (1951, 1976) reports a female-limited gene in Melipona marginata. This model of variability due to female-limited genes is supported by the genetic load studies and in more detail by allozymic studies (Contel and Mestriner, 1974). They readily found polymorphic loci of the esterase complex (Est^2; Est^3) in Meli-

pona subnita. On the other hand, only very low levels of variation in allozymic studies were reported for various other Hymenoptera (Lester and Selander, 1979; Metcalf *et al.*, 1975; Pamilo *et al.*, 1975; Snyder, 1974). Thus, it does not seem likely that there is extensive variation due to sex-limiting events in stingless bees.

The effective number (N_e) in populations of Hymenoptera is given by a formula of Kerr (1974b, 1975b), who modified the equation of Wright (1933)

$$N_e = (9Nf \times Nm)/(4Nf \times 2Nm)$$

where Nf is number of females and Nm is number of males. Since in stingless bees the queen mates only with one male, Nm will be equal to the number of colonies in the population. The number of females is not approached that easily, since males are produced by both workers and queens. Ten to 95% of all drones may be worker-produced (Beig, 1972; Contel, 1972; Contel and Kerr, 1976; Kerr, 1969b). The number of laying workers in a colony varies from four in *Melipona quadrifasciata* (Tambasco, 1971) to 23 in *Scaptotrigona postica* (Beig, 1972). Kerr (1975b) shows that the effective number of females will increase as the proportion of worker-produced males increases (with no need to increase the number of colonies in the population). Estimates of N_e in natural populations of *Apis mellifera*, *Scaptotrigona postica*, and *Melipona rufiventris* are 4524, 575, and 1064, respectively (Kerr, 1974b). Kerr and Vencovsky (1982) and Vencovsky and Kerr (1982) suggest several selection models for breeding Meliponini. Their results still have a very theoretical character since there are no experimentally derived heritabilities for *Melipona* characteristics.

C. Sex Determination

Cunha and Kerr (1957) and Kerr and Nielsen (1967) suggested a genic-balance hypothesis for sex determination in stingless bees. Kerr (1974a, 1975c) modified his gene-balance hypothesis to a more complex model. Sex determination is proposed to be the result of a balance of non- or slightly cumulative male-determining genes and totally or almost totally cumulative female-determining genes. According to Kerr and Nielsen (1967) and Kerr (1975c), the genes responsible for caste determination in *Melipona* would be female genes which have lost their capacity to determine diploid males through successive mutations. This model was severely criticized by Crozier (1971, 1977) who proposed a one-locus, multiallele model.

Camargo (1976, 1979) showed that diploid males also appear in *Melipona quadrifasciata*. He could determine 18 chromosomes in metaphase one, and, as there was no nuclear division in metaphase, two spermatozoa remained diploid. These results confirm the model that sex is determined by one multiple-allele locus. In most cases, a ratio of 1 : 1 (diploid males to females) occurs after inbreeding, which also fits the one-locus model.

However, more than one sex locus seems to be common in *Trigona* species. In natural colonies of *Trigona quadrifasciata*, a ratio of 1 : 3 (diploid drones to workers) was found (Tarelho, 1973). Assuming single mating of the queen, this suggests the presence of two independent loci, as was found in *Bombus atratus* (Garofalo, 1973; Garofalo and Kerr, 1975).

In contrast to *Apis* (Woyke, 1967), the diploid drone larvae of the stingless bees are not eaten, since each cell is mass provisioned and sealed when the egg is laid. Camargo (1979) found 97% viability in colonies producing diploid males. However, diploid drones obviously have a lower longevity than normal drones (Camargo, 1982). This seems to be due to the sex alleles, since workers of the same inbred level have a significantly higher longevity.

In *Melipona*, the X gene is epistatic to the genes determining caste (Camargo, 1977a). Thus, diploid males, which are heterozygous at two caste-determining loci, did not develop queenlike features. Nevertheless, topical applications of juvenile hormone (10^{-6} μm/prepupae) on diploid drone prepupae did lead to queen features in about 25% of them (Camargo, 1977b; Campos, 1978). These males were thought to have the double heterozygous ($X_1^a/X_2^a; X_1^b/X_2^b$) caste-determining gene combination characteristic of *Melipona* queens.

As in *Bombus*, a certain proportion of stingless-bee males are produced by workers (Beig, 1969; de Silva, 1974; Tambasco 1971). Contel and Kerr (1976) confirmed this in an electrophoretic study of *Melipona subnita*. They obtained a proportion of 0.388 worker-laid males. The origin of the males was identified due to their isozyme pattern at the Est^3 locus.

D. Caste Determination

In *Trigona*, caste seems to be determined by the amount of larval food (Darchen and Delage, 1970; Darchen and Delage-Darchen, 1971; Camargo, 1971, 1972b,c), but in the genus *Melipona* caste is probably determined genetically. A nontrophic mechanism was suspected early (Ihering, 1903), and Kerr (1946) postulated a genetic model. Kerr's original hypothesis (1950a,b) involves a two-independent-loci model. Bees heterozygous at both loci become queens. Homozygosity at one or both loci leads to workers. In succeeding papers (Kerr, 1966, 1974a,b), this genetic model was confirmed. In most of the species tested a constant ratio of queens to workers developed (1 : 3 or 1 : 8, respectively), suggesting a two- or three-loci model.

However, Kerr *et al.* (1966) and Kerr and Nielsen (1966) showed that genetically determined *Melipona* queens can become workers due to overriding environmental conditions. Poorly fed larvae never become queens regardless of their genetic disposition. Camargo *et al.* (1976) found that species-specific amounts of food must be in the cells to assure that genetically determined queens do not become workers. Queens of *Melipona quadrifasciata* with worker phenotypes could be identified either by the number of ventral ganglia [Kerr and Nielson (1966): four in queens; five in workers] or by the number of tergal glands (Cruz-Landim *et al.*, 1980a). Workers with worker genotypes only have glands at tergite 2; workers with the queen genotype have pairs of glands at tergites 3, 4, 5, and 6.

Kerr (1974a) and Kerr *et al.* (1975) suggest that the "caste" genes regulate juvenile hormone production. Campos *et al.* (1975) and Velthuis and Velthuis-Kluppel (1975) showed that juvenile-hormone applications to 1-day-old larvae induced queens. There was a positive correlation between juvenile hormone doses and queen development for several *Melipona* species (Campos, 1979). Although juvenile hormone shows a comparable effect in *Apis mellifera* (Rembold *et al.*, 1974; Wirtz, 1973; Wirtz and Beetsma, 1972), the queen substance of *A. mellifera* (9-oxodecenic acid) has no effect on caste determination in stingless bees and does not counteract juvenile hormone (Engel, 1978, 1979). Another difference from *Apis* is that *Melipona* has not been induced to produce intercastes [Velthuis (1976); however, see Cruz-Landim *et al.* (1980a)], which supports the theory of a genetic component in caste determination.

In contrast to the genetical hypothesis, Darchen (1973), Darchen and Delage (1970), Darchen and Delage-Darchen (1974, 1975, 1977), and Stejskal (1974) suggest that only trophic factors determine whether a female larva becomes queen or worker. The two models differ mainly because of the different experimental observations in overfeeding larvae. According to Camargo *et al.* (1976), supranutrition has *no* effect on the 1 : 3 ratio of queens to workers. In contrast, the Darchens found that supranourishment of larvae increases the proportion of queens.

Campos *et al.* (1979) showed that both observations might be reconciled. They also found a case of 1 : 1 segregation of queens to workers and discussed this finding on the basis of the gene-balance hypothesis of Kerr *et al.* (1975). In total, four different gene batteries were invoked to account for the regulation of sex and caste. Campos *et al.* (1979) presumed that traces of juvenile hormone in the larval food might be responsible for the increase of the proportion of queens above the expected 1 : 3 ratio.

E. Cytogenetics

As in *Apis*, polyploidy in somatic tissue of stingless bees seems to be common. Cruz-Landim and Mello (1969) and Mello (1972) showed polyploidy patterns in developing larval silk glands of *Melipona quadrifasciata*

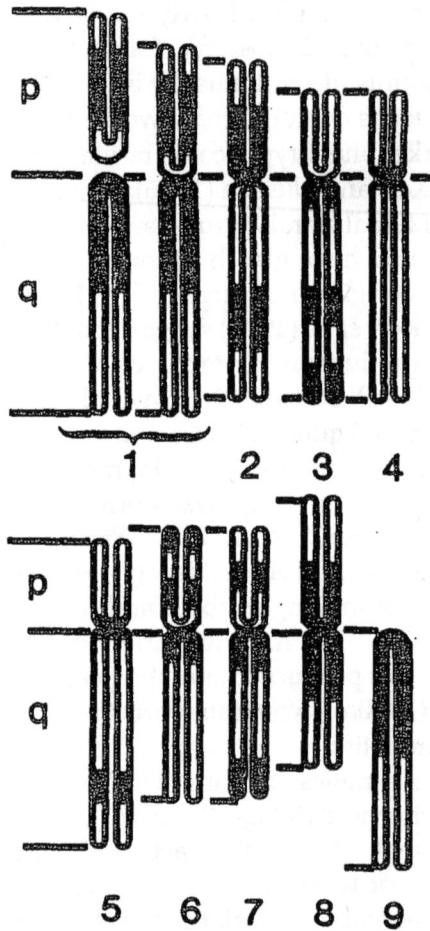

Fig. 2. G-Band pattern of chromosomes of *M. quadrifasciata*. Large chromosome arm, *q*; short chromosome arm, *p*. Each chromosome shows a characteristic G-band pattern. [From Tambasco *et al.* (1979), with permission. Copyright 1979 by *Cytologia*.]

and *M. quinquefasciata*. There is a gradient of polyploidy from the proximal to the distal end of the gland which correlates with the degree of local development and secretory activity (Mello *et al.*, 1970). Moraes *et al.* (1972) gave comparable data for midgut cells of *M. quadrifasciata* larvae. The DNA content, as well as the nuclear volumes of the salivary glands of pupae and adult workers, varies according to the stage and activity of the tissue (Moraes, 1978). Also, polyploidy was found in the seminal glands (Mello, 1969) and in the Malpighian tubules of *Melipona quinquefasciata* (Mello and Silveira, 1970) and *M. quadrifasciata* (Mello and Takahashi, 1969, 1971).

The first studies concerning the morphology of chromosomes were done on *M. quadrifasciata*. Analyzing the G-band pattern (Fig. 2), the chromosome number of *n* = 9 was determined (Tambasco *et al.*, 1979). Cruz-Landim *et al.* (1980b) studied the mechanism of meiosis and mitosis for several stingless-bee species: *M. quadrifasciata, Scaptotrigona postica, Nannotrigona testaccicornis, Plebeia droryara, Frieseomelitta varia,* and *Leurotrigona muelleri.* In all these species, a membrane was maintained around the nucleus during mitotic and meiotic cell divisions in spermatogenesis and in some myoblast cells. The centrioles remained outside this membrane. The spindle seemed to be entirely organized by the kinetochore.

Kerr (1972a), Kerr and Silveira (1972) and Silveira (1971, 1972) gave data on stingless-bee chromosome numbers. They hypothesized that in the evolution of stingless bees, polyploidy originated independently three times: twice in *Trigona* [first from an ancestor of *Leurotrigona muelleri* ($n = 8$) to *Frieseomelitta* ($n = 15$); second from an ancestral *Trigonini* ($n = 9$) to *Plebeia* ($n = 18$)] and once in *Melipona* [from an ancestor like *Melipona quadrifasciata* or *M. marginata* ($n = 9$) to *M. quinquefasciata* ($n = 18$)]. The authors suggested that the basic number of chromosomes in stingless bees was $n = 8$. Thus, species of *Melipona* with $n = 9$ are said to have their origin from polysomy. Possibly, this polysomy arose from the effect of low temperature on meiosis in *Melipona* (Kerr, 1972b), which leads to uneven divisions. According to Kerr (1974c) and Kerr and Silveira (1972), *Trigona* species having 15, 14, or 17 chromosomes could be results of Robertsonian fusions. Combining their data with those models based on morphology, ethology, and caste determination (Cruz-Landim, 1967; Kerr, 1969b; Kerr and Esch, 1965; Michener, 1944), Kerr and Silveira (1972) constructed a diagram of relationship (Fig. 3). Crozier (1977) considered this model to be highly tentative because of the small number of species (48) karyotyped and the general lack of published details on chromosome morphology.

IV. NON-*MELLIFERA* (*APIS*)

In this section the genetics of the three non-*mellifera Apis* species (*A. cerana, A. dorsata,* and *A. florea*) will be discussed. Thus, the system of Maa (1953) is not used, and species such as *A. laboriosa,* which was recently proposed by Sakagami *et al.* (1980), will not be discussed because almost no information concerning their biology is available. All non-*mellifera Apis* exist in Asia; the two open-nesting species, *A. florea* and *A. dorsata,* only appear in the tropics and subtropics.

A. Mating Behavior

Apis cerana is closely related to *A. mellifera* and has very similar mating behavior. Ruttner *et al.* (1973) and Ruttner (1973) observed matings of *A. cerana* in the same drone congregation area used by *A. mellifera* in Germany.

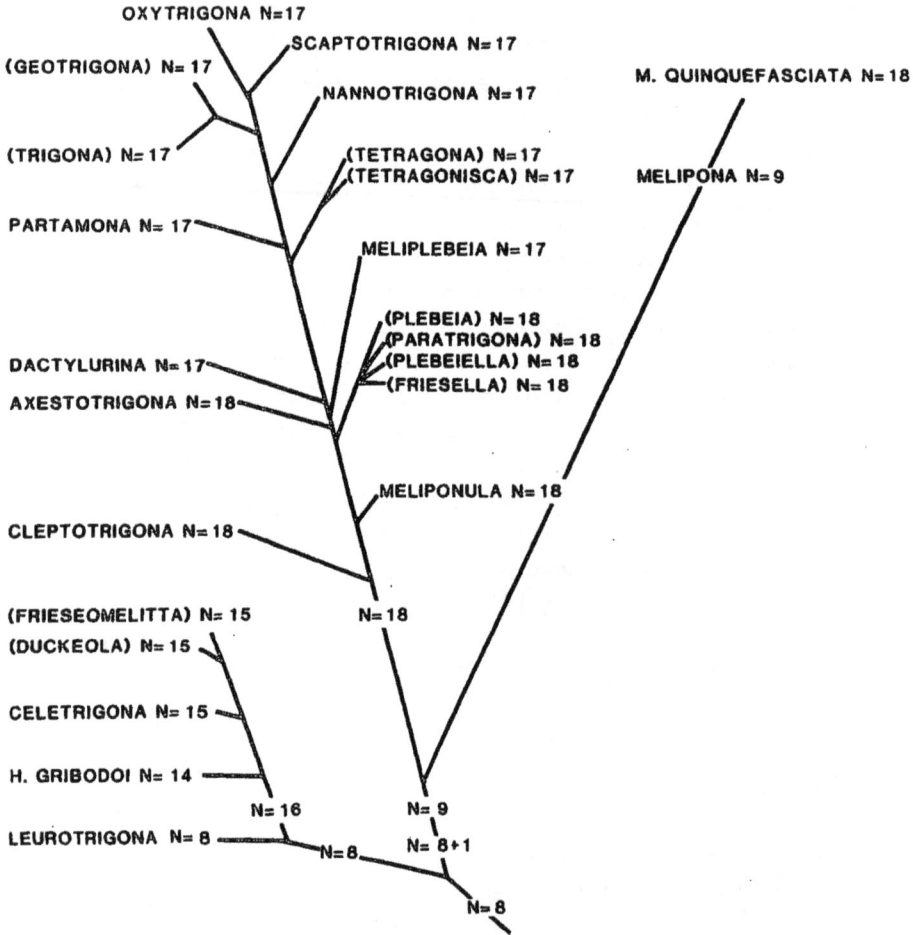

Fig. 3. Hypothetical diagram of relationship in stingless bees. Due to the small number of species, this diagram is considered highly tentative. N, number of chromosomes. [Modified from Kerr and Silveira (1972), with permission. Copyright 1972 by the Society for the Study of Evolution.]

Drones of *A. cerana* have much less semen than those of *A. mellifera,* and *A. cerana* queens mate with more males (30 males per queen). In their original habitat in India, Woyke (1975) showed that *A. cerana* queens mate with an average of 10 males during their first mating flight. To determine this he killed queens after their first successful mating flight. Larger numbers of males per queen might be common since in most cases a second mating flight is taken (Shah and Shah, 1980). From the data of Woyke (1975), I calculated the approximate flight time from hive to congregation area and the mean time per mating (Fig. 4). *Apis cerana* has a shorter mating time per drone (0.79 min) than *A. mellifera* (2.04 min; Woyke, 1956; Moritz, 1985).

Number of Drones

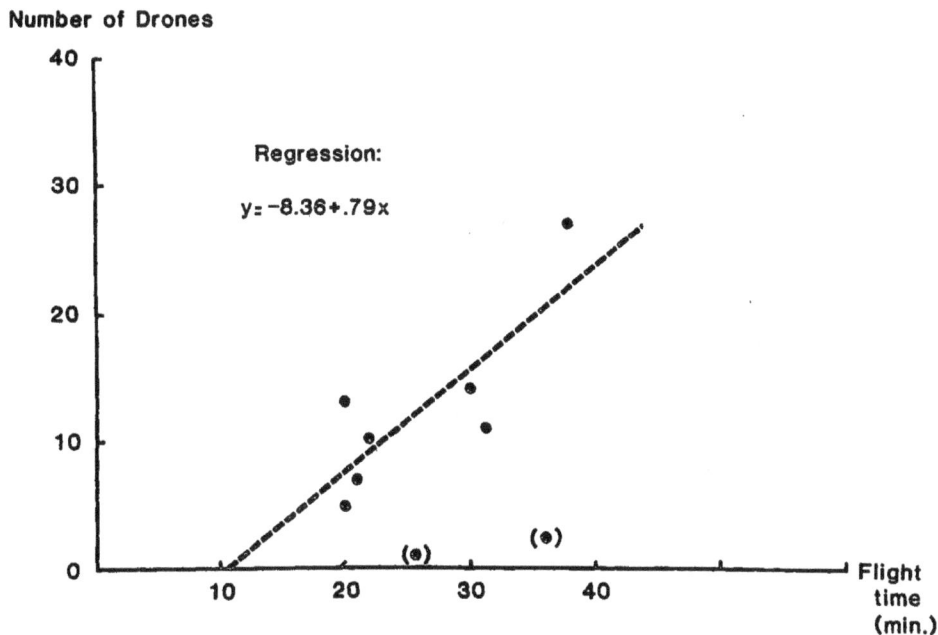

Fig. 4. Regression of flight time and number of mated drones for *A. cerana* queen mating flight. Assuming a linear regression of the queens' mating time per drone, the regression coefficient $b = 0.79$. The flight time between hive and drone congregation area is given by the intercept of the regressional line with the X axis. (•), Flights not regarded due to too few matings. [Drawing by Sandra Kleinpeter.]

The duration of the flight between the hive and the congregation area is about 10 min for both species. These calculations agree well with observations by Ruttner *et al.* (1973) of *A. cerana* and *A. mellifera* drones in the same drone congregation area.

Apis mellifera and *A. cerana* are not sympatric and have similar mating times. In their German study, Ruttner *et al.* (1972) found that *A. cerana* queens did not successfully mate when *A. mellifera* drones were abundant; however, when the *A. mellifera* drones were excluded from the congregation area, *A. cerana* queens mated successfully. Similar observations were made in Pakistan in another way. Imported *A. mellifera* queens did not successfully mate under natural conditions (Ruttner *et al.* 1973).

Little is known about the mating behavior of the other Asian *Apis* species. However, all four species of *Apis* have many similarities in behavior and anatomy (Koeniger, 1976). The same sex pheromone attracts drones of all species (Shearer *et al.*, 1970) as do queen extracts from the different species (Butler *et al.*, 1967). The three Asian species occur sympatrically in many habitats and interspecific matings, or the interference of specific matings, might easily result from such similarities. However, Koeniger and Wijaya-

% Flights/15 min

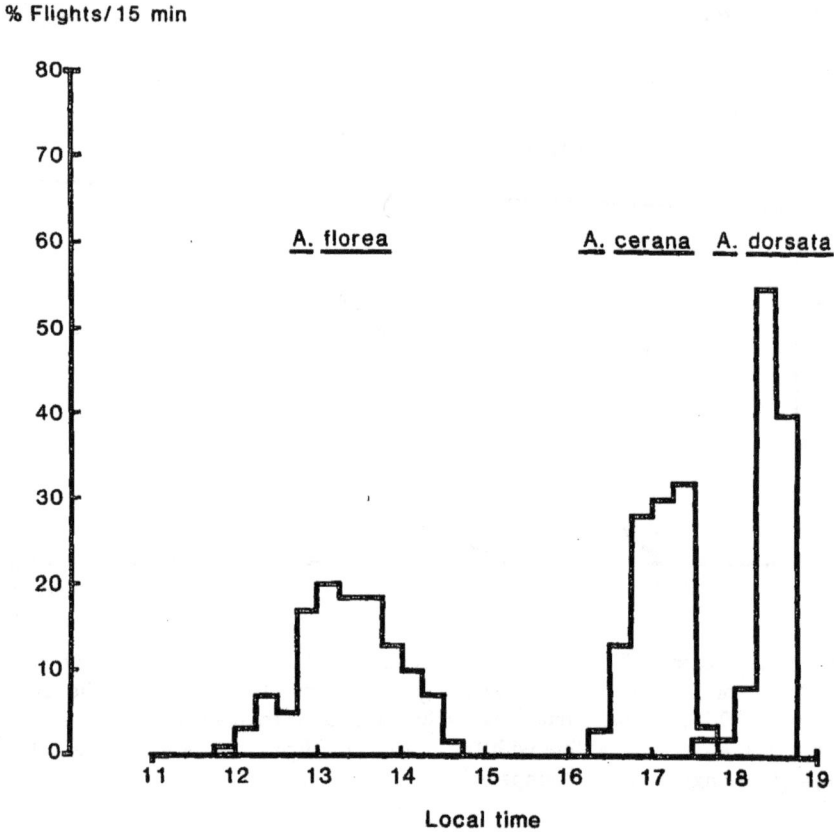

Fig. 5. Flight times of drones of three *Apis* species. The drones of the three sympatric species show well separated flight times. X axis: local time in Sri Lanka. [From Koeniger and Wijaya-gunasekera, (1976), with permission. Copyright 1976 by the International Bee Research Association.]

gunasekera (1976) showed that this is prevented by well-separated drone, and presumably queen, flight times in Sri Lanka (Fig. 5). *Apis florea* drones flew early in the afternoon; *A. cerana* made their mating flights late in the afternoon; *A. dorsata* drones had their main flight activity at dusk.

B. Sex Determination

Sex determination in *A. cerana* is nearly identical to that in *A. mellifera*. Using instrumental insemination, Woyke (1973, 1975, 1979), produced inbred lines of *A. cerana*. The drones of *A. cerana* only produced 0.2 mm³ of semen (one-sixth that of *A. mellifera* drones), and only those queens inseminated with the semen of more than 15 drones laid normal numbers of fertilized eggs. Examining 572 larvae of the offspring of inbred queens,

Woyke detected that 27.4% were diploid drones. He suggested that, as in *A. mellifera*, there is only one sex locus with several alleles. Homozygosity or hemizygosity at that locus results in males. Hoshiba *et al.* (1981) observed the same ratio of diploid drones in *A. cerana japonica* and confirmed Woyke's model.

Diploid-drone larvae of *A. cerana* are not eaten by the workers just after hatching (as in *A. mellifera*), but mostly they are eaten at the age of 1 day. Some larvae were even reared for 4 days, and Woyke (1980) suggested that under favorable natural conditions, *A. cerana* diploid drones might even develop to the adult stage. Presumably, the diploid drone larvae of *A. cerana* produce much less cannibalism substance than those of *A. mellifera* (Dietz, 1975; Woyke, 1967). However, *A. cerana* diploid-drone larvae seem to produce this substance (a type of pentane) longer.

C. Cytogenetics

The common number of chromosomes ($n = 16$) of *A. cerana* and *A. mellifera* is often considered a sign of their close relationship (Deodikar *et al.*, 1959; Hoshiba *et al.*, 1981; Sharma *et al.*, 1961). Since tetrads in male and octads in female meiosis appear often, the number of $n = 16$ is said to be the result of phylogenetic polyploid formation (Deodikar and Thakar, 1966; Kerr and Silveira, 1972). Deodikar and Thakar (1966) and Thakar and Deodikar (1966) claimed that *A. dorsata* and *A. florea* should have a chromosomal set of $n = 8$; however, Fahrenhorst (1977a,b) showed that these two species also have $n = 16$. This was previously reported for *A. dorsata* by Kumbkarni (1964). Figure 6 shows the haploid chromosomal set of all three

Fig. 6. Chromosomes of the Asian *Apis* species. The chromosomes are listed according to size for *A. cerana*, *A. dorsata*, and *A. florea*. [From Fahrenhorst, (1977b), with permission. Copyright 1977 by the International Bee Research Association.]

TABLE 2. Families, Sociality, and Distribution on Non-Apidae Bees[a]

Family	Degree of sociality	Distribution
Colletidae	Solitary	Worldwide
Halictidae	Solitary, communal, quasisocial, semisocial, eusocial	Worldwide
Andrenidae	Solitary, communal	Worldwide, except Australia
Oxaeidae	Solitary	New World
Melittidae	Solitary	Worldwide
Fidelidae	Solitary	Africa, Chile
Megachilidae	Solitary, communal, quasisocial	Worldwide
Anthophoridae	Solitary (parasitic), communal, quasisocial, subsocial, eusocial, socially parasitic	Worldwide

[a] After Michener, 1974.

Asian species of *Apis* ($n = 16$) (Fahrenhorst, 1977b). All three complements include only one large mediocentric chromosome, so the chromosome number cannot result from somatic polyploidy. Thus, the karyotype of the four species of *Apis* cannot be used to show a phylogenetic relationship in the genus.

Also, the DNA study of Jordan and Brosemer (1974) gave no clear separation between the species *A. mellifera*, *A. cerana*, and *A. florea*. They characterized the DNA of these species by the distribution of the guanine–cytosine (GC) pairs. Though *A. mellifera* and *A. cerana* DNAs were rather similar, all three species had a broad GC distribution.

Although *A. cerana* and *A. mellifera* are closely related species, they are quite distinct and there have been no reports of vital hybrids. Ruttner (1969) and Maul (1969) obtained hybrid eggs by artificial insemination. These eggs only developed to the blastoderm stage and then died. Ruttner and Maul suggested that the maternal enzyme apparatus may cease functioning at that stage and that as the hybrid genome starts its own protein synthesis, the regulation of metabolism stops.

V. OTHER SOCIAL AND SOLITARY BEES

This group contains many different families, tribes, and genera. Table 2 shows the most important families and their geographic distribution. In most families, the degree of sociality ranges from solitary to eusocial. In *Anthophoridae*, parasitic forms (e.g., *Nomadia*) appear. Due to the wide-ranging biological differences among these bees, I will not give a detailed view of their general biology. The interested reader is referred to Michener (1974).

In the last two decades the important role of solitary bees as pollinators of commercial crops has been the topic of many investigations. Leafcutter bees (Megachilidae) and sweat bees (Halictidae) have special interest because they pollinate many important crop plants that are only poorly pollinated by honey bees (Greenberg, 1982; Rank and Goerzen, 1981; Tasei, 1975, 1977). One of the main problems of domestication seems to be the overwintering of the adults. In *Osmia coeru lescens*, a leafcutter bee with a good ability to pollinate red clover, only about 50% of the overwintering females survive (Parker, 1981; Ruszkowski, 1979). Also, domestication and rearing in human-made hives is difficult. Rank and Goerzen (1981) kept *Megachile relativa* in hives and only obtained 1280 progeny cells from a starting population of 8200 cells.

A. Mating Behavior

Patterns of mating behavior are quite diverse. As in *Apis*, chemical substances and pheromones play a rather important role in courtship. In the tropical orchid bees (Euglossini), males gather substances from orchid flowers, fungus-infested wood, or other substrates (Kullenberg and Bergström, 1974, 1976; Vogel, 1966; Zucchi *et al.*, 1969). These compounds are placed in gland slits in the tibia of each hind leg and become sex attractants for the females. Michener (1974) points out that the collected substances only can act as precursors of the pheromones because the flowers themselves are not attractive to the females. Several of these substances were identified as 1,8-cineol, benzyl acetate, methyl salicylate, and methyl cinnamate (Dodson *et al.*, 1969). Kimsey (1980) showed the attractiveness of the male pheromones to females.

In Halictidae, pheromonelike substances also seem to be important for mating. The females produce similar but species-specific sex attractants in their mandibular glands (Barrows, 1974, 1975a,b). Although males pounce at anything with the same size and flight behavior as their own females, they only mate females of their species. The males even seem to have a capacity to learn the smell of the females. Smith (1982) exposed a female for 10 min to a group of males and found that the attraction of a second female was negatively correlated to the genetic relation between the two females. Smith suggested a genetically determined multicomponent chemical signal which males might be able to discriminate.

Bergström (1981) recently reviewed his work with Tengö and Kullenberg on the composition of the different compounds secreted in the Dufour and

TABLE 3. Volatile Components in Solitary Bees[a]

Genera	Dufour glands	Mandibular glands
Lasioglossum	Lactones	—
Halictus	Lactones	—
Colletes	Lactones	Linalool
Andrena	Farnesyl hexanoate; geranyl octanoate	Monoterpenes; spiroketals; straight chain ketones, alcohols, and butyrates; geraniol, neral
Mellita	Butyrates	—
Nomada	—	Farnesyl hexanoate; geranyl octanoate
Xylocopa	—	Lactones[b]

[a] A modification of Bergström (1981), which summarizes the papers of Bergström (1974, 1978, 1979), Francke et al. (1980), Bergström and Tengö (1974, 1978, 1979), Tengö (1979), and Tengö and Bergström (1975a,b, 1976, 1977a,b, 1978).
[b] Wheeler et al. (1976); Bacardit and Moreno-Mañas (1980).

mandibular glands of several bees (Table 3). Duffield et al. (1981) identified several macrocyclic lactones and isopentylester in the Dufour's gland secretion of 18 different bee species. Most mandibular gland secretions seem to be important for mating. In Andrena the compounds produced in the mandibular glands show great similarities between both sexes. The secretions are obviously attractive to both sexes. The compounds of the mandibular glands in males are also used to scent various subjects (as territory marks) in Andrena, Centris (Frankie et al., 1980), and Triepeolus (Alcock, 1978). In Centris adani, the marketing pheromone of the copulating male makes the females unattractive to other territorial males. In the cleptoparasitic Nomada, the mandibular gland secretions of the males mimic the host's Dufour's gland secretions (Tengö and Bergström, 1975b, 1977b). They may presumably aid the female's entry into the host's nest (Eickwort and Ginsberg, 1980).

Besides these sex attractants and marking compounds of the males, the queens of Lasioglossum zephyrum seem to produce a pheromone which inhibits the mating of their workers (Greenberg and Buckle, 1981). Sixty percent of the females kept in isolation from other females successfully mated, while only 7.7% of the workers from colonies with queens showed a mating receptivity. Similar observations were made in Evylaeus calceatus (Plateaux-Quénu, 1974).

There are three main strategies that male solitary bees use when finding females (Alcock et al., 1978; Barrows, 1976). Males either may mate females at the nesting site (38%), may mate at flowers or other points of resource

(41%: most common), or may mate at nonresource sites (21%: most common when nests and flowers are widely dispersed). In all three cases, males may either patrol nonaggressively (eg. *Osmia rufa*) or show territorial behavior (e.g., *Anthidium manicatum*) (Severinghaus *et al.*, 1981). In the latter case, the male occupying a territory will behave aggressively toward other males entering his territory, which ranges in size from 0.5 to 6 m² (Alcock *et al.*, 1978). Velthuis and Gerling (1980) showed that in *Xylocopa sulcatipes*, a carpenter bee, territorial behavior as well as nonaggressive behavior occur in the same species according to the environmental conditions.

B. Population Genetics

Genetic variation in populations of solitary and primitively eusocial bees seems to be as low as it is in the social species. Isoenzymes of the esterase complex in *Anthrophora quadrifasciata*, *Megachile nipponica*, and *Xylocopa* sp. were found by Tanabe *et al.*, (1970) but no polymorphisms were reported. Snyder (1974) tested 25 enzymes in *Lasioglossum zephyrum* and *Augochlora pura* and did not find any polymorphism. In *L. zephyrum*, sex-limited allozymes (Est^1; Est^2) were found. This is interesting since sex-limited genes are thought to be one of the sources of genetic variability in haplo–diploid populations (Crozier, 1977). More recently, some authors succeeded in showing polymorphisms in solitary bees. Mean heterozygosities of 0.015 for *Megachile pacifica* and of 0.041 for *Nomia melanderi* were found (Lester and Selander, 1979). A sex limitation in the esterase complex appeared in *N. melanderi*. Pamilo *et al.*, (1978) studied *Macropis labiata*, *Colletes succincta*, *Andrena clarkella*, *Andrena lapponica*, and *Andrena vaga* and obtained very low degrees of heterozygosity ranging from 0 to 0.064.

Thus, it seems to be a general rule that polymorphic loci are not common in bees. In other insects, significantly larger mean heterozygosity values are found (see Powell, 1975). However, a large intraspecific variation in bee body color seems to be common. For example, melanism clines are found in *Andrena*. Lanham (1974) reports one in *Andrena albihirta* and suggests that black color is advantageous in lower temperatures. There also seems to be a genetically determined colony odor which varies even between closely related colonies. Greenberg (1979) demonstrated that there is a positive correlation between the coefficient of relationship between an entering and a guarding bee and the frequency with which the guard allows the entering bee to pass (*Lasioglossum zephyrum*).

C. Sex Determination

Although generally males develop from unfertilized eggs and females from fertilized eggs in all of Apoidea, there is no genetical model for sex determination based on experimental data for non-Apidae bees. Greenberg

(1979) could not find any deleterious effects during an inbreeding experiment in *Lasioglossum zephyrum,* and mechanisms other than those proposed for Bombini and Apinae might exist in solitary bees.

Although there is no evidence concerning the genetic control of sex, the control by females of egg fertilization is well documented. Females of the leafcutter bee *Megachile rotunda* have a sperm pump which allows exact control of sperm release from the spermatheca (Gerber and Klostermeyer, 1970). Unmated females produce only male offspring (arrhenotkous parthenogenesis). Sweat-bee females *(Evylaeus calceatus)* also control the sex of the eggs they lay. The production of males seems to be influenced by the weight of the pollen ball deposited in the cell. Males generally develop on smaller pollen balls than females (Plateaux-Quénu, 1982).

Although the queen has control over fertilization, errors in oviposition might occur. In *Osmia rufa,* males erroneously were reared in 7% of large "female" cells (Raw and O'Toole, 1979). Such errors in oviposition also occurred in *Lasioglossum erythurum* (Knerer, 1980). In this case, the nonfertilized eggs laid on large pollen balls developed to macrocephalic males. Knerer suggested that this mechanism might be the evolutionary origin of caste determination. Those female eggs laid on small pollen balls ("male" cells) could have become infertile females; those laid on large pollen balls could have become queens.

Environmental effects on the sex ratio in *Lasioglossum* are also well documented (Knerer and Plateaux-Quénu, 1967a,b; Michener and Wille, 1961; Plateaux-Quénu and Plateaux, 1980). Similar observations were made for *Osmia lignaria* (Torchio and Tepedino, 1980). Though the overall sex ratio was rather constant in this species (2 : 1, males to females), seasonal differences resulted in more drones being produced in the second half of the season. An equal sex ratio (1 : 1) was found in *Osmia rufa* (Raw and O'Toole, 1979). Sex ratios ranging from 2 : 1 to 1 : 1 (males to females) seem to be rather common and are found in *Osmia coerulescens* (Parker, 1981), *Megachile concinna, M. zaptlana, Chalicodoma lanata,* and *C. rufipennis* (Jayasingh and Freeman, 1980).

D. Cytogenetics

Only limited cytogenetic data exist for this group. Kerr (1972a) gives the chromosome numbers of several Halictidae and Anthophoridae. The most common number of chromosomes in his study was $n = 16$, which was found also for *Xylocopa violacea, X. fenesterata* (Granata, 1909; Kumbkarni, 1965; Makino, 1951), and *Osmia rufa* (Armbruster, 1913). In a few cases, different chromosome numbers were found [*Lasioglossum rhythidophorum,* $n = 6$; *Pseudaugochloropsis graminea,* $n = 8$; *Exomalopsis aureopilosa,* $n = 9$ (Kerr, 1972)].

Jordan and Brosemer (1974) characterized the DNA of *Megachile rotunda* concerning its guanine–cytosine distribution and renaturation speed in order to separate this species from *Apis*. The DNA of *M. rotunda* had a rather narrow distribution and could be clearly distinguished from the broad distribution of *Apis*. Since no other bees were tested, a comparative examination of additional solitary bees is needed to set up a diagram of a relationship using DNA. The same is true of karyotypes. Too few species have been karyotyped to support any phylogenetic interpretation.

REFERENCES

Alcock, J. (1978). Notes on male mate locating behavior in some bees and wasps of Arizona. *Pan. Pac. Entomol.* **54**, 215–225.

Alcock, J., Barrows, E. M., Gordh, G., Hubbard, L. G., Kirkendall, L., Pyle, D. W., Ponder, T. L., and Zalom, F. G. (1978). The ecology and evolution of male reproductive behavior in the bees and wasps. *J. Linn. Soc. London Zool.* **64**, 293–326.

Alford, D. V. (1975). "Bumblebees." Davis-Poynter Ltd., London.

Almeida, M. G. (1974). Aspectos biologicos, ecologicos e genéticos da abelha *M. scutellaris* latr. 1811) Msc. Thesis. Univ. São Paulo, Ribeirão Preto.

Armbruster, L. (1913). Chromosomenverhältnisse, bei der Spermatogenese solitärer Apiden (*Osmia cornuta* latr.) *Arch. Zell-Forsch.* **11**, 242–326.

Augspurger, C. K. (1980). Mass-flowering of a tropical shrub (*Hybanthus prunifolius*): influence on pollinator attraction and movement. *Evolution* **34**, 475–488.

Awram, W. J. (1970). Flight route behaviour of bumblebees. Ph.D. Thesis. University of London.

Bacardit, R., and Moreno-Manas, M. (1980). Hydrogenations of triacetic acid lactone. A new synthesis of the carpenter bee (*Xylocopa hirsutissima*) sex pheromone. *Tetrahedron Lett.* **21**, 551–554.

Barrows, E. M. (1974). Pheromones characteristic of individuals in *Lasioglossum zephyrum*. *J. Kansas Entomol. Soc.* **47**, 540.

Barrows, E. M. (1975a). Mating behavior in Halictine bees (Hym. Halictidae) III. Copulatory behavior and olfactory communication. *Insectes Soc.* **22**, 307–332.

Barrows, E. M. (1975b). Individually distinctive odors in an invertebrate. *Behav. Biol.* **15**, 57–69.

Barrows, E. M. (1976). Mating behavior in halictine bees II. Microterritorial and patrolling behavior in ♂♂ of *Lasioglossum rohweri*. *Z. Tierpsychol.* **40**, 377–389.

Beig, D. (1969). Produção de rainhas e desenvolvinato embrionário em *Trigona* (*Scaptotrigona*) *postica* Latr. Ph.D. Thesis. Univ. São Paulo, Rio Claro.

Beig, D. (1972). The production of males in queenright colonies of *Trigona* (*Scaptotrigona*) *postica*. *J. Apic. Res.* **11**, 33–39.

Bergström, G. (1974). Macrocyclic lactones in the Dufour gland secretion of the solitary bees *Colletes cunicularius* L. and *Halictus calceatus* Scop. (Hymenoptera, Apidae). *Chem. Scr.* **5**, 39–46.

Bergström, G. (1978). Role of volatile chemicals in *Ophrys*–pollinator interactions. In "Biochemical Aspects of Plant and Animal Coevolution" (G. Harborne, ed.), pp. 207–231. Academic Press, New York.

Bergström, G. (1979). Complexity of volatile signals in hymenopteran insects. In "Proceedings of the Conference on Chemical Ecology and Odour Communication in Animals" (F. J. Ritter, ed.), pp. 187–200. Elsevier/North Holland.

Bergström, G. (1981). Chemical aspects of insect exocrine signals as a means for systematic and polygenetic discussions in aculeate Hymenoptera. *Entomol. Scand. Suppl. 15*, 173–184.

Bergström, G. (1982). Chemical composition, similarity and dissimilarity in volatile secretions: examples of indications of biological function. *Mediateur Chimique Versailles les Coll. INRA 7*, 289–296.

Bergström, G., and Tengö, J. (1974). Studies in natural odoriferous compounds. IX. Farnesyl- and geranyl esters as main volatile constituents of the secretion from Dufour's gland in 6 species of *Andrena* (Hymenoptera, Apidae). *Chem. Scr. 5*, 28–38.

Bergström, G., and Tengö, J. (1978). Linalool in mandibular gland secretion of *Colletes* bees. *J. Chem. Ecol. 4*, 437–449.

Bergström, G., and Tengö, J. (1979). C_{24}-, C_{22}-, C_{20}- and C_{18}-Macrocyclic lactones in halictide bees. *Acta Chem. Scand. 33*, 390.

Bergström, G., Kullenberg, B., and Ställberg-Stenhagen, S. (1967). Studies on natural odori- ferous compounds. II. Identification of a 2,3-dihydrofarnesol as the main component of the marking perfume of male bumble bees of the species *Bombus terrestris* L. *Ark. Kemi. 28*, 453–469.

Bergström, G., Kullenberg, B., and Ställberg-Stenhagen, S. (1973). Studies on natural odori- ferous compounds. VII. Recognition of two forms of *Bombus lucorum* L. (Hymenoptera, Apidae) by analysis of the volatile marking secretion from individual males. *Chem. Scr. 4*, 174–182.

Bergström, G., Svensson, I. G., Appelgren, M., and Groth, I. (1981). Complexity of bumble bee marking pheromones: biochemical ecological and systematical interpretations. *In* "Biosys- tematics of Social Insects" (P. E. Howse and J. L. Clement, eds.), pp. 175–183. Academic Press, New York.

Bringer, B. (1973). Territorial flight of the bumble-bee males in coniferous forest on the north- ernmost part of the island Öland. *Zoon Suppl. 1*, 15–22.

Butler, C. G., Calam, D. M., and Callow, R. K. (1967). Attraction of *Apis mellifera* drones by the odours of the queens of two other species of honeybees. *Nature (Lond.) 214*, 423–424.

Camargo, C. A. (1971). Determinação de castas em *Scaptotrigona postica* Latr. (Hymenoptera Apidae). *Proc. XXIII Reun. a. Soc. Bras. Prog. Ciênc.* p. 106.

Camargo, C. A. (1972a). Mating of the social bee *Melipona quadrifasciata* under controlled conditions. *J. Kansas Entomol. Soc. 45*, 520–523.

Camargo, C. A. (1972b). Determinação de castas em *Scaptotrigona postica* Latr. (Hymenoptera, Apidae). *Rev. Bras. Biol. 32*, 133–138.

Camargo, C. A. (1972c). Produção "in vitro" de intercastas em *Scaptotrigona postica* Latr. (Hymenoptera, Apidae). Homenagen a W. E. Kerr, *Fac. Filosof. Ciênc. Let. Rio Claro. B*, 37–45.

Camargo, C. A. (1976). Determinação do sexo e controle de reproducao em *Melipona quadri- fasciata* Lep. (Hymenoptera Apidae). Ph.D. Thesis. Univ. São Paulo, Ribeirão Preto.

Camargo, C. A. (1977a). Properties of the X° gene, sex determining in *Melipona quadrifasciata. Proc. Congr. IUSSI 8th (Wageningen)*, pp. 191–192.

Camargo, C. A. (1977b). Effects of juvenile hormone on diploid drones of *Melipona quadri- fasciata. Proc. Congr. IUSSI 8th (Wageningen)*, pp. 193–194.

Camargo, C. A. (1979). Sex determination in bees. XI. Production of diploid males and sex determination in *Melipona quadrifasciata. J. Apic. Res. 18*, 77–84.

Camargo, C. A. (1982). Longevity of diploid males, haploid males and workers of the social bee *Melipona quadrifasciata* Lep. (Hymenoptera Apidae). *J. Kansas Entomol. Soc. 55*, 8–12.

Camargo, C. A., Almeida, M. G., Parra, M. G. W., and Kerr, W. E. (1976). Genetics of sex determination in bees. IX. Frequencies of queens and workers from larvae under controlled conditions. *J. Kansas Entomol. Soc. 49*, 120–125.

Campos, L. A. O. (1978). Sex determination in bees. VI. Effect of a juvenile hormone analogue in males and females of *Melipona quadrifasciata. J. Kansas Entomol. Soc. 51*, 228–234.

Campos, L. A. O. (1979). Sex determination in bees. XIV. Role of juvenile hormone in caste differentiation in the subfamily Meliponinae. *Rev. Bras. Biol. 39*, 965–971.

Campos, L. A. O., Velthuis-Kluppell, F. M., and Velthuis, H. H. W. (1975). Sex determination in bees. VII. Juvenile hormone and caste determination in a stingless-bee. *Naturwissenschaften 62*, 98–99.

Campos, L. A. O., Kerr, W. E., and Silva, D. L. N. da (1979). Sex determination in bees. VIII. Relative action of genes X^A and X^B on sex determination in *Melipona* Bees. *Rev. Bras. Gen. 2*, 267–280.

Contel, E. P. B. (1972). Aspectos genéticos e biologicos obtidos a partir de estudos de proteinos, em duas espécies de Melipona. Ph.D. Thesis, Univ. São Paulo, Ribeirão Preto.

Contel, E. P. B., and Kerr, W. E. (1976). Origin of males in *Melipona subnitida* estimated from data of an isozymic polymorphic system. *Genetica 46*, 271–277.

Contel, E. P. B., and Mestriner, M. A. (1974). Esterase polymorphisms at two loci in the social bee. *J. Hered. 65*, 349–352.

Cortopassi-Laurino, M. (1979). Observações sobre atividades de machos de *Plebeia drorgana* Friese (Apidae, Meliponinae). *Rev. Bras. Entomol. 23*, 177–191.

Crozier, R. H. (1971). Heterozygosity and sex determination in haplo-diploidy. *Am. Nat. 105*, 399–412.

Crozier, R. H. (1977). Evolutionary genetics of the Hymenoptera. *Annu. Rev. Entomol. 22*, 263–288.

Cruz-Landim, C. da (1967). Estudo comparativo de algumas glândulas das abelhas (Hymenoptera, Apidae) e respectivas implicações evolutivas. *Arq. Zool. (São Paulo) 15*, 177–290.

Cruz-Landim, C. da, and Mello, M. L. S (1969). Development of poliploidy in silk glands of *Melipona quadrifasciata anthidioides* Lep. (Hymenoptera Apoidea) during the larval stage. *J. Exp. Zool. 170*, 149–156.

Cruz-Landim, C. da, Santos, S. M. F. dos, and Höfling, M. C. A. (1980a). Sex determination in bees XV. Identification of queens of *Melipona quadrifasciata anthidioides* (Apidae) with the worker phenotype by a study of tergal glands. *Rev. Bras. Gen. 3*, 295–302.

Cruz-Landim, C. da, Silva de Moraes, R. L. M., and Beig, D. (1980b). Membranous envelope of the spindle during the cell division in bees (Apidae Meliponinae). *Ciênc. Cult. (São Paulo) 32*, 931–936.

Cunha, A. B. da, and Kerr, W. E. (1957). A genetical theory to explain sex determination by arrhenotokous parthenogenesis. *Forma Functio 4*, 33–36.

Darchen, R. (1973). Essai d'interprétation du déterminisme des castes chez les *Trigones* et les *Melipones, C. R. Acad. Sci. Ser. D 276*, 607–609.

Darchen, R., and Delage, B. (1970). Facteur déterminant les castes chez les *Trigones. C. R. Acad. Sci. Ser. D 270*, 1372–1373.

Darchen, R., and Delage-Darchen, B. (1971). Le déterminisme des castes chez les *Trigones. Insectes Soc, 18*, 121–134.

Darchen, R., and Delage-Darchen, B. (1974). Nouvelles expériences concernant le déterminisme des castes chez les *Mélipones. C. R. Acad. Sci. Ser. D 278*, 907–910.

Darchen, R., and Delage-Darchen, B. (1975). Contribution a l'étude d'une abeille du mexique, *Melipona beecheii* B. (Hymenoptère Apide). *Apidologie 6*, 295–339.

Darchen, R., and Delage-Darchen, B. (1977). Sur le déterminisme des castes chez les *Mélipones. Bull. Biol. Fr. Belg. 111*, 91–109.

Darwin, C. R. (1886). Über die Wege der Hummelmännchen. In: "Gesammelte kleinere Schriften von Charles Darwin," Ser. 2, Vol. VII (E. Krause, ed.), pp. 84–88. Ernst Gunthers Verlag, Leipzig.

Deodikar, G. B., and Thakar, C. V. (1966). Cytogenetics of Indian honeybees and bearing on taxonomic and breeding problems. *Indian J. Genet. 26A*, 286–393.

Deodikar, G. B., Thakar, C. V., and Pushpa, N. S. (1959). Cytogenetic studies in Indian honeybees I. Somatic chromosome complement in *Apis indica* and its bearing on evolution and phylogeny. *Proc. Ind. Acad. Sci. 49*, 196–207.

Dietz, A. (1975). Influence of the "cannibalism substance" of diploid drone honey bee larvae on the survival of newly hatched worker bee larvae. *Proc. Inter. Apic. Cong. (Apimondia) 25*, 267–269.

Digby, P. S. B. (1955). Factors affecting the temperature excess of insects in sunshine. *J. Exp. Biol. 32*, 279–298.

Dodson, C. H., Dressler, R. L., Hills, H. G., Adams, R. M., and Williams, N. H. (1969). Biologically active compounds in orchid fragrances. *Science 164*, 1243–1249.

Duffield, R. M., Fernandes, A., Lamb, C., Wheeler, J. W., and Eickwort, G. C. (1981). Macrocyclic lactones and isopentyl esters in the Dufour's gland secretion of halictine bees (Hymenoptera Halictidae). *J. Chem. Ecol. 7*, 319–331.

Eickwort, G. C., and Ginsberg, H. S. (1980). Foraging and mating behavior in Apoidea. *Annu. Rev. Entomol. 25*, 421–446.

Engel, M. S. (1978). Hat *Apis*-Königinnensubstanz einen Anti-JH-Effekt auf die stachellose Biene *Melipona quadrifasciata*? *Mitt. Deutsche Ges. Allg. Angew. Entomol. 1*, 302–303.

Engel, M. S. (1979). Is caste determination in *Melipona quadrifasciata*, a stingless-bee, influenced by 9-oxo-deconoic acid. *Insectes Soc. 26*, 273–278.

Fahrenhorst, H. (1977a). Nachweis übereinstimmender Chromosomenzahlen (n = 16) bei allen 4 *Apis*-Arten. *Apidologie 8*, 89–100.

Fahrenhorst, H. (1977b). Chromosome number in the tropical honeybee species *A. dorsata* and *A. florea*. *J. Apic. Res. 16*, 56–58.

Fink, S. C., Carlson, C. W., Gurusiddaiah, S., and Brosemer, R. W. (1970). Glycerol 3-phosphate dehydrogenases in social bees. *J. Biol. Chem. 245*, 6525–6532.

Fischer, M. E. (1978). "Heterosis." VEB Gustav Fischer Verlag, Jena, GDR.

Francke, W., Reith, W., Bergström, G., and Tengö, J. (1980). Spiroketals in the mandibular glands of *Andrena* bees. *Naturwissenschaften 67*, 149–150.

Frank, A. (1941). Eigenartige Flugbahnen bei Hummelmännchen. *Z. Vergl. Physiol. 28*, 467–484.

Frankie, G. W., Coville, R. E., and Vinson, S. B. (1980). Territorial behavior of *Centris adani* and its reproductive function in the Costa Rican dry forest (Hymenoptera, Anthophoridae). *J. Kansas Entomol. Soc. 53*, 837–857.

Free, J. B. (1971). Stimuli eliciting mating behaviour of bumble bee (*B. pratorum*) males. *Behaviour 40*, 55–61.

Free, J. B. (1982). "Bees and Mankind." George Allen and Unwin, London.

Friese, H., and Wagner, F. (1912). Zoologische Studien an Hummeln II. Die Hummeln der Arktis, des Hochgebirges und der Steppe. *Zool. Jahrbuch Suppl. 15*, 155–210.

Garofalo, C. A. (1972). Ocorrência de machos des em *Bombus atratus*. *Ciênc. Cult. (São Paulo) (Suppl.) 24*, 161.

Garofalo, C. A. (1973). Occurrence of diploid drones in a neotropical bumbel bee. *Experientia 29*, 726.

Garofalo, C. A., and Kerr, W. E. (1975). Sex determination in bees 1. Balance between femaleness and maleness genes in *Bombus atratus*. *Genetica 45*, 203–209.

Gerber, H. S., and Klostermeyer, E. C. (1970). Sex control by bees: a voluntary act of egg fertilization during oviposition. *Science 167*, 82–84.

Granata, L. (1909). Le divisioni degli spermatociti di *Xylocopa violacea* L. *Biologica Torino 2*, 1–12.

Greenberg, L. (1979). Genetic component of bee odor in kin recognition. *Science 206*, 1095–1097.

Greenberg, L. (1982). Year-round culturing and productivity of a sweat bee, *Lasioglossum zephyrum* (Hymenoptera: Halictidae). *J. Kansas Entomol. Soc. 55*, 13–22.

Greenberg, L., and Buckle, G. R. (1981). Inhibition of worker mating by queens in a sweat bee, *Lasioglossum zephyrum*. *Insectes Soc. 28*, 347–352.

Haas, A. (1946). Neue Beobachtungen zum Problem der Flugbahnen bei Hummelmännchen. *Z. Naturforsch. 1*, 596–600.

Haas, A. (1949a). Arttypische Flugbahnen von Hummelmännchen. *Z. Vergl. Physiol. 31*, 281–307.

Haas, A. (1949b). Gesetzmässiges Flugverhalten der Männchen von *Psithyrus silvestris* Lep. und einiger solitären Apiden. *Z. Vergl. Physiol. 31*, 671–683.

Haas, A. (1952). Die Mandibeldrüse als Duftorgan bei einigen Hymenopteren. *Naturwissenschaften 39*, 484.

Hartl, D. L. (1971). Some aspects of natural selection in arrhenotokous populations. *Am. Zool. 11*, 309–325.

Hobbs, G. A. (1967). Ecology of species of *Bombus* (Hymenoptera: Apidae) in southern Alberta. VI. Subgenus *Pyrobombus*. *Can. Ent. 99*, 1271–1292.

Holm, S. N. (1966). The utilization and management of bumble bees for red clover and alfalfa seed production. *Ann. Rev. Entomol. 11*, 155–182.

Honk, C. G. J. van, Velthuis, H. H. W., and Röseler, P. F. (1978). A sex pheromone from the mandibular glands in bumble bee queens. *Experientia 34*, 838–839.

Hoshiba, H., Okada, I., and Kusanagi, A. (1981). The diploid drone of *Apis cerana japonica* and its chromosomes. *J. Apic. Res. 20*, 143–147.

Ihering, H. V. (1903). Biologie der stachellosen Honigbienen Brasiliens. *Zool. Jahrb. 19*, 179–287.

Jayasingh, D. B., and Freeman, B. E. (1980). The comparative population dynamics of eight solitary bees and wasps (Aculeata Apocrita; Hymenoptera) trapnested in Jamaica. *Biotropica 12*, 214–219.

Jordan, R. A., and Brosemer, R. W. (1974). Characterization of DNA from three bee species. *J. Insect Physiol. 20*, 2513–2520.

Kerr, W. E. (1946). Formação das castas no género *Melipona*. *Anais Escola Superior Arg. "Luiz de Queiroz" 3*, 299–312.

Kerr, W. E. (1950a). Evolution of the mechanism of caste determination in the genus *Melipona*. *Evolution 4*, 7–13.

Kerr, W. E. (1950b). Genetic determination of castes in the genus *Melipona*. *Genetics 35*, 143–152.

Kerr, W. E. (1951). Bases para o estudo da genética dos Hymenoptera am general e dos Apinae sociais em particular. *Anais Escola Superior Agr. "Luiz de Queiroz" 8*, 220–354.

Kerr, W. E. (1966). Determinação das castas no gênero Melipona. *In* "Reunião Anual Sociedade Brasileira de Genetica. Programas e resumos," pp. 1–5. Piracicaba, São Paolo.

Kerr, W. E. (1969a). Origiem de genes limitados ao sexo nos hymenoptera. *Ciênc Cult. (São Paulo) 21*, 652–658.

Kerr, W. E. (1969b). Some aspects of the evolution of social bees. *Evol. Biol. 3*, 119–175.

Kerr, W. E. (1972a). Numbers of chromosomes in some species of bees. *J. Kansas Entomol. Soc. 45*, 111–122.

Kerr, W. E. (1972b). Effect of low temperature on male meiosis in *Melipona marginata*. *J. Apic. Res. 11*, 95–99.

148   Robin F. A. Moritz

Kerr, W. E. (1974a). Sex determination in bees. III. Caste determination and genetic control in *Melipona. Insectes Soc.* 21, 357–367.

Kerr, W. E. (1974b). Gentik des Polymorphismus bei Bienen. *In* "Sozialpolymorphismus bei Insekten" (G. H. Schmidt, ed.), pp. 94–109. Wissenschaftl. Verlagsgesellschaft, Stuttgart.

Kerr, W. E. (1974c). Advances in cytology and genetics of bees. *Annu. Rev. Entomol.* 19, 253–268.

Kerr, W. E. (1975a). Population genetic studies in bees (Apidae, Hymenoptera). I. Genetic load. *An. Acad. Bras. Ciênc.* 47, 319–334.

Kerr, W. E. (1975b). Evolution of the population structure in bees. *Genetics* 79, 73–84.

Kerr, W. E. (1975c). Genetics of sex determination in bees. X. Proposed model of activity of genes determining sex and caste. *Anais. Cong. Bras. Apic.,* 3, 179–187.

Kerr, W. E. (1976). Population genetic studies in bees. II. Sex linked genes. *Evolution* 30, 94–99.

Kerr, W. E., and Almeida, M. G. de (1981). Genetic studies of bee populations. III. The genetic load of *Melipona scutellaris. Rev. Bras. Biol.* 41, 137–139.

Kerr, W. E., and Esch, H. (1965). Comuniçâcao entre as abelhas sociais brasileiras e sua contribução para o entendinento da sua evolucao. *Ciênc. Cult. (São Paulo)* 17, 529–538.

Kerr, W. E., and Krause, W. (1950). Contribução para o conhecimento da biomia dos Meliponini. *Dusenia* 1, 275–282.

Kerr, W. E., and Maule, V. (1964). Geographic distribution of stingless-bees and its implications (Hymenoptera: Apidae). *J. N. Y. Entomol. Soc.* 72, 2–18.

Kerr, W. E., and Nielsen, R. A. (1966). Evidences that genetically determined *Melipona* queens can become workers. *Genetics* 54, 859–866.

Kerr, W. E., and Nielsen, R. A. (1967). Sex determination in bees (Apinae). *J. Apic. Res.* 6, 3–9.

Kerr, W. E., and Silveira, Z. V. da. (1972). Karyotype evolution and corresponding taxonomic implications. *Evolution* 26, 197–202.

Kerr, W. E., and Vencovsky, R. (1982). Melhoramento genético em abelhas. I. Efeito do numero de colônias sobre o melhoramerito. *Rev. Bras. Gen.* 5, 279–285.

Kerr, W. E., Zucchi, R., Nakadaira, J. T., and Botolo, J. E. (1962). Reproduction in the social bees. *J. N. Y. Entomol. Soc.* 70, 265–276.

Kerr, W. E., Stort, A. C., and Montenegro, M. J. (1966). Importance of some environmental factors for caste determination in the genus *Melipona. An. Acad. Bras. Ciênc.* 38, 149–168.

Kerr, W. E., Akahira, Y., and Camargo, C. A. (1975). Sex determination in bees IV. Genetic control of juvenile hormone production in *Melipona quadrifasciata. Genetics* 81, 749–756.

Kimsey, L. S (1980). The behaviour of male orchid bees (Apidae, Hymenoptera, Insecta) and the question of leks. *Anim. Behav.* 28, 996–1004.

Knerer, G. (1980). Evolution of halictine castes. *Naturwissenschaften* 67, 133–135.

Knerer, G., and Plateaux-Quénu, C. (1967a). Sur la production continue ou périodique de couvain Halictinae. *C. R. Acad. Sci. Paris* 264, 651–653.

Knerer, G., and Plateaux-Quénu, C. (1967b). Sur la production de males chez les Halictines sociaux. *C. R. Acad. Sci. Paris* 264, 1096–1099.

Koeniger, N. (1976). Neue Aspekte der Phylogenie innerhalb der Gattung *Apis. Apidologie* 7, 357–366.

Koeniger, N., and Wijayagunasekera, H. N. P. (1976). Time of drone flight in the three asiatic honeybee species. *J. Apic. Res.* 15, 67–71.

Krüger, E. (1932). Über die Farbenvariationen der Hummelart *Bombus agrorum. Z. Morphol. Ökol.* 24, 148–237.

Krüger, E. (1951). Über die Bahnflüge der Männchen der Gattungen *Bombus* und *Psithyrus. Z. Tierpsychol.* 8, 61–75.

Kullenberg, B. (1956). Field experiments with chemical sexual attractants on aculeate Hymenoptera males. *Z. Morphol. Oekol. Tiere* 39, 527–545.

Kullenberg, B., and Bergström, G. (1974). The pollination of *Ophrys* orchids. *Nobel Symp.*, 1973. 25, 253–258.

Kullenberg, B., and Bergström, G. (1975). Chemical communication between living organisms. *Endeavour* 34, 59–66.

Kullenberg, B., and Bergström, G. (1976). Hymenoptera aculeata males as pollinators of *Ophrys* orchids. *Zool. Scr.* 5, 13–23.

Kullenberg, B., Bergström, G., and Ställberg-Stenhagen, S. (1970). Volatile components of the cephalic marking secretion of male bumblebees. *Acta. Chem. Scand.* 24, 1481–1483.

Kullenberg, B., Bergström, G., Bringer, B., Carlberg, B., and Cederberg, B. (1973). Observations on scent marking by *Bombus* Latr. and *Psithyrus* Lep. males (Hym. Apidae) and localization of site of production of the secretion. *Zoon Suppl.* 1, 23–30.

Kumbkarni, C. G. (1964). Cytological studies in Hymenoptera Part I: Cytology of parthenogensis in the honeybees—*Apis dorsata*. *Indian J. Exp. Biol.* 2, 65–68.

Kumbkarni, C. G. (1965). Cytological studies in Hymenoptera Part II: Cytology of parthenogenesis in the carpenter bee *Xylocopa fenesterata* Fabre. *Cytologia* 30, 222–228.

Lanham, U. N. (1974). Melanism in the bee *Andrena albihirta* in north central Colorado. *J. Kansas Entomol. Soc.* 47, 373–377.

Lester, L. J., and Selander, R. K. (1979). Population genetics of haplo–diploid insects. *Genetics* 92, 1329–1345.

Ly, Y. C. (1981). The wheat bee (*Melipona* spec). *Zhongguo Yangfeng* 2, 8.

Maa, T. C. (1953). An inquiry into the systematics of the tribus *Apidini* or honey bees (Hym.). *Treubia* 21, 525–640.

Makino, S. (1951). "An Atlas of the Chromosome Numbers in Animals." Iowa State College Press, Ames.

Maul, V. (1969). Die Ursache der Kreuzungsbarriere zwischen *Apis mellifera* L. und *Apis cerana* Fab. 2. Eibefruchtung und Embryonalentwicklung. *Proc. Inter. Apic. Cong. (Apimondia)* 22, 515–516.

Mello, M. L. S. (1969). Contribution to the study of somatic polyploidy in some insect organs. Ph.D. Thesis. Univ. São Paulo, Ribeirão Preto.

Mello, M. L. S (1972). Micro-interferometry of insect polyploid nuclei. *Cytologica* 37, 261–270.

Mello, M. L. S., and Silveira, Z. V. da. (1970). Somatic polyploidy in larval Malpighian tubes of *Melipona quinquefasciata* Lep. nucleus. *India* 13, 59–61.

Mello, M. L. S., and Takahashi, C. S. (1969). The post-embryonic changes in *Melipona quadrifasciata anthidioides* V. Polyploidy in the larval Malpighian tubules. *Cytologica* 34, 369–374.

Mello, M. L. S., and Takahashi, C. S. (1971). Occurrence of polyploidy in the Malpighian tubules of *Melipona quadrifasciata*. *Arquiros do Museu Nactional* 54, 286.

Mello, M. L. S., Takahashi, C. S., and Gagliardi, A. R. T. (1970). Polyploidy pattern in larval silk glands of *Melipona quadrifasciata*. *Insectes Soc.* 17, 295–302.

Metcalf, R. A., Marlin, J. C., and Whitt, G. S. (1975). Low levels of genetic heterozygosity in Hymenoptera. *Nature (Lond.)* 257, 792–794.

Michener, C. D. (1944). Comparative external morphology, phylogeny and a classification of the bees. (Hymenoptera). *Bull. Am. Mus. Nat. Hist.* 82, 151–326.

Michener, C. D. (1974). "The Social Behavior of the Bees." Harvard University Press, Cambridge, Mass.

Michener, C. D., and Wille, A. (1961). The bionomics of a primitively social bee, *Lasioglossum inconspicuum*. *Univ. Kansas Sci. Bull.* 42, 1123–1202.

Moraes, R. L. M. S. de. (1978). Variações de conteúdo de DNA e volume nucleares nas glândulas salivares de operárias de *Melipona quadrifasciata anthidioides* Lep. durante a diferenciação pós-embrionaria e ciclo secretor. *Papéis Avulsos de Zoologia* 31, 251–281.

Moraes, R. L. M. S. de, Cruz-Landim, C. da, and Caetano, F. H. (1972). Polyploidy in cells of the midgut of larval *Melipona quadrifasciata anthidioides*. Homenagem a W. E. Kerr, *Fac. Filosof. Ciênc. Let. Rio-Claro. B,* 237–249.

Moritz, R. F. A. (1985). The effects of multiple mating on the worker-queen conflict in *Apis mellifera* L. *Behav. Ecol. Sociobiol. 16,* 375–377.

Morton, N. E. (1975). Theory of inbreeding effect on diploid bees. *An. Acad. Bras. Ciênc. 47,* 332–334.

Morton, N. E., Crow, J. F., and Muller, H. J. (1956). An estimate of mutational damage in man from data on consanguinous marriages. *Proc. Natl. Acad. Sci. (U.S.A.) 42,* 855–863.

Nogueira-Neto, P. (1970). "A Criação de Abelhas indigenas sem Ferrão (Meliponinae)." Chácaras e Quintos, São Paulo.

Obrecht, E., and Scholl, A. (1981). Enzymelektrische Untersuchungen zur Analyse der Verwandtschaftgrade zwischen Hummel- und Schmarotzerhummelarten *(Apidae, Bombini). Apidologie 12,* 257–268.

Owen, R. E. (1980). Population genetics of social hymenoptera with worker produced males. *Heredity 45,* 31–46.

Owen, R. E., and Plowright, R. C. (1980). Abdominal pile color dimorphism in the bumble bee, *Bombus melanopygus. J. Hered. 71,* 241–247.

Owen, R. E., Rodd, F. H., and Plowright, R. C. (1980). Sex ratios in bumble bee colonies: complications due to orphaning? *Behav. Ecol. Sociobiol. 7,* 287–291.

Pamilo, P., Vepsäläinen, K., and Rosengren, R. (1975). Low allozymic variability in *Formica* ants. *Hereditas 80,* 293–296.

Pamilo, P., Varvio-Aho, S.-L., and Pekkarinen, A. (1978). Low enzyme gene variability in Hymenoptera as a consequence of haploidiploidy. *Hereditas 88,* 93–99.

Pamilo, P., Pekkarinen, A., and Varvio-Aho, S.-L. (1981). Phylogenetic relationship and the origin of social parasitism in Vespidae and in *Bombus* and *Psithyrus* as revealed by enzyme genes. *In* "Biosystematics of Social Insects" (P. E. Howse and J. L. Clement, eds.), pp. 37–48. Academic Press, New York.

Parker, F. D. (1981). A candidate red clover pollinator *Osmia coerulescens. J. Apic. Res. 20,* 62–65.

Pekkarinen. A. (1979). Morphometric, colour and enzyme variation in bumble bees (Hymenoptera, Apidae, *Bombus*) in Fennoscandia and Denmark. *Acta Zool. Fenn. 158,* 1–60.

Pekkarinen, A., Varvio-Aho, S.-L., and Pamilo, P. (1979). Evolutionary relationships in northern European *Bombus* and *Psithyrus* species (Hymenoptera, Apidae) studied on the basis of allozymes. *Ann. Entomol. Fenn. 45,* 77–80.

Pittioni, B. (1940). Analytische Untersuchungen an den Hummelfaunen des Witoscha- und Ljulin-Gebirges in Bulgarien. Eine zoogeographisch-ökologische Studie. *Mitt. Bulg. Entomol. Ges. 11,* 101–137.

Pittioni, B. (1941). Die Variabilität des *Bombus agrorum* F. in Bulgarien. Eine variationsstatistische Untersuchung unter Berücksichtigung geographischer und ökologischer Faktoren. *Mitt. Königl. Naturwiss. Inst. Sofia-Bulgarian 14,* 238–311.

Plateaux-Quénu, C. (1974). Compartement dessociétés orphelines d'*Evylaus calceatus* (Scop.) *(Hymenoptera, Halictinae). Insectes Soc. 21,* 5–12.

Plateaux-Quénu, C. (1982). Pollen ball weight and sex determination in Halictinae. *Proc. Congr. IUSSI 9th (Boulder),* p. 253.

Plateaux-Quénu, C., and Plateaux, L. (1980). Action de la température sur la taille, le sexe et le cycle des individus de première couvée chez *Evylaeus calceatus* (Scop.) (Hym., Halictinae): première étude expérimentale. *Ann. Sci. Nat. Zool. Biol. Anim. 2,* 27–33.

Plowright, R. C., and Owen, R. E. (1980). The evolutionary significance of bumble bee color patterns: a mimetic interpretation. *Evolution 34,* 622–637.

Plowright, R. C., and Pallett, M. J. (1979). Worker male conflict and inbreeding in bumble bees. *Can. Entomol. 111,* 289–294.

Pomeroy, N. (1979). Brood bionomics of *Bombus ruderatus* in New Zealand *(Hymenoptera, Apidae). Can. Entomol. 111*, 865–874.

Pouvreau, A. (1963). Observation sur l'accuplemente de *Bombus hypnorum* L. *(Hymenoptera, Apidae)* en serre. *Insectes Soc. 10*, 111–118.

Powell, J. R. (1975). Protein variation in natural populations of animals. *In* "Evolutionary Biology, Vol. 8" (T. Dobzhansky, M. K. Hecht, and W. D. Steere, eds.), pp. 79–119. Plenum Press, New York.

Radovic, I. T. (1981). Anatomy and function of the sting apparatus of stingless-bees (Hymenoptera Apinae, Meliponini). *Proc. Entomol. Soc. Wash. 83*, 269–273.

Rank, G. H., and Goerzen, D. W. (1981). Native leafcutter bee species and associated parasites in commercial hives in Saskatchewan. *Apidologie 12*, 211–220.

Raw, A., and O'Toole, C. (1979). Errors in the sex of eggs laid by the solitary bee *Osmia rufa. Behaviour 70*, 168–171.

Rembold, H., Czoppelt, Ch., and Rao, P. J. (1974). Effect of juvenile hormone treatment on caste differentiation in the honeybee *Apis mellifera. J. Insect Physiol. 20*, 1193–1202.

Roubik, D. W. (1978). Competitive interactions between neotropical pollinators and Africanized honey bees. *Science 201*, 1030–1032.

Ruszkowski, A. (1979). The search for the methods to increase population of bees pollinating alfalfa. U.S.D.A. Final Rep. Proj. PL-ARS-42 (unpublished).

Ruttner, F. (1969). The cause of the hybridization barrier between *Apis mellifera* L. and *Apis cerana* Fabr. 1. Experiments with natural and artificial insemination. *Proc. Inter. Apic. Cong. (Apimondia)* 22, 561.

Ruttner, F. (1973). Drohnen von *Apis cerana* Fabr. auf einem Drohnensammelplatz. *Apidologie 4*, 41–44.

Ruttner, F., Woyke, J., and Koeniger, N. (1972). Reproduction in *Apis cerana* I: Mating behaviour. *J. Apic. Res. 11*, 141–146.

Ruttner, F., Woyke, J., and Koeniger, N. (1973). Reproduction in *Apis cerana* II: Reproductive organs and natural insemination. *J. Apic. Res. 12*, 21–34.

Sakagami, S. F. (1976). Specific differences in the bionomic characters of bumblebees. A comparative review. *J. Fac. Sci. Hokkaido Univ. VI Zool. 20*, 390–447.

Sakagami, S. F., and Laroca, S. (1963). Additional observations on the habits of the cleptobiotic stingless-bees, the genus *Lestrimelitta friese* (Hymenoptera, Apoidea). *J. Fac. Sci. Hokkaido. Univ. Ser. 6 15*, 319–339.

Sakagami, S. F., and Zucchi, R. (1965). Winterverhalten einer neotropischen Hummel, *Bombus atratus*, innerhalb eines Beobachtungskastens: Ein Beitrag zur Biologie der Hummeln. *J. Fac. Sci. Hokkaido Univ. Ser. 6 15*, 712–762.

Sakagami, S. F., Mutsumura, T., and Ito, K. (1980). *Apis laboriosa* in Himalaya, the little known world largest honeybee. *Insecta Mastumurana 19*, 47–77.

Saraceni, N. G. (1972). Aspectos de regulacão social em colônias de *Bombus atratus*. Homenagem a W. E. Kerr, *Fac. Filosof. Ciênc. Let. Rio. Claro. B*, pp. 259–265.

Scholl, A., and Obrecht, E. (1983). Enzymelektrophoretische Untersuchungen zur Artabgrenzung im *Bombus lucorum* Komplex (Apidae, Bombini). *Apidologie 14*, 65–78.

Schwarz, H. (1948). Stingless bees *(Meliponidae)* of the western hemisphere. *Bull. Am. Mus. Nat. Hist. 90*, 1–546.

Selander, R. K. (1976). Genic variation in natural populations. *In* "Molecular Evolution" (F. J. Ayala, ed.), pp. 21–45. Sinauer, Sunderland.

Severinghaus, L. L., Kurtak, B. M., and Eickwort, G. C. (1981). The reproductive behaviour of *Anthidium manicatum (Hymenoptera Megachilidae)* and the significance of size for territorial males. *Behav. Ecol. Sociobiol. 9*, 51–58.

Shah, F. A., and Shah, T. A. (1980). Early life, mating and egg laying of *Apis cerana* queens in Kashmir. *Bee World 61*, 137–140.

Sharma, G. P., Gupta, B. L., and Kumbkarni, C. G. (1961). Cytology of spermatogenesis in the honey bee *Apis indica* F. *J. R. Microsc. Soc.* 79, 337–351.

Shearer, D. A., Boch, R., Morse, R. A., and Laigo, F. M. (1970). Occurrence of 9-oxodec-trans-2-enoic acid in queens of *Apis dorsata, A. cerana* and *A. mellifera. J. Insect Physiol.* 16, 1427–1441.

Silva, D. L. N. da. (1974). Estudos bionômicos em colônias mistas de *Meliponinae (Hymenoptera, Apoidae)*. Ph.D. Thesis. Univ. São Paulo, Ribeirão Preto.

Silva, D. L. N. da, Zucchi, R., and Kerr, W. E. (1972). Biological and behavioural aspects of the reproduction in some species of *Melipona. Anim. Behav.* 20, 123–132.

Silveira, Z. V. (1971). Número de cromosomos em Meliponideos brasileiros. *Ciênc. Cult. (São Paulo) Suppl.* 23, 105–106.

Silveira, Z. V. (1972). Número de chromosomos em Meliponideos brasileiros II. *Ciênc. Cult. (São Paulo) Suppl.* 24, 160–161.

Sladen, F. E. L. (1912). "The Humble Bee, its Life history and How to Domesticate it, with Descriptions of all the British Species of *Bombus* and *Psithyrus*." Macmillan and Co., London.

Smith, B. H. (1982). Reproductive biology in Halictine bees. *Proc. Congr IUSSI 9th (Boulder)*, p. 333.

Snyder, T. P. (1974). Lack of allozymic variability in three bee species. *Evolution* 28, 687–689.

Stein, G. (1963a). Über den Sexuallockstoff der Hummelmännchen. *Naturwissenschaften* 50, 305.

Stein, G. (1963b). Untersuchengen über den Sexuallockstoff der Hummelmännchen. *Biol. Zbl.* 82, 343–349.

Stejskal, M. (1974). La déterminisme de la formation des reines chez *Melipona fasciata merillae* Cockr. (Hymenoptère, Apide). *C. R. Acad. Sci. Paris* 278, 2647–2648.

Stephen, W. P. (1957). Bumble bees of western America. *Oreg. Agric. Exp. Stn. Techn. Bull.* 40, 2–163.

Stephen, W. P., and Cheldelin, I. H. (1973). Phenetic grouping in bees of the tribe Bombini based on the enzyme α-glycerophosphate dehydrogenase. *Biochem. Syst.* 1, 69–76.

Stiles, E. W. (1979). Evolution of color pattern and pubescence characteristics on male bumblebees. automimicry vs. thermoregulation. *Evolution* 33, 941–957.

Suomalainen, E. (1950). Parthenogenesis in animals. *Adv. Genet.* 3, 193–254.

Suomalainen, E. (1962). Significance of parthenogenesis in the evolution of insects. *Annu. Rev. Entomol.* 7, 349–366.

Svensson, B. G. (1979). Patrolling behaviour of bumble bee males *(Hymenoptera Apidae)* in a subalpine/alpine area, Swedish Lapland. *Zoon* 7, 67–99.

Svensson, B. G., and Bergström, G. (1977). Volatile marking secretions from the labial gland of north European *Pyrobombus* D. T. males *(Hymenoptera Apidae). Insectes Soc.* 24, 213–224.

Svensson, B. G., and Bergström, G. (1979). Marking pheromones of *Alpinobombus* males. *J. Chem. Ecol.* 5, 603–615.

Tambasco, A. J. (1971). Processo reprodutivo em *Melipona quadrifasciata* e seu impacto na populacao geneticamente ativa. *Ciênc. Cult. Suppl. (São Paulo)* 23, 104–105.

Tambasco, A. J., Giannoni, M. A., and Azevedo Moreira, L. M. de. (1979). Analysis of G-Bands in chromosomes of the *Melipona quadrifasciata anthidiodes* Lep. *Cytologia* 44, 21–27.

Tanabe, Y., Tamaki, Y., and Nakano, S. (1970). Variation of esterase isozymes in seven species of bees and wasps. *Japn. J. Genetics* 45, 425–428.

Tarelho, Z. V. S. (1973). Contribuição ao estudo citogenético dos Apoidea. Ph.D. Thesis, Univ. São Paulo, Rebeirão Preto.

Tasei, J. N. (1975). Le probléme de l'adaptation de *Megachile (Entricharea) pacifica* Panz. *(Megachilidae)* Américain en France. *Apidologie* 6, 1–57.

Tasei, J. N. (1977). Possibilité de multiplication du pollinisateur de la luzerne *Megachile pacifica* Panz. en France. *Apidologie 8*, 61–82.

Tengö, J. (1979). Odour released behaviour in *Andrena* male bees (*Apoidea, Hymenoptera*). *Zoon 7*, 15–48.

Tengö, J., and Bergström, G. (1975a). All-trans-farnesyl hexanoate and geranyl octanoate in the Dufour's gland secretion of *Andrena* (*Hymenoptera, Apidae*). *J. Chem. Ecol. 1*, 253–268.

Tengö, J., and Bergström, G. (1975b). Odour correspondence between *Melitta* females and males of their nest parasite *Nomada flavopicta* K. (*Hymenoptera Apidae*). *J. Chem. Ecol. 2*, 57–65.

Tengö, J., and Bergström, G. (1976). Comparative analyses of lemon-smelling secretion from the heads of *Andrena* F. (*Hymenoptera, Apidae*) bees. *Comp. Biochem. Physiol. 55B*, 179–188.

Tengö, J., and Bergström, G. (1977a). Comparative analyses of complex secretions from heads of *Andrena* bees (*Hym. Apoidea*). *Comp. Biochem. Physiol. 57B*, 197–202.

Tengö, J., and Bergström, G. (1977b). Cleptoparasitism and odor mimetism in bees: Do *Nomada* males imitate the odor of *Andrena* females? *Science 196*, 1117–1119.

Tengö, J., and Bergström, G. (1978). Identical isoprenoid esters in the Dufour's gland secretions of North American and European *Andrena* bees (*Hymenoptera: Andrenidae*). *J. Kansas Entomol. Soc. 51*, 521–526.

Thakar, C. V., and Deodikar, W. B. (1966). Chromosome number in *Apis florea* Fabr. *Curr. Sci. (Bangalore) 7*, 186.

Torchio, P. F., and Tepedino, V. J. (1980). Sex ratio, body size and seasonality in a solitary bee, *Osmia lignaria propinqua* Cresson (*Hymenoptera Megachilidae*). *Evolution 34*, 993–1003.

Velthuis, H. H. W. (1976). Environmental, genetic and endocrine influences in stingless-bee caste determination. *In* "Phase and Caste Determination in Insects" (M. Lüscher, ed.), pp. 35–53. Pergamon Press, Oxford, U.K.

Velthuis, H. H. W., and Gerling, D. (1980). Observations on territoriality and mating behaviour of the carpenter bee *Xylocopa sulcatipes*. *Entomol. Exp. Appl. 28*, 82–91.

Velthuis, H. H. W., and Velthuis-Kluppel, F. M. (1975). Caste differentiation in a stingless-bee, *Melipona quadrifasciata* L. influenced by juvenile hormone application. *Konikl-Nederl. Akad. Wetensch. Amsterdam 78*, 81–94.

Venkovsky, R., and Kerr, W. E. (1982). Melhoramento genético em abelhas II. Teoria e avaliação de alguns métodos de selecao. *Rev. Bras. Gen. 5*, 493–502.

Vogel, S. (1966). Scent organs of orchid flowers and their relation to insect pollination. *Proc. World Orchid Conf. 5*, 253–259.

Vogt, O. (1909). Studien über das Artproblem. Über das Variieren der Hummeln I. *Sitzungsber. Naturf. Freunde Berlin 1909*, 28–84.

Webb, M. C. (1961). The biology of the bumble bees of a limited area in eastern Nebraska. Ph.D. Dissertation. University of Nebraska, Lincoln.

Wheeler, J. W., Evans, S. L., Blum, M. S., Velthuis, H. H. W., and Camargo, J. M. F. de. (1976). Cis-2-methyl-5-hydroxyhexanoic acid lactone in the mandibular gland of a carpenter bee. *Tetrahedron Lett. 45*, 4029–4032.

White, M. J. D. (1954). "Animal Cytology and Evolution." University Press, Cambridge.

Wirtz, P. (1973). Differentiation in the honey bee larva. A histological, electron-microscopical and physiological study of caste induction in *Apis mellifera mellifera*. L. *Meded. Landbouwh. Wageningen. 73-5*, 1–155.

Wirtz, P., and Beetsma, J. (1972). Induction of caste differentiation in the honeybee *Apis mellifera* by juvenile hormone. *Entomol. Exp. Appl. 15*, 517–520.

Woyke, J. (1956). Anatomo-physiological changes in queenbees returning from mating flights and the process of multiple mating. *Bull. Acad. Pol. Sci. Cl. II. 4*, 81–87.

Woyke, J. (1967). Diploid drone substance. Cannibalism substance. *Proc. Inter. Apic. Cong. (Apimondia) 21*, 471–472.

Woyke, J. (1973). Instrumental insemination of *Apis cerana indica* queens. *J. Apic. Res.* 12, 151–158.

Woyke, J. (1975). Natural and instrumental insemination of *Apis cerana indica* in India. *J. Apic. Res.* 14, 153–159.

Woyke, J. (1979). Sex determination in *Apis cerana indica*. *J. Apic. Res.* 18, 122–127.

Woyke, J. (1980). Evidence and action of cannibalism substance in *Apis cerana indica*. *J. Apic. Res.* 19, 6–16.

Wright, S. (1933). Inbreeding and homozygosis. *Proc. Natl. Acad. Sci. U.S.A.* 19, 411–420.

Wynne-Edwards, V. C. (1962). "Animal Dispersion in Relation to Animal Behaviour." Oliver and Boyd, London.

Zucchi, R., Sakagami, S. F., and Camargo, J. M. F. de. (1969). Biological observations on a neotropical parasocial bee *Eulaema nigrita*, with a review of the biology of Euglossinae. A comparative study. *J. Fac. Sci. Hokkaido Univ. (VI. Zool.)* 17, 271–380.

CHAPTER 6

Behavioral Genetics

THOMAS E. RINDERER AND ANITA M. COLLINS

I. INTRODUCTION

Darwin (1859) devoted a chapter of "The Origin of Species" to the evolution of instinct. More than a quarter of that chapter discussed the nest-building behavior of honey bees and concluded with the notion that "the most wonderful of all known instincts, that of the hive bee, can be explained by natural selection." Although Darwin's chapter might have enticed scientists to study the hereditary aspects of honey-bee behavior, it was 105 years until Rothenbuhler (1964a) provided experimental evidence for the hypothesis that the tendency of honey bees to clean their nest of dead brood, called hygienic behavior, "depends upon homozygosity for two recessive genes."

Rothenbuhler's study of hygienic behavior is a widely cited classic in the field of behavioral genetics and is the founding work of honey-bee behavioral genetics. His work, indeed most behavioral genetic work, was precluded until two scientific developments occurred. After Mendel's principles were rediscovered in 1900, a very active period of scientific inquiry resulted in the rapid elucidation of the mechanisms of heredity. In this work, geneticists followed Mendel's lead and generally chose easy access to principles through study of the variation of easily measured discrete characteristics. Behavioral characteristics generally were not studied by geneticists because these were usually thought to be primarily the product of an animal's experience. This nonevolutionary view of behavior, derived from the mainstream Cuvier–Watsonian traditions of psychology (Lockard, 1971), held sway until the influence of Lorenz, Tinbergen, and von Frisch (Tinbergen, 1960) brought clear focus to the Darwinian notion that animal behavior was primarily a product of natural selection and thus had a genetic basis. Genetic studies with honey bees in the last 20 years have often explored the genic contribution to behavioral variation. Nearly all of this work has been done with the western honey bee *(Apis mellifera)*.

II. THEORETICAL CONSIDERATIONS

A. Regulators of Honey-Bee Behavior

All phenotypic characteristics, whether pysiological, morphological, or behavioral, are the products of specific inherited potential expressed in specific environmental circumstances. The current environment, both physiological (within an individual) and ecological (outside an individual), is itself dependent upon prior genetic and environmental interactions. Behavior, the functions of which are response to environmental stimuli and modification of the environment, is particularly dependent on such epigenetic considerations. Unlike many physiological and morphological traits, behavior is quite distant from the chemical nature of its underlying genes. For example, many more physiological processes are required for a honey bee to build comb than for the production of color in her eyes. In behavior where learning (previous experience) plays a part, the stimulating environment can be considered to extend through time. Moreover, honey-bee behavior is additionally complex since natural selection has shaped it in a social context. Indeed, sociality is sufficiently strong in honey bees that they cannot live long without it (Rinderer and Baxter, 1978; Rinderer and Elliot, 1977), and when they have gone astray they seek it (Free and Butler, 1955). Thus, a honey bee's behavior is the product of its genetic potentiality, its ecological and physiological environments, the social conditions of the colony, and various prior and ongoing interactions among these three (Fig. 1).

With honey-bee behavioral phenotypes so complex in origin, it may seem surprising that productive behavioral genetic work is possible with this organism. However, since behavior is a product of natural selection it is adaptive within the context of honey-bee natural history. This topic has been extensively studied by many scientists in the last century largely because the economic value of honey bees is ultimately tied to their behavior in natural settings. Thus, honey-bee behavioral geneticists have a tremendous background of information to guide the design of their experiments in order to have adequate control of environmental and social variance. Such

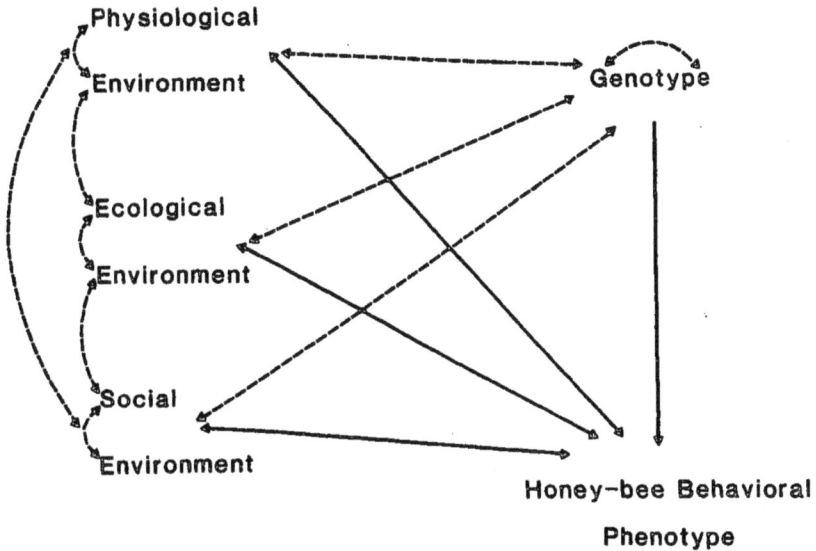

Fig. 1. The chief regulators of honey-bee behavior. Each regulator has components which interact (dashed lines) both with other components of the same regulator and with components of other regulators. This produces a complex system resulting (solid lines) in a phenotype.

designs are necessary for the productive study of behavioral genetic variance.

B. Behavioral Units of Study

Slater (1978) provides an excellent discussion of fundamental questions requiring answers before launching a research program in animal behavior. Among these questions, those involving the categorizing of behavior loom especially large for studies of the genetic aspects of honey-bee behavior. Since behavioral variation in bees has four chief interacting regulators, the kind and quality of genetic information obtained in behavioral genetic studies depends in large measure on the unit of behavior studied.

Broad units of behavior, such as nest defense or honey storage, are the products of the actions of groups of bees performed over reasonably long periods of time. Each bee performs sequences of actions which unfold according to genetic, environmental, and social regulation. The outcome of each action itself becomes a portion of the environment and greatly influences the subsequent actions of both a single bee and her hive mates. Such complexly regulated behavior will likely show a pattern of continuous variation among bee colonies. The genetic analysis of such variation is restricted to the techniques of quantitative genetics (see Chapter 11). Such analysis

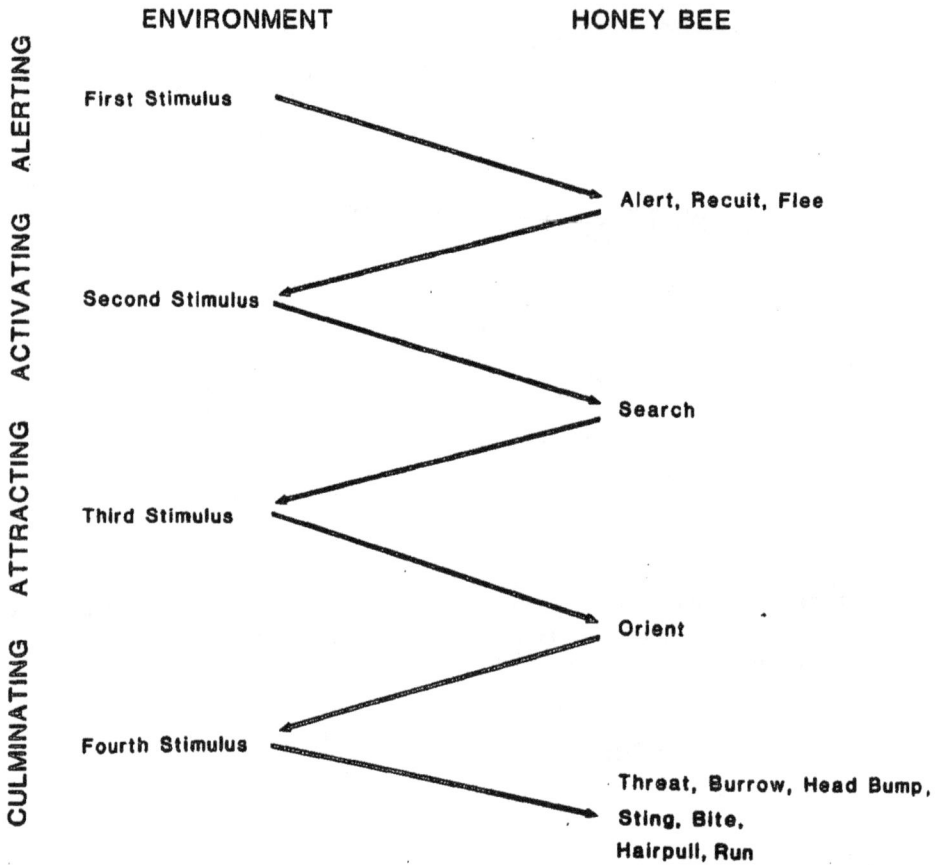

Fig. 2. Basic sequence of honey bee defensive behavior. [Redrawn from Collins *et al.* (1980). Copyright in public domain.]

will usually show that a certain portion of the variation arises from additive genetic events and that other portions of the variance are attributable to environmental and, with an appropriate experimental design, social conditions. Such information, albeit highly useful in stock improvement programs using mass selection techniques, is incomplete. Variation in early portions of the behavioral sequence arising from genetic differences may be entirely or partially masked by later events in the sequence. Also, nonadditive genetic causes of variance, as well as interactive sources of variance, are included in the variance attributed to environment (Falconer, 1960).

More information can be obtained if the behavior pattern in question is scrutinized and then described and measured in greater detail. No one method of doing so exists: both Manning (1967) and Slater (1978) suggest that units of behavior, at this stage in the development of behavioral genetics, are best resolved by an empirical approach. Nonetheless, actions resembling the fixed action patterns studied by early ethologists, which are functionally or mechanically distinct and are repeatedly recognizable, can be used to advantage in classifying the components of complex behavior. Collins *et al.* (1980) developed a model using such units to emphasize and organize variation known to exist in honey-bee defensive behavior (Fig. 2). The attributes of the model were fitted to a measurement procedure (Collins and Kubasek, 1982) which was then used in a survey of defensive behavior variation (Collins *et al.*, 1982) and in behavioral genetic studies (Collins *et al.*, 1984).

III. CONTROL OF UNDESIRABLE GENETIC VARIATION

Honey-bee behavior geneticists have the advantage of studying a organism with wide-ranging variation in nearly all of its behavioral characteristics. Details of this variation are reviewed by Ruttner (Chapter 2) and by Rothenbuhler (1967), Gonçalves and Stort (1978), Rothenbuhler *et al.* (1968), Dietz (1982), and others. In general, variation in honey-bee behavior is noted in the differences between bees adapted to different geographical conditions. While such differences occur in good measure between subspecies, they also are known to occur across more fine-grained geographic ranges containing bees assigned to the same subspecies (ecotypes, stocks, and lines). Many of these differences remain stable when bees are moved to different environments (von Frisch, 1965; Louveaux, 1969).

As well as genetic variation between bees of different subspecies, ecotypes, stocks, and lines, substantial genetic variation exists among the members of a single colony. A normal field colony of bees is composed of a heterozygous queen which has mated to several genetically different drones, parthogenically produced haploid drones which collectively reflect the heterozygosity of the queen, and sexually produced worker bees which collectively reflect both the queen's heterozygosity and the heterogeneity of the drones with which she mated. Additionally, some of the workers may be sisters of the queen, rather than her daughters. This collection of relatives has been termed a "superfamily" composed of several "subfamilies" (Rothenbuhler, 1960).

Nearly all behavioral genetic studies with bees require the elimination of genetic variation arising from "superfamily" relationships. Most of the interesting honey-bee behavior is done by worker bees, and much of this behavior is social. Thus, techniques are required which permit the composite behavior of workers in a colony to be interpreted as arising from a common genotype. Also, behavior which appears to be individual, such as a single-bee foraging or drones and queens taking mating flights, is performed in a social context that strongly influences individual behavior. For example, comb volatiles, which are products of social behavior, regulate the intensity and efficiency of an individual honey-bee's nectar foraging (Rinderer, 1982). Thus, it is desirable to assure a uniform genotype among worker bees determining the social conditions in which a behavior is measured even if that behavior is done by individual bees identified by genetic or mechanical marks.

The techniques which will control much of the "superfamily" variation are straightforward. Careful beekeeping will assure that a single queen is the mother of all the worker bees measured or contributing to the social conditions affecting measurements. Instrumental insemination, especially using only the semen from a single drone, can be used to control heterogeneity arising from the drones mated to the queen. Since drones are haploid, all the sperm from a drone are genetically identical. Thus, using the semen from a single drone assures that half the genes of worker progeny are identical. This partial identity can be enhanced when highly inbred queens are used. The technique of mating inbred queens to single drones was developed by Rothenbuhler (1960) specifically for the genetic analysis of honey-bee behavior.

By using the inbred queen–single drone technique, matings can be made which permit the behavior of entire colonies to be genetically analyzed. Worker bees from such matings will all have genotypes minimally different from the colony average. Colonies representing inbred lines differing in one or more behavioral traits can be measured as representatives of parental types. Consistent measurements suggest homozygosity for the genes regulating the traits in question. Both types of parental crosses will produce colonies of worker bees classed as F_1 progeny. An F_1 sister queen can provide genetic segregation from the F_1 generation through the drones she produces. The semen of these drones can be used to provide single drone–inbred queen backcrosses to both parental lines. The recriprocal mating of an inbred line drone to the F_1 queen is not useful since it produces a colony of heterogeneous worker bees.

One difficulty is associated with the use of the semen from a single drone for inseminations. Queens inseminated with such a small amount of semen appear to be more readily superseded and are likely to more quickly deplete their reserve of spermatozoa and cease to produce worker progeny. The selection for mating of drones that have larger amounts of sperm, careful beekeeping, and restriction of the queen to a small brood-nest area effectively reduce these difficulties. Careful insemination and beekeeping allowed Rothenbuhler *et al.* (1979) to test full-sized colonies produced by queens inseminated with the semen of single drones for an entire honey production season.

IV. APPLICATIONS OF BEHAVIORAL GENETIC TECHNIQUES

A. Behavioral Analysis of Mutants

1. Eye-Color Mutants

The detailed analysis of the effects of the mutant gene *yellow* on the mating success of *Drosophila* males (Bastock, 1956) is a major contribution to behavioral genetic literature. This study showed that the technique of mutant behavioral analysis has usefulness in revealing genetic elements in the regulation of behavior. Behavioral effects of mutants are generally maladaptive pleiotropic phenotypes attending the more apparent mutant phenotype. Detailed behavioral studies can show which precise actions in behavioral sequences are impaired and can thus also offer information on how adaptive behavior is organized and what its underlying physiological correlates are. Honey bees have several known mutants that are potentially useful in such studies.

Witherell (1972) compared the flight activity of normal drones to that of drones from 21 mutant stocks. Generally, mutant drones took fewer flights from colonies than did normal drones. This decrease was almost always associated with a shorter life span and, in some cases, with a marked reduction in the return of the mutants when they did leave colonies. One mutation, *chartreuse-red (ch')*, was exceptional. Drones carrying this mutation lived as long and returned to colonies after a flight as often as normal drones. Yet they took significantly fewer flights. This result suggests that detailed observations of normal and *ch'* drones may provide good information on the factors leading to drone flight.

Kuz'mina *et al.* (1975a,b, 1977) studied neurological and biochemical effects of *ch'* and another eye-color mutation named *snow (s)*. Pleiotropic effects of these two genes were found; they caused shorter neuromuscular excitability in homozygous mutant bees, an inability to orient in a flight-room, and an inhibition of dance communication following foraging. The biochemical changes characterizing the *s* mutation apparently have a neurogenic effect through the synthesis of serotonin-like compounds. Similar work by Neese (1969) with chartreuse (*ch*) mutants showed a reduced ability to orient to a hive or a feeding place. Also, the speed of flying and dancing was lower than in phenotypically normal bees.

2. Wing Mutants

Witherell and Laidlaw (1977) explored the effects of the recessive mutant *diminutive-wing (di)* on aspects of foraging by worker honey bees. Workers homozygous for *di* have wings with areas only 62.9% of normal. Generally, these bees displayed foraging activities adjusted to the reduction in wing surface area. The wing-beat frequency of *di* workers was 22.2% more than normal. Outgoing foraging flights required 5.4% more time while returning foraging flights with nectar loads required 35.6% more time. Nectar foragers left hives carrying less food reserves and returned with smaller nectar loads. Pollen loads weighed less. Details of dance communication were the same with normal and *di* bees.

Work on a similar wing mutation, *short* (*sh*), by Kuz'mina (1977) and Lopatina *et al.* (1977) revealed effects of the mutation on a dance rhythm (depressed), and neuromuscular excitement levels (also lower). By making a phenocopy (artificial shortening of the wings) they showed that the pleiotropic effects were not simply a result of the morphological defect.

B. F_1 and Backcross Experiments

1. Nest-Cleaning Behavior

Rothenbuhler's work (1964a) with nest-cleaning behavior is the most commonly referenced work in honey-bee behavior genetics and perhaps in the larger field of behavior genetics. The work is elegantly designed and provides a classic example of genes regulating behavior. The genetic work was part of a larger study of the mechanisms of honey-bee resistance to a disease of brood, American foulbrood, caused by a sporulating bacteria, *Bacillus larvae* White. Two inbred lines were developed which were resistant to American foulbrood through several mechanisms, including one which was behavioral (Rothenbuhler *et al.* 1968). Resistant bees removed dead larvae and pupae from the brood nest at a high rate and were termed hygienic. Contrasting susceptible lines were also developed which removed dead brood only slowly or not at all and were termed nonhygienic (Rothenbuhler, 1964b).

Rothenbuhler (1964a) experimentally assessed an inbred resistant line (Brown), an inbred susceptible line (Van Scoy), and appropriate crosses. Three Brown colonies were uniformly hygienic and four Van Scoy colonies were uniformly nonhygienic. This contrast permitted the hypothesis that the lines differed genetically in their regulation of nest-cleaning behavior. Furthermore, five F_1 colonies, produced by queens inseminated with the semen of single drones and reared concurrently, showed nest-cleaning behavior very close to that of the nonhygienic Van Scoy colonies. This result suggested that the difference in nest-cleaning behvior was due to recessive genes at one or more loci.

Drones were reared from two F_1 queens and 29 single-drone inseminations were made with Brown line queens. The frequency of hygienic nest

TABLE 1. Genetic Hypothesis Explaining Differences in Nest-Cleaning Behavior among Inbred Lines F_1 Hybrids, and Backcrosses[a]

Hygienic Brown inbred line × nonhygienic Van Scoy inbred line			$uu,rr \times ++;++$	
Nonhygienic F_1			$+u,+r$	
F_1 gametes as drones from F_1 queens occur in equal frequency			$u,r;\ u,+;\ +,r;\ +,+$	
Progeny of F_1 and recessive Brown line in equal frequency	uu,rr Completely hygienic, uncap and remove	$uu,+r$ Uncap but do not remove	$+u,rr$ Do not uncap but remove	$+u,+r$ Completely nonhygienic, neither uncap nor remove

[a] Data from Rothenbuhler (1964a).

cleaning among these backcrosses to the line carrying the recessive genes provided information concerning both the underlying genetic mechanism and also the behavioral mechanisms of nest cleaning. Half (14 of 29) of the backcross colonies were nonhygienic and half were hygienic to some degree. Some of the hygienic colonies (six of 15) were completely hygienic while some (nine of 15) removed the caps of cells containing dead brood but did not remove the dead remains. This curious result prompted further testing of the nonhygienic colonies. Combs containing cells with foul-brood-killed brood were uncapped by the investigator and placed in the brood nests of the 14 nonhygienic colonies. Six of these colonies removed dead remains from the uncapped cells at a high rate and were classified as removers. Eight of the colonies removed dead remains at a lower rate or not at all and were classified as nonremovers. Thus, the backcross colonies fell into four classifications: $\frac{1}{4}$ uncapped cells and quickly removed the dead remains, $\frac{1}{4}$ uncapped but only slowly removed remains, $\frac{1}{4}$ did not uncap but quickly removed the remains, and $\frac{1}{4}$ did not uncap and only slowly removed remains.

These results permitted a genetical hypothesis to account for the differences in nest-cleaning behavior between the two inbred lines (Table 1). The hypothesis suggested that the difference may be due to genetic differences at two loci, with one regulating uncapping and the other regulating removal. The hygienic line was hypothesized to be homozygous for recessive alleles for uncapping (u) and for recessive alleles for removal (r). The nonhygienic line was hypothesized to be homozygous for dominant wild-type alleles ($+$) at both loci. F_1 hybrids, heterozygous for both loci, display the nonhygienic phenotype due to the effects of dominance. Backcrosses to the hygienic line from the F_1 as it segregates through drones from F_1 queens produce four classes of colonies in equal frequency: those with worker bees homozygous recessive at both loci (uu,rr), those with bees homozygous recessive for uncapping but not removal ($uu,+r$), those with bees homozygous recessive for removal but not uncapping ($+u,rr$), and those with bees that are homozygous at neither locus ($+u,+r$).

This hypothesis explained most, but not all, of the results. Backcrosses were also made to the nonhygienic Van Scoy line. All of the resulting colonies were expected to be nonhygienic, yet one of eight contained bees that both uncapped and removed at a good rate. The measurements of dead brood removed by the non-uncapping backcrosses when given experimentally uncapped dead brood ranged from 0 to 92%. Generally, the distinction between fast removal and slow removal was clear but one or two colonies might be viewed as intermediate. Among the backcross colonies that uncapped but did not remove the contents, one colony had opened only half the cells by the end of the test. However, the distinction between uncapped and non-uncapped was more decisive than the distinction between removed and nonremoved.

Rothenbuhler (1964a) identified these anomalous data, but felt they were not significant enough to negate the hypothesis which explains the bulk of the data. Although the Van Scoy line is named a nonremover, it does, in fact, remove dead brood, but at a very slow rate. This rate is strongly influenced by environmental factors. Young bees up to 28 days of age show hygienic behavior and then become nonhygienic foragers (Thompson, 1964). However, incoming liquid food enhances hygienic behavior (Thompson, 1964) and will even cause foragers of a hygienic line to return to nest cleaning (Palmquist-Momot and Rothenbuhler, 1971). Because hygienic behavior is measured in field colonies, the natural differential of incoming nectar between colonies, possible occasional drifting of hygienic foragers to nonhygienic colonies, and colony age-structure differences may account for the anomalous data.

2. Defensive Behavior

a. European Bees. In conjunction with his analysis of hygienic behavior, Rothenbuhler (1964a) also observed differences in stinging behavior. The susceptible Van Scoy line almost never stung the experimenter during visits to the colonies, while the resistant Brown bees stung often. A common belief that disease resistance and defensive behavior were due to the same underlying character, vigor, was disproved by observations on the 29 backcrosses of F_1 to Brown line queens. Both colonies that were hygienic and those that were nonhygienic showed various levels of stinging. Analysis of the distribution of stinging behavior in these colonies indicated that more than one or two loci were involved in this behavioral difference and that the tendency to sting was recessive.

Three other measures of defensive behavior were later studied using the Brown and Van Scoy lines, as was a related physiological character, production of isopentyl acetate (IPA), a honey-bee sting alarm pheromone (Boch and Rothenbuhler, 1974). Six Brown colonies, seven Van Scoy colonies and eight F_1 crosses were measured for their response to human breath at the hive entrance, opening of the hive without smoke, and IPA presented on a cork next to the entrance. Also, 25 workers from each colony were sampled to determine the amount of IPA present.

The defensive Brown line was more responsive than the Van Scoy line in all tests and produced more IPA. The F_1 hybrids resembled the Van Scoy parent in response to breath and IPA, which indicated dominance of the gene or genes for mild response. The intermediate responses of the F_1s in the opening test indicated a lack of dominance in this character. Extreme variation in IPA production among the F_1 colonies was attributed to a complex genetic situation which probably included effects of a polygenic determination, heterozygosity in the parental lines, dominance, epistasis, and heterosis.

Two of Boch and Rothenbuhler's behavioral tests and a third, response to a moving leather target, were used by Farrell (1977) to analyze backcrosses from an inbred queen – single drone mating scheme between the Brown line and a different gentle line named YD. Backcrosses to both parental types were tested. The backcrosses to the Brown line had more bees responding to the IPA and to opening of the colony (a characteristic that Farrell calls recruitability). They were also faster to sting the leather target; yet, they delivered fewer stings than the YD backcrosses. This demonstrates the complex nature of composite colony defense behavior. Some components of this behavior are inherited differently, while others have similar modes of inheritance and probably have some common genes.

A different component of defensive behavior was studied by Collins (1979). She measured the response of caged worker honey bees to a component of their alarm pheromone (IPA) as time to react and intensity of the initial response. Initial activity level of the bees was also measured, since it affected the expression of the other characters. For all these components the expression associated with the defensive phenotype was dominant (partial, full, or over) to the less defensive phenotype. Each character difference was determined to be due to approximately two genes which were different from the genes regulating the other characteristics.

b. Africanized Bees. A series of papers by Stort (1975a,b,c, 1976, 1980) presented the results of a study of defensive behavior which measured the progeny of queens from an Italian colony mated to drones from an Africanized colony. Backcrosses of drones from F_1 queens were made to each parental colony. This follows the Rothenbuhler (1960) scheme except that the parentals were not inbred; they were single representatives of two types of bee. Stort measured five components of the behavior: (1) time at which the first sting reached a leather ball, (2) time taken for the colony to become aggressive, (3) number of stings in the gloves of the observer, (4) number of stings in the leather ball, and (5) observer persecution behavior (or distance followed while walking away).

For two of the components, Stort proposed specific genetic systems since the distribution of behavior in the two types of backcrosses followed simple Mendelian segregation patterns. Number of stings in the gloves was inferred to be controlled by a pair of genes, F_1 and F_2: F is the dominant gentle Italian behavior type (few stings) and f is the recessive aggressive type. Only bees with f_1f_1, f_2f_2 genotypes were aggressive. A different two-gene system (A and B) with m alleles for gentleness and br alleles for aggressiveness was proposed to account for the differences seen in the number of stings in the leather ball. When an $A^mA^m,B^{br}B^{br}$ genotype is found, or there are more m alleles than br, the behavior expressed is gentle (few stings). Otherwise the bees are aggressive.

In the time to the first sting and the time to become aggressive the F_1 colonies were as aggressive as the Africanized type. The variation shown by the backcross colonies for these measures did not follow a simple segregation pattern, and their genetic bases were assumed to be complex and involve at least two genes for each behavior. The fifth character, persecution, was also complex. The F_1 was gentle, which indicated dominance of the gentle Italian type. However, the backcrosses to both parentals had both gentle and aggressive colonies. Stort suggested at least three genes controlling the differences here.

The disagreement between conclusions from several studies of defensive behavior serves to demonstrate the difficulty of studying a complex behavior. The different ways that the behavior was quantified caused the measurement of different components of the behavior that were determined by distinct sets of genes. Only a few of the measures were specific enough to clearly show the underlying modes of inheritance. It is possible that fine tuning of the processes of actual measurement could more clearly show the genotypes involved. However, some characteristics may be so polygenic in regulation that only quantitative approaches are possible.

3. Flight Activity

The daily flight patterns of Africanized, European, and F_1 hybrid bees were investigated by Kerr et al. (1970) and reviewed by Gonçalves and Stort (1978). Small but consistent differences were observed in the time of day at which foraging flights began in the morning and ended in the afternoon. Africanized bees flew both earlier and later. Additionally, the time of peak flight by European bees occurred in the morning and by Africanized bees in the afternoon. The daily flight pattern of hybrids was similar to that of Africanized bees and suggests dominance effects in the genetic systems regulating flight activity.

4. Learning

Several simple experiments on genetic components of learning in honey bees have been done. Five colonies of different genetic origin, some of which were F_1s of some of the others, were measured by Kerr et al. (1975) and found to differ in learning ability measured as discrimination and extinction during feeding at an artificial source. Ott and Brückner (1980) found that inbred lines learned color choice more slowly and had lower percentages of correct choices than did non-inbred bees. Their conclusion was that these behavioral traits are affected by inbreeding depression and are, therefore, partially genetic in origin. All of these results imply genetic influences on the behavior because the bees used differed in their genotypes. As such, these studies set the stage for more refined genetic studies of learning.

5. Hoarding Behavior

Laboratory hoarding behavior was independently reported several years ago by investigators at two laboratories (Free and Williams, 1972; Kulinčević and Rothenbuhler, 1973). Small numbers of bees in laboratory cages will take sucrose solution from a gravity feeder and place it in comb provided in the cage. This laboratory behavior, at least in certain instances, is correlated to the amount of nectar collected by bees of the same genotype in field experiments (Kulinčević and Rothenbuhler, 1973; Kulinčević et al., 1974). Partially because of its economic possibilities, hoarding behavior has received considerable experimental attention.

An inbred-line F_1 experiment (Brückner, 1980) was used to determine whether or not heterotic effects were involved in the genetic regulation of hoarding. Worker bees from three unrelated inbred lines had slower hoarding rates than non-inbred workers, even though the inbred lines were not selected for reduced hoarding rate. Hybrid (F_1) workers from a mating of bees from two of the inbred lines hoarded significantly faster than bees of either parental line. These results indicate that heterozygosity at some or all of the several loci involved (Rinderer and Sylvester, 1978) enhances hoarding.

C. Selection Experiments

1. Pollen Collection

Pollen collection is a fundamentally important activity of honey bees, since pollen is their exclusive protein source. Plants producing pollen collected by bees vary in floral morphology and hence in the difficulties confronting bees during pollen collection. Alfalfa (Medicago sativa) flowers have a tripping mechanism which causes the sexual column to strike the underside of the head of bees collecting pollen. This mechanism, although assuring effective pollination, seems to cause bees to collect pollen from other floral sources if they are available (McGregor, 1976).

Nye and Mackensen (1965) surveyed foragers returning to colonies in an alfalfa seed-production area and found substantial variation in the percentage of pollen collecting bees that collected alfalfa pollen. From this survey they identified three colonies that had collected a high percentage of alfalfa pollen and three colonies that had collected a low percentage. These colonies were used to start a breeding program which continued through seven selected generations (Mackensen and Nye, 1966, 1969; Nye and Mackensen, 1968, 1970). Selection was effective in producing two lines differing in their pollen collection behavior within four generations (Table 2). Fifth-generation backcrosses were generally intermediate in the percentage of alfalfa pollen they collected. The variability of the response of backcrosses indicated that the character was genetically regulated by several genes with additive effects. The reduction of progress in the later generations, in combination with a substantial reduction in within-line variance, provided evidence that a selection plateau had been reached in both lines. The underlying behavioral changes that resulted in increased collection of alfalfa pollen

TABLE 2. Change in Percentage of Pollen
Collectors Carrying Alfalfa Pollen Resulting
from Bidirectional Selection[a]

Selected generation	High line (%)	Low line (%)
2	40	26
3	50	15
4	66	8
5	85	18
6	86	8
7	87	36

[a] Values are group averages. Base stock: 32%.

also affected the percentage of pollen collected by the two lines from several different sources (Mackensen and Tucker, 1973).

2. Hoarding Behavior

The feasibility of bidirectional artificial selection on hoarding behavior to develop stocks producing greater and lesser amounts of honey in the field was studied by Rothenbuhler *et al.* (1979). A mating design styled after the one presented in Fig. 3 was organized to reduce problems associated with inbreeding and still use an experimentally manageable number of colonies.

Results in the first selected generation show that hoarding rate increased in the fast line but did not decrease in the slow line (Fig. 4). Thereafter, good progress was made in the slow line in the second, third, and fifth generations, but little progress was made at increasing hoarding rate in the fast-hoarding line. However, the selection pressure applied to the slow-hoarding line was considerably less than the selection pressure applied to the

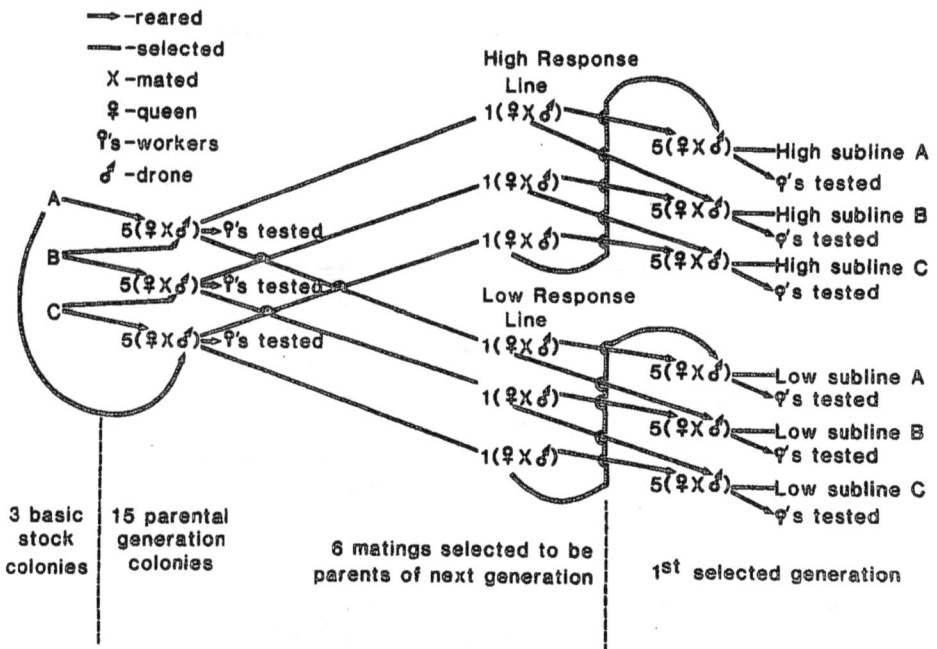

Fig. 3. A mating design reducing problems associated with inbreeding. Arrows indicate the origin of individuals; lines (without arrowheads) indicate that a selection was made among matings. Drones are haploid, and each queen was mated by artificial insemination to one drone. Progeny tests of each mating were made by testing worker bees for response. Results of these tests indicated which matings should be chosen to produce each new (selected) generation. [Redrawn after Kulinčević and Rothenbuhler (1975), with permission. Copyright 1975 by Academic Press.]

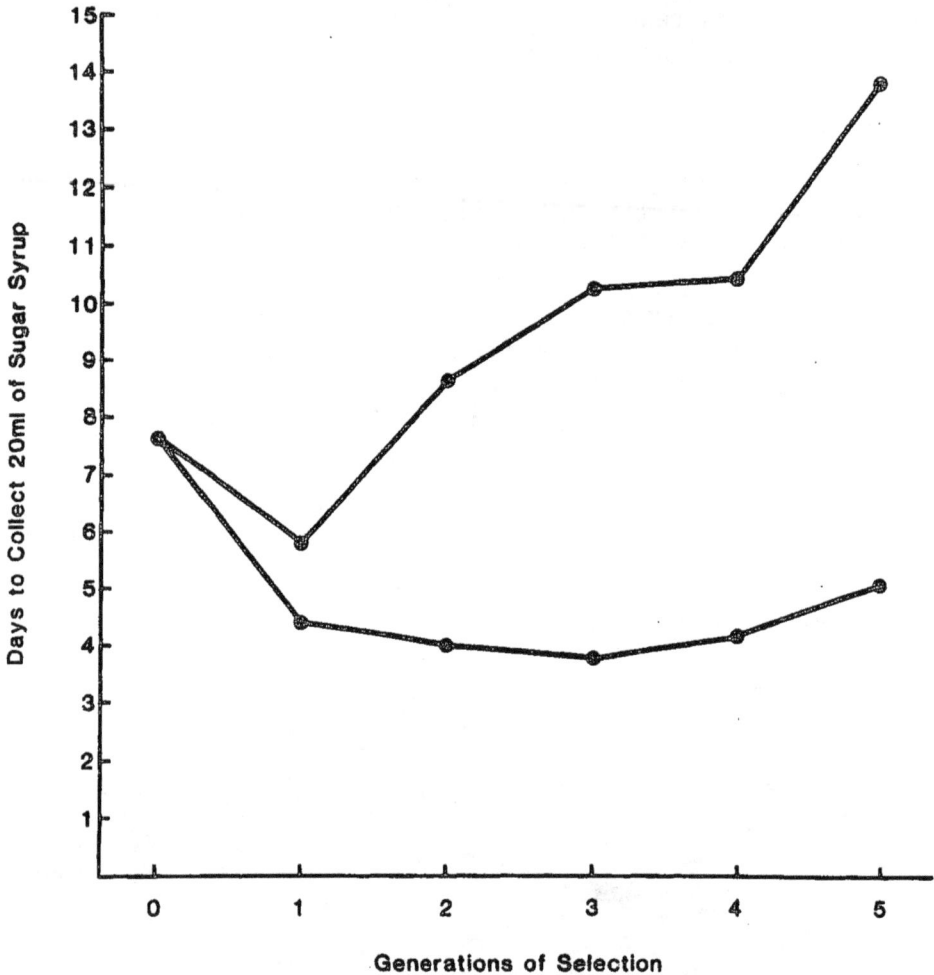

Fig. 4. Progress in selecting for fast and slow hoarding behavior in laboratory cages expressed as the time taken by caged samples of bees in successive generations to collect 20 ml of sugar syrup. Connected points are means from about 20 matings tested in each generation of each line. [From Rothenbuhler *et al.* (1979), with permission. Copyright 1979 by International Bee Research Association.]

fast-hoarding line. The hoarding rate of the fast line (4 days to hoard 20 ml of sucrose solution) is slower by at least 2 days than several other colonies tested in Rothenbuhler's laboratory. Thus, the fast selected line did not plateau in its response because a physiological limit was reached. Changes in response in the slow-hoarding selected stock and survey data from Rinderer and Sylvester (1978) suggest that additive genetic events contribute substantially to the regulation of hoarding. The plateau reached in the fast-hoarding line may indicate that genetic variation available in the base stock was generally exhausted in the development of the first selected generation. This is especially likely since the colonies of the base population were products of another selection experiment and thus were already related.

The expression of desired correlated responses in honey production was ambiguous. Three separate field tests of the second generation showed that the fast-hoarding line stored more honey. Yet field tests of the fourth and fifth selected generations did not show such differences. Interactions of the behavioral genotypes with the various environments during the field tests of each generation probably produced these results. The effects of a piece of that environment, empty comb, which stimulates hoarding and honey production, have been shown to interact strongly with nectar flow conditions (Rinderer, 1982).

D. Correlated Behavior

1. Correlated Flight Speed

Drescher and Gonçalves conducted a bidirectional selection program on a morphological character, the number of winghooks or hamuli joining fore and hind wings, which resulted in a correlated behavioral difference between lines [reported in Gonçalves and Stort (1978)]. After 22 selected generations, the line selected for fewer hamuli averaged 10.6, while the line selected for more hamuli averaged 28.6. A collaborator of Gonçalves, M. C. O. Campos, measured the flight speed of both workers and drones of these lines and found that the bees with more hamuli flew faster.

2. Defensive Behavior

Stort (1978) calculated correlations between all his measured components of defensive behavior and correlations between each measure and abdominal color. In the Africanized backcrosses, time to first sting and time to become aggressive were positively correlated. Both were negatively correlated with the number of stings both in the ball and the gloves and with the distance bees followed the observer. The number of stings and distance were positively correlated. This is expected since shorter times, more stings, and following for longer distances are all aggressive behavior. In the Italian backcross colonies, number of stings in the gloves was correlated with distance followed; time to the first sting was not correlated to number of stings in the ball; number of stings in the ball was not correlated with the distance followed. All of these relationships reflect the differences in the way colony defense is expressed by the two genetically different groups.

There was no correlation between behavioral components and abdomen color, which was a discrete trait controlled by one gene.

The Brown and YD backcross colonies used by Farrell (1977) to test colony defense were the same colonies used by Collins (1979) to study worker response to isopentyl acetate (IPA). Rank correlations were calculated for seven characters: number of bees responding to IPA at the hive entrance, number of bees responding to opening the hive, number of stings in a leather target, time to first sting (Farrell), initial activity level, time to react, and initial intensity of the response (Collins). A fast and strong response to IPA in the cage was seen in bees that responded in large numbers to IPA at the colony entrance and to opening of the hive, but there were no significant correlations between the response by the caged bees and the stinging behavior of the colony.

A. M. Collins and H. A. Sylvester (unpublished data) found that time to respond to IPA and duration of response by caged workers were significantly correlated with hoarding of sucrose solution in those same cages. This indicates that some common basis for the two activities is likely; possibly they are related to the sensory perception of both IPA and sucrose solution.

E. Heritability Estimates and Genetic Correlations

Heritability (h^2) is a genetic parameter that represents the proportion of the variation of a phenotype that can be attributed to additive genetic variance. It is frequently used in conjunction with selection programs to predict results and to assess the success of selection. A genic correlation is a parameter estimating the covariance between two characteristics arising from common additive genes. A more detailed discussion of these parameters is presented by Collins (Chapter 11).

Pirchner et al. (1962), Soller and Bar-Cohen (1967), el-Banby (1969), and Bar-Cohen et al. (1978) reported estimates of h^2 for honey production from calculations of the regression of responses of open-mated offspring with those of the female parents or from variance components. These estimates ranged from $h^2 = 0.23$ to $h^2 = 0.75$ and collectively indicate that honey production, which is in part dependent on a variety of behavioral traits, is strongly regulated by additive genetic events. As such, selection programs designed to increase honey production have a remarkably high chance of success when compared to the prospects of genetically improving traits in other livestock.

Several of these authors (Soller and Bar-Cohen, 1967; el-Banby, 1969; Bar-Cohen et al., 1978) also reported h^2 values for brood rearing. While the range of estimates was from $h^2 = 0.10$ to $h^2 = 0.90$, five of the seven estimates were above 0.30. Again, selection programs designed to improve the trait have good chances of success. These authors also calculated genetic and phenotypic correlations between brood rearing and honey production and generally found quite high genic correlations ($r = 0.77 - 1.12$) and moderate phenotypic ones ($r = 0.34 - 0.51$). Nonetheless, since the variance of the regression was larger than the predicted improvement from a reasonable selection differential, Bar-Cohen et al. (1978) do not view a selection for number of brood cells as a reliable means of increasing honey production.

Collins (1979) used a different calculation system (regression of offspring on midparent from colonies produced by single-drone insemination) to calculate h^2 for the time to respond to IPA by caged worker bees. The estimate of 0.68 was high for a behavioral character. A more sophisticated sibling analysis approach was used by Rinderer et al. (1983) and estimated $h^2 = 0.03$ for the same character. Longevity of caged worker bees was also measured and had a $h^2 = 0.32$.

Quantitative genetic studies on the honey bee are in their infancy. The theoretical groundwork for adapting existing extensive methodology to the haplo–diploid social honey bee is limited. Investigators are also hampered by the biology of the organism that currently precludes the large sample sizes of thousands that are desirable for more accurate estimation of parameters. Nonetheless, this is an important area of honey-bee behavioral genetic investigation.

ACKNOWLEDGMENT

The preparation of this chapter was done in cooperation with the Louisiana Agricultural Experiment Station.

REFERENCES

Banby, M. A. el (1969). Heritability estimates and genetic correlation for brood-rearing and honey production in the honeybee. *Proc. Arab. Sci. Cong. 6*, 517–526.

Bar-Cohen, R., Alpern, G., and Bar-Anan, R. (1978). Progeny testing and selecting Italian queens for brood area and honey production. *Apidologie 9*, 95–100.

Bastock, M. (1956). A gene mutation which changes a behavior pattern. *Evolution 10*, 421–439.

Boch, R., and Rothenbuhler, W. C. (1974). Defensive behaviour and production of alarm pheromone in honeybees. *J. Apic. Res. 13*, 217–221.

Brückner, D. (1980). Hoarding behaviour and life span of inbred, non-inbred and hybrid honeybees. *J. Apic. Res. 19*, 35–41.

Collins, A. M. (1979). Genetics of the response of the honeybee to an alarm chemical, isopentyl acetate. *J. Apic. Res. 18*, 285–291.

Collins, A. M., and Kubasek, K. J. (1982). Field test of honey bee (Hymenoptera: Apidae) colony defensive behavior. *Ann. Entomol. Soc. Am. 75*, 383–387.

Collins, A. M., Rinderer, T. E., Tucker, K. W., Sylvester, H. A., and Lackett, J. J. (1980). A model of honeybee defensive behaviour. *J. Apic. Res.* 19, 224–231.

Collins, A. M., Rinderer, T. E., Harbo, J. R., and Bolten, A. B. (1982). Colony defense by Africanized and European honey bees. *Science* 218, 72–74.

Collins, A. M., Rinderer, T. E., Harbo, J. R., and Brown, M. A. (1984). Heritabilities and correlations for several characters in the honey bee. *J. Hered.* 75, 135–140.

Darwin, C. (1859). "The Origin of Species." Murray Press, London.

Dietz, A. (1982). Honey bees. *In* "Social Insects, Vol. III" (H. Hermann, ed.), pp. 323–360. Academic Press, New York.

Falconer, D. S. (1960). "Introduction to Quantitative Genetics." Ronald Press, New York.

Farrell, K. R. (1977). Some differences in the defensive behavior of genetically different stocks of *Apis mellifera* (the honeybee). Master's Thesis. The Ohio State University.

Free, J. B., and Butler, G. C. (1955). An analysis of the factors involved in the formation of a cluster of honeybees. *Behaviour* 7, 304–316.

Free, J. B., and Williams, I. H. (1972). Hoarding by honey bees (*Apis mellifera* L.). *Anim. Behav.* 20, 327–334.

Frisch, K. von. (1965). "Tanzsprache und Orientierung der Bienen." Springer-Verlag, Berlin.

Gonçalves, L. S., and Stort, C. A. (1978). Honey bee improvement through behavioral genetics. *Annu. Rev. Entomol.* 31, 197–213.

Kerr, W. E., Gonçalves, L. S., Blotta, L. F., and Maciel, H. B. (1970). Biologia comparada de abelhas italianas (*Apis mellifera ligustica*) abelhas africanizadas (*Apis mellifera adansonii*) e suas hibridas. *Proc. Congr. Bras. Apic.* 1, 151–185.

Kerr, W. E., Duarte, F. A. M., and Oliveira, R. S. (1975). Genetic component in learning ability in bees. *Behav. Genet.* 5, 331–337.

Kulinčević, J. M., and Rothenbuhler, W. C. (1973). Laboratory and field measurements of hoarding behaviour in the honeybee (*Apis mellifera*). *J. Apic. Res.* 12, 179–182.

Kulinčević, J. M., and Rothenbuhler, W. C. (1975). Selection for resistance and susceptibility to hairless-black syndrome in the honeybee. *J. Invertebr. Pathol.* 25, 289–295.

Kulinčević, J. M., Thompson, V. C., and Rothenbuhler, W. C. (1974). Relationship between laboratory tests of hoarding behavior and weight gained by honey-bee colonies in the field. *Am. Bee J.* 114, 93–94.

Kuz'mina, L. A. (1977). Effect of the mutation *short* on signaling behaviour and neurological characteristics of the honeybee. (In Russian, English summary) *Genetika* 13, 1552–1560.

Kuz'mina, L. A., Lopatina, N. G., Nikitina, I. A., Ponomarenko. V. V., and Saifutdinova, Z. N. (1975a). Effect of mutant genes snow and chartreuse-red on neurological features of the honeybee. *Doklady Acad. Sci. U.S.S.R. Biol. Sci.* 221, 111–112.

Kuz'mina, L. A., Lopatina, M. G., Nikitina, I. A., Ponomarenko, V. V., and Saifutdinova, Z. N. (1975b). Effect of mutant genes snow and chartreuse red, controlling tryptophan metabolism *in vivo* on signal behavior of the honey bee. *Doklady Acad. Sci. U.S.S.R. Biol. Sci.* 222, 206–207.

Lockard, R. B. (1971). Reflections on the fall of comparative psychology: is there a message for us all? *Am. Psychol.* 26, 168–179.

Lopatina, N. G., Kuz'mina, L. A., Nikitina, I. A., Ponomarenko, V. V., and Chesnokova, E. G. (1977). Effect of hereditary changes in the flying system on a signaling behaviour and on nervous system activity in honeybees. *Proc. Inter. Apic. Cong. (Apimondia)* 26, 308–313.

Louveaux, J. (1969). Ecotype in honey bee. *Proc. Inter. Apic. Cong. (Apimondia)* 22. 499–501.

Mackensen, O., and Nye, W. P. (1966). Selecting and breeding honeybees for collecting alfalfa pollen. *J. Apic. Res.* 5, 79–86.

Mackensen, O., and Nye, W. P. (1969). Selective breeding of honeybees for alfalfa pollen collection: sixth generation and out-crosses. *J. Apic. Res.* 8, 9–12.

Mackensen, O., and Tucker, S. C. (1973). Preference for some other pollens shown by lines of honeybees selected for high and low alfalfa pollen collection. *J. Apic. Res.* 12, 187–190.

Manning, A. (1967). Genes and the evolution of social behavior. In "Behavior-Genetic Analysis" (J. Hirsch, ed.), pp. 44–60. McGraw-Hill, New York.

McGregor, S. E. (1976). "Insect Pollination of Cultivated Crop Plants." U.S.D.A. Agric. Handbook No. 496, U.S. Government Printing Office, Washington, D.C.

Neese, V. (1969). The behaviour of the chartreuse eye mutant of *Apis mellifera* L. *Proc. Congr. 6th IUSSI (Bern)*, p. 195–200.

Nye, W. P., and Mackensen, O. (1965). Preliminary report on selection and breeding of honeybees for alfalfa pollen collection. *J. Apic. Res.* 4, 43–48.

Nye, W. P., and Mackensen, O. (1968). Selective breeding of honeybees for alfalfa pollen: fifth generation and backcrosses. *J. Apic. Res.* 7, 21–27.

Nye, W. P., and Mackensen, O. (1970). Selective breeding of honeybees for alfalfa pollen collection with tests in high and low alfalfa pollen collection regions. *J. Apic. Res.* 9, 61–64.

Ott, I., and Brückner, D. (1980). Genetisch bedingte Unterschiede im Lernverhalten der Honigbiene *(Apis mellifica carnica)*. *Apidologie* 11, 3–15.

Palmquist-Momot, J., and Rothenbuhler, W. C. (1971). Behaviour genetics of nest cleaning in honeybees. VI. Interactions of age and genotype of bees, and nectar flow. *J. Apic. Res.* 10, 11–21.

Pirchner, F., Ruttner, F., and Ruttner, H. (1962). Erbliche Unterschiede zwischen Ertragseigenschaften von Bienen. *Proc. Int. Congr. Entomol.* 11, 510–516.

Rinderer, T. E. (1982). Regulated nectar harvesting by the honeybee. *J. Apic. Res.* 21, 74–87.

Rinderer, T. E., and Baxter, J. R. (1978). Honey bees: the effect of group size on longevity and hoarding in laboratory cages. *Ann. Entomol. Soc. Am.* 71, 732.

Rinderer, T. E., and Elliott, K. D. (1977). The effect of a comb on the longevity of caged adult honey bees. *Ann. Entomol. Soc. Am.* 70, 365–366.

Rinderer, T. E., and Sylvester, H. A. (1978). Variation in response to *Nosema apis*, longevity, and hoarding behavior in a free-mating population of the honey bee. *Ann. Entomol. Soc. Am.* 71, 372–374.

Rinderer, T. E., Collins, A. M., and Brown, M. A. (1983). Honey bee response to *Nosema apis*, longevity, and alarm response to isopentyl acetate: heritabilities and correlations. *Apidologie* 14, 79–85.

Rothenbuhler, W. C. (1960). A technique for studying genetics of colony behavior in honey bees. *Am. Bee J.* 100, 176, 198.

Rothenbuhler, W. C. (1964a). Behvior genetics of nest cleaning in honey bees. IV. Responses of F₁ and backcross generations to disease-killed brood. *Am. Zool.* 4, 111–123.

Rothenbuhler, W. C. (1964b). Behaviour genetics of nest cleaning in honey bees. I. Response of four inbred lines to disease-killed brood. *Anim. Behav.* 12, 578–583.

Rothenbuhler, W. C. (1967). Genetic and evolutionary considerations of social behavior of honeybees and some related insects. In "Behavior-Genetic Analysis" (J. Hirsch, ed.), pp. 61–106. McGraw-Hill, New York.

Rothenbuhler, W. C., Kulinčević, J. M., and Kerr, W. E. (1968). Bee genetics. *Annu. Rev. Genet.* 2, 413–438.

Rothenbuhler, W. C., Kulinčević, J. M., and Thompson, V. C. (1979). Successful selection of honeybees for fast and slow hoarding of sugar syrup in the laboratory. *J. Apic. Res.* 18, 272–278.

Slater, P. J. B. (1978). Data collection. In "Quantitative Ethology" (P. W. Colgan, ed.), pp. 7–24. J. Wiley and Sons, New York.

Soller, M., and Bar-Cohen, R. (1967). Some observations on the heritability and genetic correlation between honey production and brood area in the honeybee. *J. Apic. Res.* 6, 37–43.

Stort, A. C. (1975a). Genetic study of aggressiveness of two subspecies of *Apis mellifera* in Brazil. 2. Time at which the first sting reached a leather ball. *J. Apic. Res.* 14, 171–175.

Stort, A. C. (1975b). Genetic study of aggressiveness of two subspecies of *Apis mellifera* in Brazil. IV. Number of stings in the gloves of the observer. *Behav. Genet.* 5, 269–274.

Stort, A. C. (1975c). Genetic study of aggressiveness of two subspecies of *Apis mellifera* in Brazil. V. Number of stings in the leather ball. *J. Kansas Entmol. Soc.* 48, 381–387.

Stort, A. C. (1976). Genetic study of aggressiveness of two subspecies of *Apis mellifera* in Brazil. III. Time taken for the colony to become aggressive. *Ciênc. Cult. (São Paulo)* 28, 1182–1185.

Stort, A. C. (1978). Genetic study of aggressiveness of two subspecies of *Apis mellifera* in Brazil. VII. Correlation of the various aggressiveness characters among each other and with the genes for abdominal color. *Ciênc. Cult. (São Paulo)* 30, 491–496.

Stort, A. C. (1980). Genetic study of aggressiveness of two subspecies of *Apis mellifera* in Brazil. VI. Observer persecution behavior. *Rev. Bras. Genet.* 8, 285–294.

Thompson, V. C. (1964). Behaviour genetics of nest cleaning in honeybees. III. Effect of age of bees of a resistant line on their response to disease-killed brood. *J. Apic. Res.* 3, 25–30.

Tinbergen, N. (1960). Behavior, systematics, and natural selection. *Ibis* 101, 318–330.

Witherell, P. C. (1972). Flight activity and natural mortality of normal and mutant drone honeybees. *J. Apic. Res.* 11, 65–75.

Witherell, P. C., and Laidlaw, H. H., Jr. (1977). Behavior of the honey bee (*Apis mellifera* L.) mutant, diminutive-wing. *Hilgardia* 45, 1–30.

CHAPTER 7

Biochemical Genetics

H. ALLEN SYLVESTER

I. INTRODUCTION

The most striking characteristic of the biochemical genetics of honey bees and other Hymenoptera is the lack of variation which is found when their allozymes are analyzed. Allozyme variability in most organisms studied has yielded heterozygosity (H) estimates ranging from about 0.05 to about 0.20 with a median of 0.11 (Lewontin, 1974). In large outbreeding populations of invertebrates, the H is about 0.15 (Selander and Kaufman, 1973). Marine organisms studied at many loci have H of about 0.01 (Schopf and Murphy, 1973) to 0.22 (Ayala *et al.*, 1973). The work of Mestriner (1969) and Mestriner and Contel (1972) with honey bees, Contel and Mestriner (1974) with *Melipona*, and Crozier (1973), Johnson *et al.* (1969), and Tomaszewski *et al.* (1973) with ants indicated that the haplo–diploid Hymenoptera would conform to the high-variability pattern. However, these latter reports dealt with only two or three loci. In more extensive surveys, very low levels of variability have been reported in most of the Hymenoptera.

To save space, only the abbreviations for the commoner enzymes and corresponding loci are used in this chapter. The full names are listed as a footnote to Table 1. "Systems" means "enzyme staining systems," which may be controlled by one or more loci. Unless shown otherwise, each different stained band position revealed in a system is assumed to be controlled by a different locus. In discussing both the locus coding for a particular enzyme and the enzyme itself, it can become unclear whether the locus

177

TABLE 1. Enzyme Systems Which Have Been Assayed[a] in Electrophoretic Surveys of the Hymenoptera

| | Organism and reference(s)[b] | | | | | | | | | | | |
Enzyme[c]	Apis mellifera 1	2	3	4	5	Bombus spp.: 2,6,7	Other bees: 2,6,8,9	Polistes spp.: 8,10	Parasitic wasps: 8,11	Other wasps: 2,9	Ants: 12,13	Pont (saw
Gp						+	+	+			+	
Acph	+	−	+			+	+*		+*	−		+
Acon	+		+	+								
Ak		+	+	+		+	+			−		−
Adh			+									
Ao	+	+		+		+*	+*			+	+*	−
Aldo	+		+	+								
Aph	+		+						+*			
Amy											+*	
Cat			+									
Est	+	+	+	+	+*	+*	+*	+*	+*	+*	+*	+
F1-6dp			+									
Fum			+	+					+		+	
Galdh	+		+									
G6pdh	+		+	+		+	+*	+	+	+*	+	
Gdh	+	−	+		+	+	+		+	+		−
Got			+			−	+		+		+	
Gr	+											
Gapdh	+	+	+		+	+	+	+		+		−
αGpdh	+	+	+		+	+*	+*	+*	+*	+*	+	+
Hk	+	+	+		+	+	+*	+		+	+	+
H6pdh						−	+					
Hbdh	+	−			+	+	+	+		+*		+
Idh	+	+	+	+	+	+*	+*	+*	+	+		+
Idhn	+											
Ldh	+		+	+	+	−	+*	+	+*	+*	+	
Lap	+	+			+	+*	+	+*	+	+	+	+
Mdh	+*	+*	+*	+*	+*	+*	+*	+	+	+*	+*	+
Me	+	+		+	+*	+*	+		+	+	+	+
M6pi					+							
Pep	+						+*	+*				
6Pgdh	+	−	+			+	+*	+*	+*	+*	+	−
Pgi	+	−		+	+	+	+*	+*	+*	+		+
Pgm	+	−	+	+	+	+*	+*	+*	+*	+*	+*	+
Pk			+									
Rdh	+	−										
Rn			+									
Sodh	+	−	+		+	+	−		+*	−		−
Sudh			+									
Sdh						−	+					
To		−				−	+	+	+	+	+*	+
Tre	+		+									
Tpi	+	−		+								
Ty	+											
Xdh	+	+	+			+	+			−	+	+

or the enzyme is being discussed at a particular point. A useful convention to avoid this problem is to write both the name of the locus and its abbreviation in italics (e.g., *malate dehydrogenase* or *Mdh*) and write the enzyme name in standard type (malate dehydrogenase or Mdh). Mean heterozygosity is abbreviated as H in all cases in this chapter, regardless of the abbreviation used by the authors cited. These authors may not always be referring to exactly the same quantity.

II. VARIABILITY IN HYMENOPTERA

A. Results in *Apis*

For *Apis mellifera*, Sylvester (1976) reported one polymorphic locus (*Mdh*) and an H of 0.01 for 39 loci studied using horizontal starch gel electrophoresis (Table 1). Pamilo *et al.* (1978) reported $H = 0.012$ for 16 loci in 11 systems (Table 1). In *A. mellifera*, Nunamaker (1980) found only *Mdh* polymorphic in an analysis of 30 systems using isoelectric focusing (Table 1). Nunamaker and Wilson (1980) reported 28 to 30 tested systems exhibited isozymes, but did not discuss variation (Table 1). Badino *et al.* (1983) reported only *Mdh-1* polymorphic in an analysis of 8 systems (Table 1) in honey bees of the Piedmont of Italy. Sheppard and Berlocher (1985) assayed Italian honey bees for 21 systems (number of loci not reported) and found 3 polymorphic loci, *Mdh, Me,* and an *Est,* and a new *Mdh* allele (Table 1).

Tanabe *et al.* (1970) found that *A. mellifera* and *A. cerana* Est were differ-

---

[a] +, Staining activity detected at one or more loci in one or more species; —, no staining activity detected at one or more loci in one or more species; *, genetic variation detected at one or more loci in one or more species.

[b] Reference key: 1, Sylvester (1976); 2, Pamilo *et al.* (1978); 3, Nunamaker (1980), Nunamaker and Wilson (1980); 4, Badino *et al.* (1983); 5, Sheppard and Berlocher (1985); 6, Snyder (1974); 7, Pekkarinen (1979); 8, Lester (1975), Lester and Selander (1979); 9, Metcalf *et al.* (1975); 10, Metcalf *et al.* (1984); 11, Shaumar *et al.* (1978); 12, Pamilo *et al.* (1975); 13, Ward (1980a).

[c] Gp, general protein; Acph, acid phosphatase; Acon, aconitase; Ak, adenylate kinase; Adh, alcohol dehydrogenase; Ao, aldehyde oxidase; Aldo, aldolase; Aph, alkaline phosphatase; Amy, amylase; Cat, catalase; Est, esterase; F1-6dp, fructose-1,6-diphosphatase; Fum, fumarase; Galdh, galactose dehydrogenase; G6pdh, glucose-6-phosphate dehydrogenase; Gdh, glutamate dehydrogenase; Got, glutamate oxaloacetate transaminase; Gr, glutathione reductase; Gapdh, glyceraldehyde-3-phosphate dehydrogenase; αGpdh, α-glycerophosphate dehydrogenase; Hk, hexokinase; H6pdh, hexose-6-phosphate dehydrogenase; Hbdh, 3-hydroxybutyrate dehydrogenase; Idh, isocitrate dehydrogenase; Idhn, isocitrate dehydrogenase (NADP cofactor); Ldh, lactate dehydrogenase; Lap, leucine aminopeptidase; Mdh, malate dehydrogenase; Me, malic enzyme; M6pi, mannose-6-phosphate isomerase; Pep, peptidase; 6Pgdh, 6-phosphogluconate dehydrogenase; Pgi, phosphoglucose isomerase; Pgm, phosphoglucomutase; Pk, pyruvate kinase; Rdh, retinol dehydrogenase; Rn, ribonuclease; Sodh, sorbitol dehydrogenase; Sudh, succinate dehydrogenase; Sdh, ?, unspecified by author (Snyder, 1974); To, tetrazolium oxidase (Sod); Tre, trehalase; Tpi, triose phosphate isomerase; Ty, tyrosinase; Xdh, xanthine dehydrogenase.

ent but did not find differences between *A. cerana* "flocks" from three locations. Nunamaker *et al.* (1984) detected *Mdh* and *Est* in *A. florea, A. dorsata,* and *A. cerana,* from Pakistan. *Est* could be used to distinguish the examined populations of all three species, while *Mdh* could not be used to distinguish *A. dorsata* from *A. cerana.* None of the populations examined exhibited intraspecific genetic variability.

B. Results in Other Hymenoptera

Pamilo *et al.* (1978) reported estimates as follows: $H = 0.017, 0.005,$ and 0.007 for 18, 15, and 15 loci in 12, 11, and 11 systems for 3 populations of *Bombus lucorum;* $H = 0.037$ for 15 loci in 10 systems for *B. terrestris;* $H = 0.048$ for 12 loci in 8 systems for *B. hypnorum;* $H = 0.007$ for 16 loci in 10 systems for *B. lapidarius;* $H = 0.003, 0$ and 0 for 14, 8 and 9 loci in 10, 7, and 7 systems for 3 populations of *B. pascuorum;* $H = 0$ for 11 loci in 7 systems in *B. hortorum;* $H = 0.033$ for 10 loci in 8 systems in *Macropis labiata;* $H = 0.064$ for 8 loci in 7 systems in *Colletes succincta;* $H = 0.037$ for 15 loci in 11 systems in *Andrena clarkella;* $H = 0.007$ for 10 loci in 10 systems in *A. lapponica;* $H = 0$ for 9 loci in 7 systems in *A. vaga* (all bees); $H = 0$ for 13 loci in 11 systems in *Vespula vulgaris* (wasp); $H = 0$ for 10 loci in 6 systems in *Mimesa equestris* (wasp), and $H = 0.021$ for 18 loci in 13 systems for *Pontania vesicator* (sawfly)(Table 1).

Gorske and Sell (1976) reported four polymorphic loci ($\alpha Gpdh$, *Lap, Pgm,* and *Mdh*) in purslane sawflies (*Schizocerella pilicornis* Holmgren). They did not report how many other systems were assayed nor how many were monomorphic, so no estimate of H can be made. However, the presence of this many polymorphic loci is suggestive of a higher level of heterozygosity than that reported for other Hymenoptera.

Snyder (1974) reported no variation for 24 loci in *Lasioglossum zephyrum,* 13 loci in *Augochlora pura,* and 12 loci in *Bombus americanorum* (Table 1). *Lasioglossum zephyrum* may have a female-limited *Est.*

Pekkarinen (1979) reported estimates of H of $0.00–0.14$ for 23 populations in 8 species of *Bombus,* with the mean for up to 12 loci (in 9 systems) over all species of 0.03 (Table 1). "The total number of loci studied was 24, but information on many of these loci was so scant that they have been excluded."

Obrecht and Scholl (1981) found sporadic polymorphisms for 3 loci (*Pgi, Acon, Est*) out of 10 studied in 12 species of *Bombus* and 4 species of *Psithyrus,* plus more frequent polymorphism for *Pgm* in *Bombus lucorum.* However they did not report allele frequencies, etc.

Lester (1975; Lester and Selander, 1979) reported estimates of mean heterozygosity as follows: $H = 0.043$ for 13 loci in 7 enzyme systems in

Opius juglandis; $H = 0.033$ for 19 loci in 12 systems in *Megachile rotundata* (*M. pacifica*); $H = 0.066$ for 13 loci in 11 systems in *Nomia melanderi;* $H = 0.057$ for 15 loci in 11 systems in *Polistes annularis;* $H = 0.122$ for 13 loci in 9 systems in *P. apachus;* $H = 0.071$ for 13 loci in 9 systems in *P. bellicosus;* $H = 0.061$ for 16 loci in 10 systems in *P. exclamans* (Table 1).

Metcalf *et al.* (1975) reported estimates as follows: $H = 0.056$ for 17 loci in 12 systems in *Stictia carolina;* $H = 0.073$ for 16 loci in 9 systems in *Chalybion californicum;* $H = 0.078$ for 12 loci in 8 systems in *Sceliphron caementarium;* $H = 0.051$ for 15 loci in 8 systems in *Scolia dubia dubia;* $H = 0.059$ for 19 loci in 11 systems in *Trypargilum politum* (all wasps); $H = 0.070$ for 15 loci in 10 systems in *Nomia heteropoda;* and $H = 0.038$ for 16 loci in 11 systems in *Savastra obliqua* (bees) (Table 1). Metcalf *et al.* (1984) reported $H = 0.065$ for 20 loci in 10 systems in *Polistes metricus,* and $H = 0.073$ for 20 loci in 10 systems in *P. variatus.* Fewer wasps were analyzed for *P. apachus, P. rubiginosis,* and *P. exclamans,* especially the latter two, so H was not calculated; however, 5 out of 17 loci in 7 systems, 0 out of 19 loci in 9 systems, and 0 out of 13 loci in 5 systems, respectively, were found to vary (Table 1).

In a study of genetic relatedness and social organization of *Polistes* wasp colonies, Lester and Selander (1981) found polymorphic *Lap, Pgm, Pep, Est,* and *Idh.*

Shaumar *et al.* (1978) reported 6 polymorphic loci (*Est-3, Acp-2, Ldh-1, Ldh-2, Sodh, Pgi*) out of 22 analyzed in 14 systems in the parasitic hymenopteran *Diadromus pulchellus* WSM (Table 1). However, the most common allele at *Pgi* had an allele frequency of 0.992 in females and 0.974 in males in a laboratory population, so it does not meet the usual requirements to be termed polymorphic. Nevertheless, this is a high level of polymorphism for the Hymenoptera. They also found a monomorphic *Est-4* whose expression was limited to males.

Crozier (1977) reported on variation for *Est, Mdh,* and *Amy* between populations of the ant *Aphaenogaster rudis.*

Pamilo *et al.* (1975) studied up to 10 loci in 6 systems in 10 species of *Formica* ants and found only one polymorphic *Mdh* in each of only 2 species (Table 1). In a study of genetic population structure in polygynous *Formica* ants, Pamilo (1982) reported polymorphisms in *Mdh, Est,* and *Pgi.*

Halliday (1981) surveyed 8 color forms of the *Iridomyrmex purpureus* group in Australia. Levels of genetic variation were estimated in 84 populations for the 15 loci surveyed. Variation was found for *Me, Est-1, Est-2,* and *Amy,* while the other 11 loci did not show any variation. "The average level of heterozygosity was 3.8% and an average of 11.8% of loci was polymorphic in each population" (Halliday, 1981).

Ward (1980a) reported estimates of H per population of $0.000-0.072$ (mean 0.036) for 35 populations (5 species) of the *Rhytidoponera impressa* group of ants. This was in an electrophoretic survey of 22 loci in 16 systems (Table 1). Ward and Taylor (1981) reported one polymorphic locus, *Amy*, among 16 studied with $H = 0.032$ for the Australian ant *Nothomyrmecia macrops* Clark.

C. Discussion

While these results indicate that honey bees as well as most other Hymenoptera are very uniform in their allozymes, researchers in bee genetics and behavior have found a significant amount of variation in characters which have been shown to be genetically controlled. This then leads to some question as to how accurately allozyme variability reflects or assays the variability of an organism's genome.

One possible explanation for this apparent conflict is that these allozymes are part of the honey bee's metabolism, and thus of its basic interaction with the environment. Honey bees live in perennial colonies, even though each worker only lives a few weeks. There is no reasonable mechanism to allow the queen, which usually lives 1 or more years, to produce workers with a genotype correlated with a particular season. Therefore, every worker must be reasonably well adapted genetically to support the colony at any season. The seasonal changes in the environment would then be dealt with through phenotypic plasticity based on an adapted, generalized genotype. This genotype would show very little variation in basic metabolic characteristics, such as most of the enzymes detected by electrophoresis. There are other genes involved with important but more peripheral characteristics, such as disease resistance, body color, and morphology. These characteristics would vary among locations but would probably be more stable over time at any one location. Selection would then be expected to lead to genetically determined variants adapted to a location but differing from populations at different locations. For example, bees from more northerly regions tend to be larger with relatively smaller appendages and darker (Ruttner, 1975). Such adaptations to a colder climate leave unchanged the basic social structure and metabolism. However, Johnson *et al.* (1969) and Tomaszewski *et al.* (1973) did find a correlation between allozyme polymorphism and environmental patterns in ants. This is expectable since some allozymes may be involved in the more peripheral characters (Ayala and Powell, 1972).

Levins (1968) discusses this relationship between genotype and environment in terms of the grain of the environment. A coarse-grained environment will produce selection pressure favoring balanced polymorphisms, while a fine-grained environment will produce selection pressure favoring monomorphism, with the challenges of environmental variation being met by phenotypic plasticity. The relative grain of the environment must be defined as it is perceived by the organism. The biology and behavior of the Hymenoptera in these reports, particularly the eusocial species, are such that they "perceive" the environment as being more fine-grained than it at first appears. This is primarily due to the following set of factors which applies particularly to honey bees and to a varying extent to other Hymenoptera: (1) adult care of immatures, (2) communication in social species, (3) limited food source, (4) modification of the environment by the organism, (5) food storage, (6) limited temperature range for activity, (7) reduced competition in immatures, (8) foraging area, (9) perennial colonies, and (10) mass mating and drone congregating areas (Sylvester, 1976).

While none of these environment-related factors are exclusive to the Hymenoptera, the interacting complex is rare. This phenotypic plasticity enables them to respond genetically as if they lived in a much more uniform, i.e., fine-grained, environment than that to which most other insects and probably most other organisms are exposed.

Ayala *et al.* (1975) propose that large changes in food availability over time lead to habitat and food-use flexibility and low genetic variability. This model is based on marine organisms facing high instability in one or more environmental characteristics. Availability of food exhibits the most obvious trend. Tropical and deep-sea species, which should have the most stable food supply, are highly polymorphic. Species from environments with an unstable food supply are the least polymorphic.

Over most of the Earth, the availability of food for bees varies greatly through the year. This is particularly noticeable in the temperate zone, where no flowers are available in the winter while an abundance is available for a more or less limited period at other times. Dry season–rainy season changes in the tropics produce similar effects. Bees have met this challenge of an unstable food supply with extensive phenotypic or behavioral plasticity but very limited allozyme variability. Bees then perceive their environment as fine-grained.

Selander and Kaufman (1973) argue that increased mobility and homeostatic control lead to decreased genic heterozygosity. Bees forage over a comparatively large area. Bee colonies can have foragers simultaneously over a large area, giving the colony an effective mobility much greater than any individual forager. The set of environment-related factors discussed above has the effect of significantly increasing the degree of homeostatic control exerted by bees. This supports the argument of Selander and Kaufman. However, this argument is not in opposition to that of Ayala and Valentine, nor is it even an alternative, at least in the case of bees. Mobility and homeostatic control are simply two factors involved in an organism's ability to cope with the environment and in particular with variations in availability of food. More mobile and more highly homeostatic organisms are less affected by food and other instabilities and so perceive the environment as more fine-grained.

Haplo–diploidy is the other obvious explanation for low allozyme variability in these Hymenoptera, since they all have haploid males and diploid females. With the exception of lethal alleles, there have been two schools of thought about the effects of haplo–diploidy on genetic variability. Crozier (1970) stated that "there is no a priori reason to believe haplo–diploidy reduces the likelihood of balanced polymorphism." However, Hartl (1971) stated, "one should expect to find fewer polymorphisms maintained by overdominance in a male haploid population than in a comparable diplo-diploid population." Since their arguments were based on different assumptions, "Which is in fact the better assumption is a matter that can be decided only by experiment" (Hartl, 1971).

The electrophoretic work reported so far in the Hymenoptera does not allow a decision to be made about haplo–diploidy, since its effects can not yet be separated from those of phenotypic plasticity in responding to environmental variability. However, these reports are sufficient to indicate some studies which should be particularly rewarding. These fall into two classes: studies of other haplo–diploid Hymenoptera, and studies of diplo–diploid, social, non-Hymenoptera. The object of these studies would be to assay the relative contributions of various factors to genetic variability.

Haplo–diploid Hymenoptera exhibiting various aspects of the phenotypic plasticity discussed above do exist. By making appropriate choices, it should be possible to separate the effects of haplo–diploidy from those of phenotypic plasticity. The generally plant-feeding Symphyta should be particularly interesting. Of these, the external feeding sawflies would probably be the best choice since they are so similar to Lepidoptera. Pamilo et al. (1978) found $H = 0.021$ for the sawfly Pontania vesicator (Table 1), which was similar to the other Hymenoptera he studied (average $H = 0.018$). The results of Gorske and Sell (1976) with the purslane sawfly may be similar, but they cannot be compared here since they did not list how many monomorphic loci they discovered. The parasitic Hymenoptera should also be interesting, particularly those with a narrow host range. The report by Shaumar et al. (1978) of higher variability in the parasitic hymenopteran Diadromus underscores the potential of such research.

The most obvious and possibly only choice for a eusocial, diplo–diploid nonhymenopteran is the Isoptera or termites. The similarity of their biology to that of ants indicates that they should be a good choice for the separation of the effects of haplo–diploidy from those of the set of environment-related factors. The report by Clement (1981) that European Reticulitermes species are polymorphic for 13 out of 25 loci supports the influence of haplo–diploidy in causing low variability in the Hymenoptera.

Thus, if the reduction in allozyme variability in bees and ants is due primarily to haplo–diploidy, other Hymenoptera should have the same reduced variability regardless of their biology. If the reduction in allozyme variability is due primarily to environmental stability effected by the biology (phenotypic plasticity) of these assayed Hymenoptera, then termites, with a similar biology, should have a similar reduction in allozyme variability. If, as is likely, the reduction is due in part to both causes, choosing Hymenoptera with various combinations of environment-related factors should give some insight into the relative importance of the various factors.

Honey bees are reported as being first in North America in 1622 and in California in the 1850s (Oertel, 1980). Because of the difficulties of transporting live honey bees, these introductions certainly involved only small numbers of queens and their attendant workers. Several subspecies were eventually introduced, and there have been many subsequent importations, but again these probably involved fairly small numbers of queens and workers, at least until recently. It might be argued that the bees in Europe are variable, but this variability was lost in the "bottleneck" when a few bees were introduced to the Americas. Since the introduced bees were from various locations and subspecies, were imported to several locations, and yet are still virtually identical electrophoretically, this does not seem likely to have affected common alleles. Rare alleles may very well have been lost. However, Sheppard and Berlocher (1984, 1985) feel that such a bottleneck has occurred.

Another perspective on the low variability in Hymenoptera is that stated by Crozier (1980): "Rather than Hymenoptera having unusually low levels of genic variation, it seems likely that *Drosophila* has unusually high levels and that this has biased previous surveys."

However, Berkelhamer (1983) analyzed mean heterozygosity values for 101 insect species and found 50 hymenopteran species have significantly lower H values than do 51 diplo–diploid nonhymenopterans. This difference remains significant when *Drosophila* are eliminated from the analysis. Graur (1985), Reeve *et al.* (1985), and Owen (1985) point out classification errors within the Hymenoptera which were made by Berkelhamer and which affect her conclusions about variation within the Hymenoptera. Graur also discusses other problems with Berkelhamer's analysis and concludes that haplo–diploidy per se does not reduce genetic variability. However, the data he presents do not support such a strong conclusion.

Simon and Archie (1985) discussed the effect of choice of enzymes sampled on heterozygosity estimates. They showed that small differences in the selection and resolution of loci can cause large differences in H.

Genetic variation in male haploids and in sex-linked loci has been examined through models by Pamilo (1979), Curtsinger (1980), and Pamilo and Crozier (1981), among others.

Moritz *et al.* (1986) applied DNA restriction enzyme technology to

TABLE 2. Electrophoretic Allele Frequencies Reported for *Apis mellifera*

Researcher	Country	Bee[a] type	Number of colonies	Malate dehydrogenase (Mdh-1)			
				0.50 (S,C,65)[b]	0.63 (M,B,80)[b]	87	1.00 (F,A,100)[b]
Sylvester (1976)	United States	I	24	0.70	0.11		0.20
	Brazil	Az	34	0.01	0.16		0.84
	Colombia	E	13	0.39	0.37		0.24
	Trinidad	E	10	0.50	0.13		0.37
Contel *et al.* (1977)	Brazil	I	34	0.71	0.15		0.14
	Brazil	Az	78	0.03	0.20		0.77
Gartside (1980)[c]	Australia	Cn	9	0.31	0.32		0.37
	Australia	Cc	2	0.7	0.1		0.2
	Australia	I	4	0.7	0.1		0.2
	Australia	F	4	0.2	0.5		0.3
Nunamaker (1980)[d]	Brazil	Az	4	A: 0.1	B: 0.2		C: 0.7
	Brazil	Az	6		0.008		0.992
	South Africa	A	18				1.0
	South Africa	Ac	2[c]	0.3	0.3		0.4
Nunamaker and Wilson (1981a)	South Africa	A	10				1.0
	Brazil	Az	12	0.04	0.03		0.93
Cornuet and Louveaux (1981)	Italy	I	8	0.75			0.25
	France	M	14		1.00		
Badino *et al.* (1983)	Italy	I	412	0.77	0.03		0.21
	France	M	?[e]		1.00		
Martins *et al.* (1977)	Brazil	I	28				
	Brazil	Az	78				
Mestriner and Contel (1972)	Brazil	I	7				
	Brazil	Az	68				
Bitondi and Mestriner[c] (1983)	Brazil	Az	183				
Sheppard and Berlocher (1984)	Norway	M	6	0.04	0.85		0.11
Sheppard and Berlocher[c] (1985)	Italy	I	5	0.64		0.06	0.30

[a] A, African (*A. m. scutellata*); Ac, Cape (*A. m. capensis*); Az, Africanized; Cc, Caucasian (*A. m. caucasica*); Cn, Carniolan (*A. m. carnica*); E, European; F, "feral swarms;" I, Italian (*A. m. ligustica*); M, *A. m. mellifera*.

[b] Allele designations: Some authors have used designations different from those first published. In these cases, the later designations are given in parentheses below the original designation.

[c] Means calculated by present author.

[d] *Mdh-1* allele frequencies for small numbers of colonies from several other locations around the world are also presented in this thesis.

[e] For 68 bees from several colonies in one apiary.

Alcohol dehydrogenase (Adh-1)			Esterase				Malic enzyme (Me)			General protein (P-3)	
			(Est-1)		(Est-3)						
3	2 (S)[b]	1 (F)[b]	S	F	S (100)[b]	F (130)[b]	79	100	106	S	F
	0.44	0.56			0.95	0.05					
	0.2	0.8			1.00						
	0.3	0.7			1.00						
	0.3	0.7			0.98	0.02					
	0.10	0.90									
0.05	0.70	0.25									
					0.91	0.09				0.53	0.47
					0.98	0.02				0.995	0.005
			0.02	0.98	0.98	0.02					
							0.14	0.86			
					0.94	0.06		0.94	0.06		

honey-bee mitochondrial DNA and found polymorphism, but at a low level relative to other organisms. Since mitochondrial DNA is maternally inherited, its low level of variability would not be due to haplo–diploidy. However, selection for a generalized genotype with little variation in basic metabolic characteristics would account for this low variability. These results must be regarded with caution since only three determinations were made.

III. KNOWN VARIABLE LOCI IN *APIS*

A. *Malate Dehydrogenase (Mdh)*

Sylvester (1976) reported the presence of 3 alleles at this polymorphic locus in adult worker bees. Allele frequencies were presented for Italian bees from California, Africanized bees from Venezuela, and European bees from Trinidad and Colombia (Table 2). He reported that workers display either one or three bands while drones display only one band, which can be explained if it is assumed that the Mdh enzyme is a dimer composed of any two monomers.

Contel *et al.* (1977) reported the Mendelian inheritance pattern of the three alleles at the *Mdh-1* locus and that *Mdh-1* and *Adh-1* are not linked. They also observed an extra band in pupae, compared to larvae and adults. They reported allele frequencies for Africanized and Italian bees in Brazil (Table 2).

Pamilo *et al.* (1978) list 2 polymorphic loci, *Mdh-1* and *Mdh-2*, in a table but do not discuss them further.

Cornuet (1979) found a preponderance of the mid allele *(0.63)* on Guadeloupe, with the fast *(1.00)* and slow *(0.50)* alleles found only where queens have been recently imported and where imported genes would be most likely to diffuse. This indicates the naturalized population was monomorphic until recent importations from Europe introduced the other two alleles. Cornuet and Louveaux (1981) reported that 8 colonies of *Apis mellifera ligustica* (origin unspecified) showed only Mdh^A *(1.00)* and Mdh^C *(0.50)* alleles, with frequencies of 0.25 and 0.75, respectively. Only the Mdh^B *(0.63)* allele was present in 14 colonies of *A. m. mellifera* from France.

Gartside (1980) reported *Mdh* allele frequencies for bee stocks in Australia (Table 2), based on worker larvae.

Nunamaker (1980) described allele frequencies from small numbers of colonies for several races from many countries around the world. A few of these values are presented in Table 2. Nunamaker and Wilson (1981a) reported that African bees *(A. m. adansonii)* from 10 colonies from two locations in South Africa were all homozygous for the fast *(1.00)* Mdh allele. Africanized bees from 12 colonies from Brazil had an $Mdh\text{-}1^{1.00}$ allele fre-

quency of 0.93, with 8 of the 12 colonies displaying only *1.00/1.00* genotypes (Table 2).

Badino *et al.* (1982, 1983) reported that *A. m. ligustica* in a wide area of the
Piedmont of Italy shows a homogenous *Mdh-1* allelic frequency distribution
(Table 2). *Apis m. mellifera* from one apiary in France is homozygyous for the
mid allele, and in two areas where these varieties hybridize the hybrid
populations show intermediate frequencies. They also found a new fourth
allele which they designated "S_1," which is slower than the 3 previously
reported alleles and was rare and present in few samples (Badino *et al.*,
1983).

Sheppard and Berlocher (1984) reported that Norwegian *A. m. mellifera*
are polymorphic (see Section IV,A.). They (1985) also reported that *A. m.
ligustica* from Italy are polymorphic and have a new *Mdh* allele (Table 2).

Snyder *et al.* (1979) developed a system for the isolation of cytoplasmic
Mdh from honey-bee larvae. They showed the enzyme is a dimer with a
molecular weight of 34,000.

B. *Alcohol Dehydrogenase (Adh)*

Martins *et al.* (1977) found a polymorphic system with 3 *Adh* alleles in
drone and worker pupae. The allele frequencies were different in Africanized and European bees (Table 2). Adh activity was not detected in young
larvae, increased to a maximum in prepupae and white-eyed pupae, and
declined to total absence in emerging bees.

Gartside (1980) reported *Adh* allele frequencies for bee stocks in Australia
(Table 2), based on analyses of worker larvae.

C. *Esterase (Est)*

Mestriner (1969) determined the inheritance of an *Est* polymorphism
with 2 alleles, based on analyses of worker and drone pupae. The esterase
patterns are constant during the entire cycle of development. Mestriner and
Contel (1972) determined the fast allele of *Est* has an allele frequency of 0.02
in Africanized and 0.09 in Italian bees in Brazil (Table 2). *Est* and *P-3* are not
linked.

Tanabe *et al.* (1970) reported that *Apis mellifera* esterase is an acetylcholine esterase and that activity was much higher in female tissue than in male
tissue, based on agar gel electrophoresis. They did not find any variation
within *A. mellifera* but found differences among 7 species of bees and wasps.

Gartside (1980) assayed *Est* allele frequencies for bee stocks in Australia
(Table 2), based on analyses of worker larvae.

Bitondi and Mestriner (1983) detected six esterases that differ in electrophoretic mobility, substrate specificity, inhibition properties, and profiles during ontogenetic development. *Est-1, Est-3, Est-5,* and *Est-6* showed variation, with *Est-3* being the same as discussed above (Table 2). *Est-6* variants had migratory rates which were so similar that they were not separable, while *Est-5^F* was detected only in workers from a single colony. Also, the commonest alleles for the other 2 loci both had allele frequencies of 0.98 over the 4 populations analyzed, so none of these loci meet the conservative criterion, of the commonest allele <0.95, to be counted as polymorphic (Ayala *et al.,* 1973).

Sheppard and Berlocher (1985) assayed adults from *A. m. ligustica* colonies from Italy and reported finding *Est^{130}* and *Est^{100}* in one colony while the other 4 were monomorphic for *Est^{100}*. *Est^{130}* did not occur in Norwegian samples of *A. m. mellifera* but does occur in United States *A. mellifera.*

D. *Malic Enzyme (Me)*

Sheppard and Berlocher (1984) discovered two allozymes in 6 Norwegian colonies of *A. m. mellifera* with allele frequencies of 0.14 for *Me 79* and 0.86 for *Me 100* (Table 2). Sheppard and Berlocher (1985) also reported that *A. m. ligustica* from Italy are polymorphic with allele frequencies of 0.94 for *ME^{100}* and 0.06 for *ME^{106}*, for the 5 colonies assayed (Table 2).

Variation in *Me* was not detected by Sylvester (1976), Pamilo *et al.* (1978), or Badino *et al.* (1982, 1983).

E. *Peptidase*

Del Lama and Mestriner (1984) reported a *Pep* polymorphism with a null allele ($f = 0.84$) but did not provide details in their publication.

F. *General Protein (P-3)*

Mestriner (1969) determined the inheritance of a *P-3* polymorphism with 2 alleles which appears only during the pupal stage. Workers show only a single band for either heterozygotes or homozygotes. Mestriner and Contel (1972) reported the *P-3^F* allele had an allele frequency of 0.47 in Italian bees and 0.005 in Africanized bees in Brazil (Table 2). *P-3* and *Est* are not linked.

Gartside (1980) was unable to detect *P-3* in Australian stocks. This is not surprising since this study analyzed larvae and not pupae.

IV. USES OF VARIATION

A. Taxonomy

One of the promising uses for the identifiable biochemical variation in honey bees is as an additional tool for the taxonomist. Sylvester (1982) used the published data for three loci (*Mdh-1, Adh-1,* and *P-3*) and the method of Ayala and Powell (1972; Ayala, 1983) to demonstrate an efficient method for separating Africanized bees of Brazil from the sampled European bees of Brazil. This method combines the analysis of these three loci to yield a probability of more than 99% of correct identification of an individual worker bee. Since separately these three loci yield probabilities of 93, 87, and 85%, respectively, the combined analysis is clearly more reliable.

Nunamaker and Wilson (1981a) reported that samples of *A. m. adansonii* from Africa were homozygous for the *C (Mdh-1$^{1.00}$)* allele, while Africanized bees from Brazil had a *C* allele frequency of 0.86. This suggested that the high level of the *C* allele in Africanized bees is from their African ancestors.

Gartside (1980), however, found that *Mdh, Adh,* and *Est* were not absolutely diagnostic of different commercial stocks present in Australia.

Cornuet (1979) found that most bees on the islands of Guadeloupe had only or predominantly one *Mdh-1* allele. The presence of the other two alleles was correlated with recent importations of queens. Cornuet and Louveaux (1981) reported that Mdh seems to discriminate between *A. m. mellifera* and *A. m. iberica* versus *A. m. caucasica, A. m. ligustica,* and *A. m. carnica,* with *mellifera* and *iberica* showing only the *B (0.63)* allele while *lingustica* showed only the *A (1.00)* and *C (0.50)* alleles. Badino *et al.* (1982, 1983) reported *A. m. mellifera* French populations are monomorphic for the *M (0.63)* allele while *A. m. ligustica* populations show homogeneous allelic frequency distributions with the *M* allele absent or at very low frequencies. Hybrid populations in very limited alpine areas show intermediate *M* frequencies. Nunamaker (1980) reported that bees he sampled of *A. m. mellifera* populations showed high frequencies of the *B (0.63)* allele. Sheppard and Berlocher (1984) reported that 6 colonies of Norwegian *A. m. mellifera* had *Mdh* allele frequencies of 0.04 (*Mdh^{65}* = 0.50), 0.85 (*Mdh^{80}* = 0.63), and 0.11 (*Mdh^{100}* = 1.00), while *A. m. ligustica* from Italy (Sheppard and Berlocher, 1985) had frequencies of 0.64 (*Mdh^{65}*), 0.06 (*Mdh^{87}*), and 0.30 (*Mdh^{100}*).

This shows that biochemical variation can be useful in separating at least some populations within *A. mellifera.* Discovery of other variable loci, particularly those detectable in adult bees, should significantly increase the discriminatory power of this method.

Biochemical variation would be even more useful in taxonomy if sample collection problems could be solved. The present requirement for live or very-low-temperature frozen samples presents difficulties in obtaining samples from remote areas, particularly in areas where frozen carbon dioxide is not readily available for sample maintenance. Possible solutions include the analysis of different sources of variation, such as more stable proteins or improved methods of storing samples to maintain the enzymes without degradation.

Pekkarinen *et al.* (1979), Obrecht and Scholl (1981), and Pamilo *et al.* (1981) demonstrated the use of enzyme gene variation in studies of phylogenetics in Vespidae, *Bombus*, and *Psithyrus*. Metcalf *et al.* (1984) demonstrated the use of isozyme data in analyzing speciation within the *Polistes fuscatus* species complex. Ward (1980b) used electrophoretic (allozyme) differences in addition to morphological differences in a systematic revision of the *Rhytidoponera impressa* group of ants. Crozier (1981) further discussed the use of allozymes in ant systematics.

B. Genetic Markers

Another promising use for the identified biochemical variation in honey bees is as genetic markers in research. As far as is known, this identified variation has no effects on any characteristic of the bees, including their general fitness. In contrast, most of the other known genetic markers do have other effects on the bees. The eye-color mutants affect vision when the phenotype is displayed, and the wing mutants prevent flight (*short, truncate,* and *wrinkled*) or significantly affect flight (*diminutive*). The body-color mutants, *major factor black* and *cordovan,* may affect body temperature or affect general fitness. Therefore allozymes (allelic forms of isozymes) offer the unique advantages of being reasonably easily identified while still causing little or no change in the bees. The allozymes which are likely to be most useful are those which are rare or absent at a particular location.

V. OTHER KNOWN ENZYME SYSTEMS

Several researchers have studied honey-bee enzymes and proteins using electrophoresis or other techniques. While those studies had different aims than assessing population variability, a brief discussion of their results is nevertheless appropriate as a guide to other enzymes or proteins which might be included in a future survey or as further background on those which have already been surveyed.

Galuszka and Kubicz (1968), using paper electrophoresis, reported dif-

ferences between the protein patterns of seminal plasma and drone hemolymph, while spermathecal fluid showed the same pattern as queen hemolymph. Differences were found between the hemolymph protein patterns
of queens, drones, and workers.

Tanabe et al. (1970) compared esterases in adults of seven species of bees,
including A. mellifera and A. cerana, and wasps using electrophoresis.

Giebel et al. (1971) studied endopeptidases from the midgut of adult
worker honey bees.

Marquardt and Brosemer (1966) purified and crystallized αGpdh. Brosemer and Marquardt (1966) reported the enzymic properties and amino
acid composition of αGpdh. Using cellulose acetate electrophoresis, Brosemer et al. (1967) studied αGpdh in honey bees and five other species of
Hymenoptera, and reported that honey-bee thoracic extracts showed one
major and one very minor band. Tomimatsu and Brosemer (1972) reported
that honey-bee αGpdh is a dimer. Fink et al. (1970), using cellulose acetate
electrophoresis, found that the honey bee has only one major αGpdh band
while all bumble-bee species show several constant bands. Storey and Hochachka (1975) studied the kinetic requirements of αGdph from flight
muscle.

In a spectrophotometric survey of oxidases in 79 animal species, Wurzinger and Hartenstein (1974) reported that honey bees did not display
peroxidase or aldehyde oxidase (Ao) activity. Nevertheless they may still be
correct in a sense since their assay used vanillin as the substrate and honey-
bee Ao may not react with vanillin. Sylvester (1976) noted that bees displayed two bands for Ao.

Huber and Thompson (1973) reported transglucolytic activity and unusual kinetics for honey-bee sucrase (invertase).

Using a spectrophotometric method, Alumot et al. (1969) demonstrated
the presence of trehalase (Tre) in honey bees. Lefebvre and Huber (1970)
studied the solubilization, purification, and properties of honey-bee Tre.
Talbot et al. (1975) reported the purification and properties of a free and a
bound Tre from thoraces. Talbot and Huber (1976) reported the electrophoretic and pH characteristics of a thoracic and an abdominal Tre. Brandt and
Huber (1979) found that thoracic Tre is totally mitochondrial. They also
studied the distribution of activities of cytochrome C oxidase, adenyl kinase,
and Mdh after dispersion treatments. Brandt et al. (1979) reported the kinetic parameters of Tre.

Kubicz and Galuszka (1971) reported on polyacrylamide gel electrophoresis of proteins and acid phosphatase of hemolymph from queens, drones,
and workers, and found differences between sexes and castes but did not
discuss genetic polymorphisms.

Plantevin and Nardon (1972) used the method "Auxotab" to detect

Acph, Aph, Est, lipase, aminopeptidase, protease, and β-glycosidase activities in the gut of honey bees.

Metcalf and March (1950) and Metcalf et al. (1955) reported properties of an acetylcholinesterase. Metcalf et al. (1956) separated a specific cholinesterase, an aliphatic esterase, and an aromatic esterase based on reactions with selective inhibitors. Substrate specificities are also noted. Kunkee and Zweig (1963) reported the purification and substrate specificity of acetylcholinesterase.

Marquardt et al. (1968) studied the crystallization, quantitative immunochemistry, and electrophoresis of Gapdh from thoraxes.

Gilliam and Jackson (1972a) detected Est, Mdh, Ldh, and αGpdh activity in polyacrylamide gel disc electrophoresis of adult worker honey-bee hemolymph but did not detect any electrophoretic differences (polymorphisms).

Arnold and Delage-Darchen (1978) reported on a survey to detect the presence of various enzymes in the different salivary glands of honey bees. They tested for activity of the following enzymes: Aph, Est (C4), Est-lipase (C8), lipase (C14) (none), aminopeptidase, trypsin and chymotrypsin (none), Acph, phosphoamidase, α-galactosidase, β-galactosidase, β-glucuronidase (none), α-glucosidase, β-glucosidase, β-glucosaminidase, α-mannosidase, and α-fucosidase.

Gilbert and Wilkinson (1974) reported on epoxidase, hydroxylase, and o-demethylase activities in larval and adult workers and drones. Huber and Thompson (1973) reported unusual kinetics for a sucrase (invertase) isolated from whole honey bees. Huber (1975) reported the purification and properties of a honey-bee abdominal sucrase. Huber and Mathison (1976) reported the properties of a different sucrase found mainly in the head.

Blum and Taber (1965) reported on the activities of 13 dehydrogenases in washed honey-bee spermatozoa; NADH₂ dh, NADPH₂ dh, Sudh, Mdh, NADP-Idh, NAD-Idh, αGpdh, Ldh, Adh, αGpdh flavoprotein, Gdh, Hbdh (none), and G6pdh.

Metcalf et al. (1966) reported that no soluble Ty activity could be detected in homogenates of the honey bee.

In a study of the effect of inbreeding on drones, Moritz (1982) reported on the volume-activities of the enzymes cholinesterase, Acph, Aph, Got, Gpt, Idh, Mdh, and G6pdh.

Martin (1965) reported on an electrophoretic study of hemolymph albumins.

Barker et al. (1966) reported the separation of hyaluronidase and phospholipase from venom and their interactions with human serum. Allalouf et al. (1975) reported characteristics of a testicular hyaluronidase. Owen (1979) reported on the hyaluronidase activity in the venom of queens and workers.

Lensky (1971a) separated and characterized the hemolymph proteins of worker larvae using several techniques.

Baars and Driessen (1984) reported on aryl hydrocarbon hydroxylase and glutathione S-transferase activity in honey bees and *Varroa* mites.

Del Lama and Mestriner (1984) described and compared zones of exopeptidase activity after starch gel electrophoresis of 14 species of bees (including *A. mellifera*).

Turner *et al.* (1979) reported on protease, hexosaminidase, hyaluronidase, and aryl sulfatases A and B from seminal plasma or sperm extracts.

VI. DEVELOPMENTAL VARIATION IN ENZYMES AND PROTEINS IN *APIS*

Tripathi and Dixon (1968) studied hemolymph Est patterns at eight ages of queen and worker larvae, by starch-gel electrophoresis. They found quantitative and qualitative differences in the patterns with age within and between castes.

Bitondi and Mestriner (1983) reported that three of six esterases studied vary during ontogenetic development.

Tripathi and Dixon (1969) studied hemolymph dehydrogenase isozymes at seven ages of queen and worker larvae. By starch-gel electrophoresis, they found differences within a caste and between castes for G6pdh, 6Pgdh, Mdh, Ldh, αGpdh, and Gapdh. By spectrophotometric assays, they found quantitative differences over time within a caste and between castes for G6pdh, 6Pgdh, and Mdh.

Liu and Dixon (1965) studied the patterns of hemolymph proteins during the larval life of queen and worker larvae. They found the total protein concentration drops to its lowest level in both castes during the third day of larval life, with queens having the lower concentration. Starch-gel electropherograms showed differences over time for both castes as well as differences between castes.

Using disc, double diffusion, and immunoelectrophoresis, Lensky and Alumot (1969) reported female specific proteins in hemolymph. Lensky (1971b) studied worker hemolymph proteins during development. He found only slight changes during the larval stage and found three main patterns of hemolymph proteins — larval, adult, and common to all stages.

Gilliam and Jackson (1972b) reported on the changes during development in the fluid proteins of workers, as shown by disc electrophoresis. They did not observe any differences in the patterns of mature adult workers. Gilliam and Valentine (1973) reported they were unable to separate adult worker hemolymph proteins on cellulose acetate membranes.

Using isoelectric focusing, Nunamaker and Wilson (1981b) found that the number of Mdh and Est isoenzymes in eggs increased as development proceeded. Nunamaker and Wilson (1982) found multiple molecular forms of Mdh, αGpdh, Adh, 6Pgdh, Xdh, G6pdh, Acph, and Est in larval hemolymph. In general, activity declined during the later stages of larval life.

VII. SAMPLING, ANALYSIS, AND INTERPRETATION

Assessing biochemical variation through electrophoresis or other methods is simply a way of measuring a phenotype in order to study the genetics of an organism. Therefore the usual rules of genetic analysis apply. In particular, statements made about populations or subspecies of bees must be based on proper sampling and interpretation. Because the workers in a colony of bees are generally only descended from one queen, sampling a large number of workers from one colony is not the same as sampling the same number of individuals in a population of a non-colonial species. The queen carries her own two haploid genomes plus the genomes of the drones with which she mated. Since a queen is usually estimated to mate with an average of 10 drones (Koeniger, Chapter 10), a colony of bees represents, on average, 12 genomes. Furthermore, each succeeding bee sampled is not independent and has a decreasing probability of sampling even those 12 genomes. It is clear that there is a very rapidly decreasing probability that each subsequent worker or drone sampled will contribute new information about the queen's two genomes. Similarly, but more slowly, each subsequent worker sampled will have a lower probability of sampling an unsampled drone genome. Also, the 10 drones will probably not produce equal numbers of progeny (Moritz, 1983), increasing the error. Therefore, at least with regard to population genetics, the usefulness of repeated samples from one colony is greatly reduced after only a very few workers have been sampled: probably about three or four, but the calculations have not yet been made. Thus exhaustive sampling of a few colonies, regardless of the number of bees sampled, only provides information about a few genomes. It is much more informative to sample one or a few workers from each of several colonies in as many apiaries and locations as possible, rather than to exhaustively sample a few colonies. Unfortunately some authors have chosen the latter approach and, in so doing, greatly reduced the value of their studies in answering questions about biochemical variation in bees.

Statements about the variation present in a population or subspecies of bees are best supported by allocating resources to extensive sampling of as many locations as possible. Next in priority is to sample as many apiaries per location as possible. Colonies should be regarded as families, not as populations, and sampled accordingly, i.e., very few bees per colony.

Exhaustive sampling would be appropriate for some studies, such as mating behavior or sperm clumping. Therefore, the aims of the research must be considered in deciding on the proper sampling method.

It is incorrect to calculate allele and genotype frequencies for a colony (family) and present them as if the numbers represented a population. They should at least be calculated for units consisting of several colonies and to the mathematically correct number of decimal places.

Another useful convention is designating alleles by a numerical super-script in relation to an allele, usually the most common or the fastest, arbitrarily designated as *100* or *1.00*. The other alleles are then designated by a number calculated as the ratio of the distance they travel from the origin relative to the designated allele; e.g., if a second allele travels $\frac{1}{2}$ as far it is designated *0.50* or *50 (Mdh-1^{50})*, if it travels 1.3 times as far it is designated *1.30* or *130 (Est-3^{130})*.

The alternatives of fast (*F*) versus slow (*S*) or alphabetic (*A, B, C*) super-scripts are simpler to assign initially but are much less flexible in allow-ing simple and logical naming of additional alleles discovered later—i.e., what do you call an allele that is faster than *"Fast"* or migrates between *"A"* and *"B"*?

The detection of biochemical variation is not only dependent on proper sampling in the field and proper storage of unprocessed samples, but is also dependent on the choice of suitable laboratory methods. For example, Brückner (1974) reported being unable to detect any variation for *Mdh* using electrophoresis. However, when Hung and Vinson (1977) compared her method with another electrophoretic method, they also found no variation under her conditions but did find variation in the same bees when analyzed under other conditions.

Another problem which has appeared is the use of data on biochemical variation to make improper inferences about the genetics or taxonomy of populations where baseline data are insufficient or absent. Examining pop-ulations of bees and finding a characteristic, such as homozygosity for a particular allele, which seems to be unique to a particular population, does not allow one to infer that an unknown population which shows this char-acteristic was necessarily derived from the previously examined population. Without proper baseline data showing that the population did not show this characteristic prior to some presumed introduction of bees, such a discovery only supports the conclusion that the two populations share a characteristic. Further research on other characteristics or other evidence for an introduc-tion is necessary to support statements that an introduction of a particular type of bee took place.

Likewise, demonstrating that a method of analyzing biochemical varia-tion such as that discussed by Sylvester (1976) will separate particular populations does not allow one to infer that it will separate unknown populations for which no baseline data are available. It may very well be capable of doing so, but this must be determined for each population before claims of the value of the method are made for such populations or types of bees.

ACKNOWLEDGMENTS

This chapter was prepared in cooperation with Louisiana Agricultural Experiment Station. R. H. Crozier, R. F. A. Moritz, T. E. Rinderer, and H. Shimanuki made useful suggestions for improvements on earlier manuscript drafts.

REFERENCES

Allalouf, D., Ber, A., and Ishay J. (1975). Properties of testicular hyaluronidase of the honey bee and oriental hornet: comparison with insect venom and mammalian hyaluronidases. *Comp. Biochem. Physiol. 50B*, 331–337.

Alumot, E., Lensky, Y., and Holstein P. (1969). Sugars and trehalase in the reproductive organs and hemolymph of the queen and drone honey bees (*Apis mellifica* L. var. *ligustica* Spin.). *Comp. Biochem. Physiol. 28*, 1419–1425.

Arnold, G., and Delage-Darchen, B. (1978). Nouvelles données sur l'équipment enzymatique des glandes salivaires de l'ouvrière d'*Apis mellifica* (Hyménoptère Apidé). *Ann. Sci. Nat. Zool. Paris 20*, 401–422.

Ayala, F. J. (1983). Enzymes as taxonomic characters. *In* "Protein Polymorphism: Adaptive and Taxonomic Significance" Systematics Assn. Special Vol. 24 (G. S. Oxford and D. Rollinson, eds.), pp. 3–26. Academic Press, London and New York.

Ayala, F. J., and Powell, J. R. (1972). Allozymes as diagnostic characters of sibling species of *Drosophila. Proc. Natl. Acad. Sci. U.S.A. 69*, 1094–1096.

Ayala, F. J., Hedgecock, D., Zumwalt, G. S., and Valentine, J. W. (1973). Genetic variation in *Tridacna maxima;* an ecological analog of some unsuccessful evolutionary lineages. *Evolution 27*, 177–191.

Ayala, F. J., Valentine, J. W., DeLaca, T. E., and Zumwalt, G. S. (1975). Genetic variability of the antarctic brachiopod *Liothyrella notorcadensis* and its bearing on mass extinction hypotheses. *J. Paleontol. 49*, 1–9.

Baars, A. J., and Driessen, O. M. J. (1984). Aryl hydrocarbon hydroxylase and glutathione S-transferase activity in the *Varroa* mite and the honeybee. *J. Apic. Res. 23*, 37–39.

Badino, G., Celebrano, G., and Manino A. (1982). Genetic variability of *Apis mellifera ligustica* Spin. in a marginal area of its geographical distribution. *Experientia 38*, 540–541.

Badino, G., Celebrano, G., and Manino, A. (1983). Population structure and *Mdh-1* locus variation in *Apis mellifera ligustica. J. Hered. 74*, 443–446.

Barker, S. A., Mitchell, A. W., Walton, K. W., and Weston, P. D. (1966). Separation and isolation of the hyaluronidase and phospholipase components of bee-venom and investigation of bee-venom–human serum interactions. *Clin. Chim. Acta 13*, 582–596.

Berkelhamer, R. C. (1983). Intraspecific genetic variation and haplodiploidy, eusociality, and polygyny in the Hymenoptera. *Evolution 37*, 540–545.

Bitondi, M. M. G., and Mestriner, M. A. (1983). Esterase isozymes of *Apis mellifera*: substrate and inhibition characteristics, developmental ontogeny, and electrophoretic variability. *Biochem. Genet.* 21, 985–1002.

Blum, M. S., and Taber, S., III. (1965). Chemistry of the drone honey bee reproductive system —III. Dehydrogenases in washed spermatozoa. *J. Insect Physiol.* 11, 1489–1501.

Brandt, N. R., and Huber, R. E. (1979). The localization of honey bee thorax trehalase. *Can. J. Biochem.* 57, 145–154.

Brandt, N. R., Hurlburt, K. L., and Huber, R. E. (1979). The kinetic parameters of trehalase in whole and disrupted mitochondrial preparations from two insects with asynchronous muscle. *Can. J. Biochem.* 57, 1210–1215.

Brosemer, R. W., and Marquardt, R. R. (1966). Insect extramitochondrial glycerophosphate dehydrogenase. II. Enzymatic properties and amino acid composition of the enzyme from honeybee (*Apis mellifera*) thoraces. *Biochim. Biophys. Acta* 128, 464–473.

Brosemer, R. W., Grosso, D. S., Estes, G., and Carlson, C. W. (1967). Quantitative immunochemical and electrophoretic comparisons of glycerophosphate dehydrogenases in several insects. *J. Insect Physiol.* 13, 1757–1767.

Brückner, D. (1974). Reduction of biochemical polymorphism in honeybees (*Apis mellifica*). *Experientia* 30, 618–619.

Clement, J. L. (1981). Enzymatic polymorphism in the European populations of various *Reticulitermes* species (Isoptera). In "Biosystematics of Social Insects" Systematics Assn. Special Vol. No. 19 (P. E. Howse and J. L. Clement, eds.), pp. 49–62. Academic Press, London and New York.

Contel, E. P. B., and Mestriner, M. A. (1974). Esterase polymorphisms at two loci in the social bee. *J. Hered.* 65, 349–352.

Contel, E. P. B., Mestriner, M. A., and Martins, E. (1977). Genetic control and developmental expression of malate dehydrogenase in *Apis mellifera*. *Biochem. Genet.* 15, 859–876.

Cornuet, J. M. (1979). The MDH system in honeybees of Guadeloupe. *J. Hered.* 70, 223–224.

Cornuet, J. M., and Louveaux, J. (1981). Aspects of genetic variability in *Apis mellifera* L. In "Biosystematics of Social Insects" Systematics Assn. Special Vol. No. 19 (P. E. Howse and J. L. Clement, eds.), pp. 85–94. Academic Press, London and New York.

Crozier, R. H. (1970). On the potential for genetic variability in haplo–diploidy. *Genetica* 41, 551–556.

Crozier, R. H. (1973). Apparent differential selection at an isozyme locus between queens and workers of the ant *Aphaenogaster rudis*. *Genetics* 73, 313–318.

Crozier, R. H. (1977). Genetic differentiation between populations of the ant *Aphaenogaster 'rudis'* in the southeastern United States. *Genetica* 47, 17–36.

Crozier, R. H. (1980). Genetical structure of social insect populations. In "Evolution of Social Behavior: Hypotheses and Empirical Tests" (H. Markl, ed.), pp. 129–146. Dahlem Konferenzem 1980, Weinheim: Verlag Chemie Gmbh.

Crozier, R. H. (1981). Genetic aspects of ant evolution. In "Evolution and Speciation. Essays in Honor of M. J. D. White" (W. R. Atchley and D. Woodruff, eds.), pp. 356–370. Cambridge Univ. Press, New York and Cambridge.

Curtsinger, J. W. (1980). On the opportunity for polymorphism with sex-linkage or haplodiploidy. *Genetics* 96, 995–1006.

Del Lama, M. A., and Mestriner, M. A. (1984). Starch gel electrophoretic patterns of exopeptidase phenotypes in 14 different species of bees. *Rev. Bras. Genet.* 7, 9–20.

Fink, S. C., Carlson, C. W., Gurusiddaiah, S., and Brosemer, R. W. (1970). Glycerol 3-phosphate dehydrogenase in social bees. *J. Biol. Chem.* 245, 6525–6532.

Galuszka, H., and Kubicz, A. (1968). Proteins from spermatheca fluid and seminal plasma of the honey bee (*Apis mellifica* L.). *Zool. Pol.* 18, 239–248.

Gartside, D. F. (1980). Similar allozyme polymorphism in honeybees *(Apis mellifera)* from different continents. *Experientia 36,* 649–650.

Giebel, W., Zwilling, R., and Pfleiderer, G. (1971). The evolution of endopeptidases. XII. The proteolytic enzymes of the honey bee *(Apis mellifica* L.). *Comp. Biochem. Physiol. 38B,* 197–210.

Gilbert, M. D., and Wilkinson, C. F. (1974). Microsomal oxidases in the honey bee, *Apis mellifera* (L.). *Pest. Biochem. Physiol. 4,* 56–66.

Gilliam, M., and Jackson, K. K. (1972a). Enzymes in honey bee *(Apis mellifera* L.) hemolymph. *Comp. Biochem. Physiol. 42B,* 423–427.

Gilliam, M., and Jackson, K. K. (1972b). Proteins of developing worker honey bees, *Apis mellifera. Ann. Entomol. Soc. Am. 65,* 516–517.

Gilliam, M., and Valentine, D. K. (1973). Unusual electrophoretic behavior of proteins from honey bee hemolymph. *Comp. Biochem. Physiol. 45B,* 463–466.

Gorske, S. F., and Sell, D. K. (1976). Genetic differences among purslane sawfly biotypes. *J. Hered. 67,* 271–274.

Graur, D. (1985). Gene diversity in Hymenoptera. *Evolution 39,* 190–199.

Halliday, R. B. (1981). Heterozygosity and genetic distance in sibling species of meat ants *(Iridomyrmex purpureus* group). *Evolution 35,* 234–242.

Hartl, D. L. (1971). Some aspects of natural selection in arrhenotokous populations. *Am. Zool. 11,* 309–325.

Huber, R. E. (1975). The purification and study of a honey bee abdominal sucrase exhibiting unusual solubility and kinetic properties. *Arch. Biochem. Biophys. 168,* 198–209.

Huber, R. E., and Mathison, R. D. (1976). Physical, chemical, and enzymatic studies on the major sucrase of honey bees *(Apis mellifera). Can. J. Biochem. 54,* 153–164.

Huber, R. E., and Thompson, D. J. (1973). Studies on a honey bee sucrase exhibiting unusual kinetics and transglucolytic activity. *Biochemistry 12,* 4011–4020.

Hung, A. C. F., and Vinson, S. B. (1977). Electrophoretic techniques and genetic variability in Hymenoptera. *Heredity 38,* 409–411.

Johnson, F. M., Schaffer, H. E., Gillaspy, J. E., and Rockwood, E. S. (1969). Isozyme genotype–environment relationships in natural populations of the harvester ant, *Pogonomyrmex barbatus,* from Texas. *Biochem. Genet. 3,* 429–450.

Kubicz, A., and Galuszka, H. (1971). Polyacrylamide gel electrophoresis of proteins and the acid phosphatase isoenzymes from hemolymphs of the honey bees, *Apis mellifica* L. *Zool. Pol. 21,* 51–57.

Kunkee, R. E., and Zweig, G. (1963). Substrate specificity studies on bee acetylcholinesterase purified by gradient centrifugation. *J. Insect Physiol. 9,* 495–507.

Lefebvre, Y. A., and Huber, R. E. (1970). Solubilization, purification, and some properties of trehalase from honeybee *(Apis mellifera). Arch. Biochem. Biophys. 140,* 514–518.

Lensky, Y. (1971a). Haemolymph proteins of the honey bee—I. Separation and characterization of haemolymph proteins of worker larvae. *Comp. Biochem. Physiol. 38B,* 129–139.

Lensky, Y. (1971b). Haemolymph proteins of the honey bee—II. Differentiation during the development of bee workers. *Comp. Biochem. Physiol. 39B,* 335–341.

Lensky, Y., and Alumot, E. (1969). Proteins in the spermathecae and hemolymph of the queen bee *(Apis mellifera* L. var *ligustica* Spin.). *Comp. Biochem. Physiol. 30,* 569–575.

Lester, L. J. (1975). Population genetics of the Hymenoptera. Ph.D. dissertation. Univ. Texas, Austin, Tex.

Lester, L. J., and Selander, R. K. (1979). Population genetics of haplodiploid insects. *Genetics 92,* 1329–1345.

Lester, L. J., and Selander, R. K. (1981). Genetic relatedness and the social organization of *Polistes* colonies. *Am. Nat. 117,* 147–166.

Levins, R. (1968). "Evolution in Changing Environments." Princeton Univ. Press, Princeton, N.J.

Lewontin, R. C. (1974). "The Genetic Basis of Evolutionary Change." Columbia Univ. Press, New York.

Liu, T. P., and Dixon, S. E. (1965). Studies in the mode of action of royal jelly in honey bee development. VI. Haemolymph protein changes during caste development. *Can. J. Zool.* 43, 873–879.

Marquardt, R. R., and Brosemer, R. W. (1966). Insect extra-mitochondrial glycerophosphate dehydrogenase. I. Crystallization and physical properties of the enzyme from honeybee (*Apis mellifera*) thoraces. *Biochim. Biophys. Acta* 128, 454–463.

Marquardt, R. R., Carlson, C. W., and Brosemer, R. W. (1968). Glyceraldehyde phosphate dehydrogenase: crystallization from honeybees; quantitative immunochemical and electrophoretic comparisons of the enzyme in other insects. *J. Insect Physiol.* 14, 317–333.

Martin, P. (1965). Elektrophoretische Untersuchungen an der Hämolymphe von. Bienen. *Berl. Münch. Tierärztl. Wochenschr.* 78, 16–17.

Martins, E., Mestriner, M. A., and Contel, E. P. B. (1977). Alcohol dehydrogenase polymorphism in *Apis mellifera*. *Biochem. Genet.* 15, 357–366.

Mestriner, M. A. (1969). Biochemical polymorphisms in bees (*Apis mellifera ligustica*). *Nature (Lond.)* 223, 188–189.

Mestriner, M. A., and Contel, E. P. B. (1972). The *P-3* and *Est* loci in the honeybee *Apis mellifera*. *Genetics* 72, 733–738.

Metcalf, R. L., and March, R. B. (1950). Properties of acetylcholine esterases from the bee, the fly and the mouse and their relation to insecticide action. *J. Econ. Entomol.* 43, 670–677.

Metcalf, R. L., March, R. B., and Maxon, M. (1955). Substrate preferences of insect cholinesterases. *Ann. Entomol. Soc. Am.* 48, 222–228.

Metcalf, R. L., Maxon, M., Fukuto, T. R., and March, R. B. (1956). Aromatic esterase in insects. *Ann. Entomol. Soc. Am.* 49, 274–279.

Metcalf, R. L., Fukuto, T. R., Wilkinson, C., Fahmy, M. H., El-Aziz, S. A., and Metcalf, E. R. (1966). Mode of action of carbamate synergists. *J. Agr. Food. Chem.* 14, 555–562.

Metcalf, R. A., Marlin, J. C., and Whitt, G. S. (1975). Low levels of genetic heterozygosity in Hymenoptera. *Nature (Lond.)* 257, 792–794.

Metcalf, R. A., Marlin, J. C., and Whitt, G. S. (1984). Genetics of speciation within the *Polistes fuscatus* species complex. *J. Hered.* 75, 117–120.

Moritz, R. F. A. (1982). Inzuchteffekte auf den Stoffwechsel von Drohne (*Apis mellifera carnica*). *Z. Tierz. Züchtungsbiol.* 99, 69–80.

Moritz, R. F. A. (1983). Homogeneous mixing of honeybee semen by centrifugation. *J. Apic. Res.* 22, 249–255.

Moritz, R. F. A., Hawkins, C. F., Crozier, R. H., and Mackinlay, A. G. (1986). A mitochondrial DNA polymorphism in honeybees (*Apis mellifera* L.). *Experientia* submitted.

Nunamaker, R. A. (1980). Subspecies determination in the honey bee (*Apis mellifera* L.) based on isoelectric focusing of malate dehydrogenase. Ph.D. dissertation. University of Wyoming, Laramie, Wyo.

Nunamaker, R. A., and Wilson, W. T. (1980). Some isozymes of the honey bee (*Apis mellifera* L.). *Isozyme Bull.* 13, 111–112.

Nunamaker, R. A., and Wilson, W. T. (1981a). Comparison of MDH allozyme patterns in the African honey bee (*Apis mellifera adansonii* L.) and the Africanized populations of Brazil. *J. Kansas Entomol. Soc.* 54, 704–710.

Nunamaker, R. A., and Wilson, W. T. (1981b). Malate dehydrogenase and non-specific esterase isoenzymes of eggs of the honey bee (*Apis mellifera* L.). *Comp. Biochem. Physiol.* 70B, 607–609.

Nunamaker, R. A., and Wilson, W. T. (1982). Isozyme changes in the honeybee, *Apis mellifera* L., during larval morphogenesis. *Insect Biochem. 12,* 99–104.

Nunamaker, R. A., Wilson, W. T., and Ahmad, R. (1984). Malate dehydrogenese and non-specific esterase isoenzymes of *Apis florea, A. dorsata,* and *A. cerana* as detected by isoelectric focusing. *J. Kansas Entomol. Soc. 57,* 591–595.

Obrecht, E., and Scholl, A. (1981). Enzymelektrophoretische Untersuchungen zur Analyse der Verwandtschaftsgrade zwischen Hummel- und Schmarotzerhummelarten *(Apidae, Bombini). Apidologie 12,* 257–268.

Oertel, E. (1980). History of beekeeping in the United States. *In* "Beekeeping in the United States." U.S.D.A. Agriculture Handbook 335, pp. 2–9, U.S. Government Printing Office, Washington, D.C.

Owen, M. D. (1979). Relationship between age and hyaluronidase activity in the venom of queen and worker honey bees *(Apis mellifera* L.) *Toxicon 17,* 94–98.

Owen, R. E. (1985). Difficulties with the interpretation of patterns of genetic variation in the eusocial Hymenoptera. *Evolution 39,* 201–205.

Pamilo, P. (1979). Genic variation at sex-linked loci: quantification of regular selection models. *Hereditas 91,* 129–133.

Pamilo, P. (1982). Genetic population structure in polygynous *Formica* ants. *Heredity 48,* 95–106.

Pamilo, P., and Crozier, R. H. (1981). Genic variation in male haploids under deterministic selection. *Genetics 98,* 199–214.

Pamilo, P., Vepsalainen, K., and Rosengren, R. (1975). Low allozymic variability in *Formica* ants. *Hereditas 80,* 293–296.

Pamilo, P., Varvio-Aho, S.-L., and Pakkarinen, A. (1978). Low enzyme gene variability in Hymenoptera as a consequence of haplodiploidy. *Hereditas 88,* 93–99.

Pamilo, P., Pekkarinen, A., and Varvio-Aho, S. L. (1981). Phylogenetic relationships and the origin of social parasitism in Vespidae and in *Bombus* and *Psithyrus* as revealed by enzyme genes. *In* "Biosystematics of Social Insects" Systematics Assn. Special Vol. No. 19 (P. E. Howse and J. L. Clement, eds.), pp. 37–48. Academic Press, London and New York.

Pekkarinen, A. (1979). Morphometric, colour and enzyme variation in bumblebees (Hymenoptera, Apidae, *Bombus)* in Fennoscandia and Denmark. *Acta Zool. Fenn. 158,* 1–60.

Pekkarinen, A., Varvio-Aho, S.-L., and Pamilo, P. (1979). Evolutionary relationships in northern European *Bombus* and *Psithyrus* species (Hymenoptera, Apidae) studied on the basis of allozymes. *Ann. Entomol. Fenn. 45,* 77–80.

Plantevin, G., and Nardon, P. (1972). Utilisation d'une microméthode de détection (Auxotab) pour la recherche qualitative d'activités enzymatiques dans les tissues d'insectes. *Ann. Zool. Ecol. Anim. 4,* 229–248.

Reeve, H. K., Reeve, J. S., and Pfennig, D. W. (1985). Eusociality and genetic variability: a re-evaluation. *Evolution 39,* 200–201.

Ruttner, F. (1975). Races of bees. *In* "The Hive and the Honey Bee" (Dadant & Sons, eds.), pp. 19–38. Dadant & Sons, Hamilton, Ill.

Schopf, T. J. M., and Murphy, L. S. (1973). Protein polymorphism of the hybridizing seastars *Asterias forbesi* and *Asterias vulgaris* and implications for their evolution. *Biol. Bull. 145,* 589–597.

Selander, R. K., and Kaufman, D. W. (1973). Genic variability and strategies of adaptation in animals. *Proc. Natl. Acad. Sci. U.S.A. 70,* 1875–1877.

Shaumar, N., Rojas-Rousse, D., and Pasteur, N. (1978). Allozyme polymorphism in the parasitic hymenoptera *Diadromus pulchellus* WSM. (Ichneumonidae). *Genet. Res. Camb. 32,* 47–54.

Sheppard, W. S., and Berlocher, S. H. (1984). Enzyme polymorphism in *Apis mellifera* from Norway. *J. Apic. Res. 23,* 64–69.

Sheppard, W. S., and Berlocher, S. H. (1985). New allozyme variability in Italian honey bees. *J. Hered.* 76, 45–48.

Simon, C., and Archie, J. (1985). An empirical demonstration of the lability of heterozygosity estimates. *Evolution* 39, 463–467.

Snyder, T. P. (1974). Lack of allozymic variability in three bee species. *Evolution* 28, 687–689.

Snyder, T. P., Chambers, G. K., and Ayala, F. J. (1979). Isolation of the cytoplasmic form of malate dehydrogenase from honey bee (*Apis* mellifera) larvae. *Biochem. Biophys. Res. Commun.* 88, 668–675.

Storey, K. B., and Hochachka, P. W. (1975). The kinetic requirements of cytoplasmic alpha-glycerophosphate (α-GP) dehydrogenase in muscles with active α-GP cycles. *Comp. Biochem. Physiol.* 52B, 175–178.

Sylvester, H. A. (1976). Allozyme variation in honeybees (*Apis mellifera* L.). Ph.D. dissertation. University of California, Davis, Calif.

Sylvester, H. A. (1982). Electrophoretic identification of Africanized honeybees. *J. Apic. Res.* 21, 93–97.

Talbot, B. G., and Huber, R. E. (1976). An electrophoretic and pH comparison of the soluble trehalases of several insect species. *Comp. Biochem. Physiol.* 53B, 367–369.

Talbot, B. G., Muir, J. G., and Huber, R. E. (1975). Properties of a free and a solubilized form of bound α,α-trehalase purified from honey bee thorax. *Can. J. Biochem.* 53, 1106–1117.

Tanabe, Y., Tamaki, Y., and Nakano, S. (1970). Variations of esterase isozymes in seven species of bees and wasps. *Jpn. J. Genet.* 45, 425–428.

Tomaszewski, E. K., Schaffer, H. E., and Johnson, F. M. (1973). Isozyme genotype-environment associations in natural populations of the harvester ant, *Pogonomyrmex badius*. *Genetics* 75, 405–421.

Tomimatsu, Y., and Brosemer, R. W. (1972). The molecular weights of glycerol-3-phosphate dehydrogenases from rabbit, honeybee and bumblebee muscle. *Comp. Biochem. Physiol.* 43B, 403–407.

Tripathi, R. K., and Dixon, S. E. (1968). Haemolymph esterases in the female larval honeybee, *Apis mellifera* L., during caste development. *Can. J. Zool.* 46, 1013–1017.

Tripathi, R. K., and Dixon, S. E. (1969). Changes in some hemolymph dehydrogenase isozymes of the female honeybee, *Apis mellifera* L., during caste development. *Can. J. Zool.* 47, 763–770.

Turner, R. B., Chavez-Arnold, R. A. M., and Holguin, T. A. (1979). Preliminary report of some hydrolytic-enzyme activities of honey bee semen. *Int. J. Invertebr. Reprod.* 1, 267–270.

Ward, P. S. (1980a). Genetic variation and population differentiation in the *Rhytidoponera impressa* group, a species complex of Ponerine ants (Hymenoptera: Formicidae). *Evolution* 34, 1060–1076.

Ward, P. S. (1980b). A systematic revision of the *Rhytidoponera impressa* group (Hymenoptera: Formicidae) in Australia and New Guinea. *Aust. J. Zool.* 28, 475–498.

Ward, P. S., and Taylor, R. W. (1981). Allozyme variation, colony structure and genetic relatedness in the primitive ant *Nothomyrmecia macrops* Clark (Hymenoptera: Formicidae). *J. Aust. Entomol. Soc.* 20, 177–183.

Wurzinger, K., and Hartenstein, R. (1974). Phylogeny and correlations of aldehyde oxidase, xanthine oxidase, xanthine dehydrogenase and peroxidase in animal tissues. *Comp. Biochem. Physiol.* 49B, 171–185.

CHAPTER 8

Cytology and Cytogenetics

CHARLES P. MILNE, JR.

I. INTRODUCTION

Cytogenetics is a blend of cytology, genetics, and molecular biology. There is, therefore, a large body of information which could be included in a review of honey-bee cytology and cytogenetics. However, the objective of this chapter is restricted to the examination of the centerpiece of cytogenetics, the chromosome, and also molecular cytogenetics, cell division, and developmental cytogenetics.

If the worker bee were kinder to researchers, undoubtedly more would be known than will be reviewed here. Honey bees have several advantages for cytogenetic research, including their large colony populations, their ability to take care of themselves, and their physiological, morphological, and genetic diversity. Honey-bee cytology and cytogenetics, until now, have never been the exclusive subjects of a detailed review, although several have touched on these areas (Ruttner and Mackensen, 1952; Kerr and Laidlaw, 1956; Rothenbuhler, 1958a; Fyg, 1959; DuPraw, 1967; Rothenbuhler et al., 1968; Rothenbuhler, 1975).

II. CHROMOSOMES

A. Number

Dzierzon (1845) advanced the idea that males (drones) develop from unfertilized eggs and females (workers and queens) originate from fertilized ones. Much classical cytological work was done between 1900 and 1920 to

determine the haploid chromosome number from diploid females and hap
loid males. The honey-bee chromosomes proved to be small and difficult t
count. Petrunkewitsch (1901, 1903) first arrived at the correct chromosom
number. He reported 32 chromosomes in diploid oogonia. This was con
firmed when consistent chromosome numbers were found in various stage
of oogenesis, spermatogenesis, and early cleavage mitoses by Meves (190:
1907), Mark and Copeland (1906), Doncaster (1906, 1907), Nachtsheir
(1913), Armbruster (1913), and Jegen (1920). Recent work, employing ad
vanced techniques, verified the earlier reports (Sanderson and Hall, 194£
1951; Hachinohe and Onishi, 1952; Hoshiba and Kusanagi, 1978; Verma є
al., 1982). Fahrenhorst (1977a,b) and Hoshiba et al. (1981) showed that th
haploid chromosome number is 16 for each of the four species of Apis.

Woyke et al. (1966) and Woyke and Skowronek (1974) determined tha
laboratory-reared diploid drones have 32 chromosomes in spermatogenesi
and in the blastoderm stage. Recently, Chaud-Netto (1980a,b) inseminate
normal diploid queens with diploid semen from diploid drones. Worke
ovarian tissue was cytogenetically triploid ($3n = 48$).

B. Structure

The techniques employed before 1948 to observe honey-bee chromo
somes rendered them round and somewhat clumped. Sanderson and Hal
(1948) first reported that honey-bee chromosomes are rod-shaped. The)
mention that one chromosome is larger than the rest and appears hooked o1
bent. This conspicuous chromosome is present once in haploid cells anc
twice in diploid cells (Sanderson and Hall, 1951).

Hoshiba and Kusanagi (1978) published excellent photographs of honey·
bee chromosomes (Fig. 1). Colchicine was used to prevent spindle formation
and arrest cells in meiosis or mitosis. Cells were examined from queen
ovaries, drone testes, drone head ganglia, and blood cells from queens,
workers and drones. Sufficient detail was obtained from 30 drone cells tc
describe a karyogram based on the chromosome length, and arm length
ratio. Eight metacentric and 8 submetacentric chromosomes were identified. The chromosomes varied from 1.3 to 4.3 μm long. One chromosome
with a secondary constriction was considerably longer than the rest. It is
likely that this chromosome is the large hooked one seen by Sanderson and
Hall 30 years earlier. Hoshiba and Kusanagi also looked for any obvious sex
chromosome or sex chromatin region in the mitotic chromosomes. Sex
chromatin usually condenses and stains differently (allocycly) than autoso-
mal chromatin. None was found, although its existence cannot be excluded.

Hoshiba (1979) compared the chromosomes of diploid and haploid
drones. The chromosome pairs from the diploid and the corresponding
chromosome from the haploid appeared identical. It should be mentioned

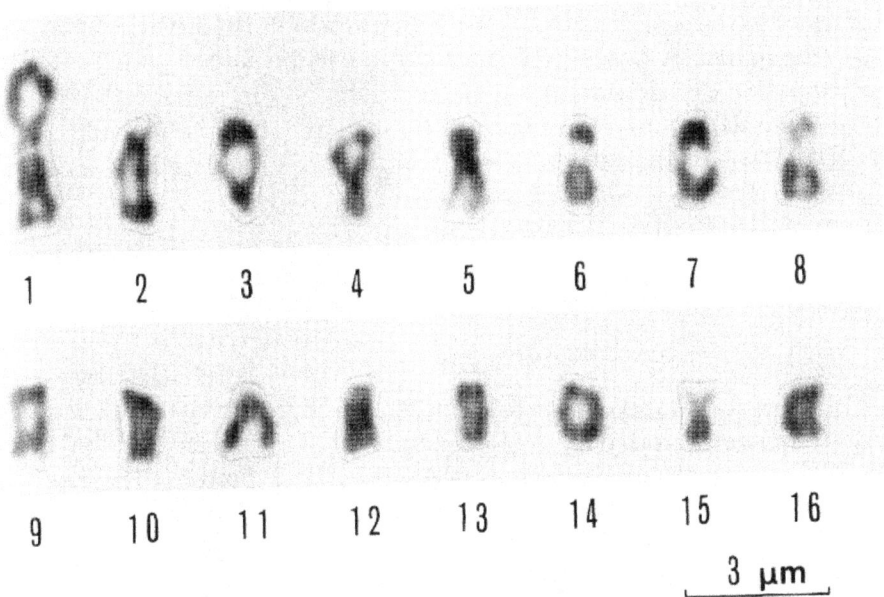

Fig. 1. Karyogram of *Apis mellifera* showing eight metacentric chromosomes (1, 2, 4, 7, 9, 11, 13, 16) and eight submetacentric ones. The serial alignment of these male meiotic chromosomes after colchicine treatment was based on their length. A secondary constriction can be seen in chromosome number 1. [From Hoshiba and Kusanagi (1978), with permission. Copyright 1978 by International Bee Research Association.]

that the chromosome arm length ratios listed in this publication are quite muddled and should be ignored. The correct arm ratios are tabulated in Hoshiba and Kusanagi (1978). Finally, Hoshiba *et al.* (1981) reported that the haploid and diploid drone chromosome complements of *Apis cerana* are identical to those of *Apis mellifera*: 8 metacentric chromosomes and 8 submetacentric chromosomes. A similar karyogram was prepared, but neither chromosome lengths nor arm ratios were given.

Hoshiba (1984) presented more detailed studies of the honey-bee karyotype. Cells were colchicine-treated spermatogonial cells in mitosis from haploid drones. The use of G- and C-banding techniques has enabled identification of each of the 16 chromosomes. These two banding techniques are different staining methods using Giemsa which produce different, reproducible banding patterns on chromosomes. Each chromosome has unique banding patterns by this staining procedure and can be more reliably identified by the banding than by length and arm ratio. Hoshiba reported 4 metacentric and 12 submetacentric or subtelocentric chromosomes. The difference between this and the earlier work by Hoshiba and Kusanagi (1978) was attributed to the source of the material.

Lee (1958) exposed drones to a range of gamma radiation doses. The sperm from those drones was used to instrumentally inseminate queens. Frames of eggs from the queens were checked later for larvae, pupae, and emerging adults. A dosage response curve was produced which reflected the production of dominant lethal mutations by the radiation. Although they were not examined cytologically, these dominant lethals should consist of chromosomal aberrations. Nearly all the dominant lethals produced death before the egg hatched.

III. MOLECULAR CYTOGENETICS

A. DNA

Cytogenetics extends beyond the understanding of the particulate nature of chromosomes and their mode of transmission in cell division to the primary molecules of heredity: DNA, RNA, and protein. The study of honey-bee DNA has revealed an interesting picture. Jordan and Brosemer (1974) examined the DNA from three species of *Apis* (*mellifera*, *cerana*, and *florea*). First, the double-stranded DNA was centrifuged in a cesium chloride (CsCl) density gradient. The buoyant densities of the major DNA peak for the *Apis* species revealed a guanine–cytosine (GC) base composition of 37, 38, and 35%, respectively. They also possessed differently shaped DNA peaks in the density gradients. However, none showed a satellite peak of DNA—a distinct major species of DNA banding at a location different from the major peak because of different base composition. This lack of satellite DNA is uncommon in the animal kingdom. *Apis florea* DNA in the density gradient appeared symmetrical and broad in shape. The DNA peaks of *A. mellifera* and *A. cerana* were broad and asymmetrical. Both were skewed to a lower GC content, with the shoulder most prominent in *A. cerana*. The shoulder fraction of *A. mellifera* and *A. cerana* represents over 30% of the total DNA. This indicates that for the size of DNA fragments examined, ca. 10^6 daltons, the *Apis* species were unusually heterogeneous in base composition. Also, there were more sections of DNA with a significantly lower percentage of GC in *A. cerana* and *A. mellifera* than in *A. florea*.

Jordan and Brosemer also examined the reassociation kinetics of the DNA from *A. mellifera* and *A. cerana*. The double-stranded DNA molecules were first sheared to a certain length and then disassociated into single strands by heating them to about 70°C. Lowering the temperature to about 60°C allowed reassociation of complementary DNA stretches. From the kinetics of reassociation, a genome size of 1.14×10^{11} daltons was calculated for the two species. As expected from the absence of a satellite DNA peak in the density gradient, the reassociation kinetics showed little repetitious DNA, only 3–5% of the genome. A high 89% of the genome of these bees is composed of unique DNA sequences. Although the function of highly repetitious DNA is unknown, its loss has severe consequences in *Drosophila* (Swanson *et al.*, 1981). The honey bee, unusual in its lack of repetitious DNA, may help to elucidate its function.

Most eukaryotic organisms have their DNA sequences arranged in the chromosomes in a regular fashion termed the *Xenopus* pattern, after the animal in which it was first identified (Davidson *et al.*, 1973). The DNA is arranged in alternating unique DNA sequences (about 1000 nucleotides in length) and spacer DNA of about 300 nucleotides. Few organisms have been found to deviate from this pattern. Both *Drosophila melanogaster* (Manning *et al.*, 1975) and the honey bee (Crain *et al.*, 1976) do so. Both genomes contain long stretches of unique DNA. Crain *et al.* (1976) propose that the *Drosophila* pattern occurs in organisms with small genomes and the *Xenopus* pattern in those with large genomes.

B. RNA

Translation of DNA to protein is mediated by messenger RNA and is carried out by ribosomes which are composed of both RNA (rRNA) and proteins. Several eukaryotic organisms have a low-molecular-weight RNA of unknown function hydrogen-bonded to the larger rRNA (Pene *et al.*, 1968). It is believed to arise by posttranscriptional processing of the rRNA precursor molecule.

Honey-bee RNA follows this general pattern. The structure of the honey-bee rRNA (sedimentation coefficient = 26S) was examined by De Lucca *et al.* (1974) and Giorgini (1977). When this rRNA was heated briefly to break hydrogen bonding between complementary RNA strands, it produced an 18S rRNA molecule and a 5.8S RNA molecule. This release can also be effected by agents known to disrupt hydrogen bonding, such as urea and dimethyl sulfoxide. The 5.8S RNA has about 150 nucleotides and a molecular weight of 50,000 daltons. It is a homogeneous RNA hydrogen bonded to a specific region of the 18S RNA in a 1:1 ratio. This 5.8S RNA closely resembles that from other eukaryotic cells and can be isolated from larvae and pupae.

C. Protein

Studies of two honey-bee protein chemistries have special cytogenetic interest. The first involves the nuclear histones of honey-bee spermatozoa. During spermiogenesis, the histones of many organisms are modified from the lysine-rich histone found in somatic cells, to a highly arginine-rich one, and finally to a protamine form. However, honey-bee histones are not changed during spermiogenesis (Verma, 1972). Both mature sperm and somatic cells contain similar lysine-rich histones. A change in histones is not responsible for the long viability of sperm in the queen's spermatheca.

The second study involves melittin, a small protein of 26 amino acids, which is the major component of queen or worker venom (Habermann and Jentsch, 1967). Upon injection, this protein lyses phospholipid membranes and disrupts protein synthesis. Partial Edman degradation showed a primary structure having an unequal distribution of hydrophobic neutral amino acids and hydrophilic basic residues. Kreil (1973a, 1975) examined the sequence of melittin from all 4 species of *Apis*. Only a few differences were noted in the primary structure. Each of these differences appears to be conservative: the unequal distribution of hydrophobic and hydrophilic amino acid residues is maintained, as is, presumably, the biological activity. It is reasonable to expect that this pernicious protein, whose activity may be explained by its amino acid sequence, would be synthesized in an inactive form by the ribosomes of the cells in the venom gland, and would appear in active form in the aqueous solution of the poison sac. Posttranslational modification is characteristic of many secreted proteins (Steiner *et al.*, 1980).

Kreil and Bachmayer (1971) examined extracts of venom glands after feeding newly emerged workers radioactively labeled amino acids. The label first appears in a protein other than melittin. This other peptide, termed promelittin, is a precursor of melittin. Its loss of label with time was accompanied by the appearance of label in melittin. Promelittin contains the entire 26 amino acid sequence of melittin with additions to the amino end (Kreil, 1973b). These additional amino acids are rich in acidic residues and presumably render the peptide inactive by affecting either its solubility properties or secondary structure. A further proof of the precursor nature of promelittin is indicated by studies of promelittin synthesis and its conversion into melittin in queens and workers (Bachmayer *et al.*, 1972). Promelittin synthesis begins during worker development, days before any conversion to melittin can be detected. In virgin queens, which must often fight rivals, both the synthesis and conversion processes are functioning at high levels upon emergence.

Kindås-Mügge *et al.* (1974, 1976) focused their attention on the messenger RNA (mRNA) directing the synthesis of promelittin. Unfractionated RNA from the venom glands of queens injected into *Xenopus laevis* oocytes directed the synthesis of promelittin. It is remarkable that an insect mRNA was translated well in a frog cell. Since the conversion of promelittin was not detected, the posttranslational modification of promelittin is specific to the honey-bee venom gland. Characterization of the mRNA revealed several interesting features. The most surprising is the large size of the mRNA. A sedimentation coefficient of 8–9S corresponds to a molecular weight of 150,000 daltons. This mRNA of about 450 nucleotides is over four times as long as necessary to code for promelittin. It is, nonetheless, one of the smallest mRNAs known.

Suchanek *et al.* (1975) translated the promelittin mRNA in a mammalian cell-free system. The polypeptide produced had a higher molecular weight than promelittin. Analysis of fragments from pepsin digests and Edman degradation revealed sequences characteristic of promelittin. This larger precursor is called prepromelittin. Most of the additional amino acids are on the amino end of promelittin. It also has an extra glycine residue at the carboxyl end (Suchanek and Kreil, 1977). They examined prepromelittin produced by a wheat-germ cell-free system and suggested that in the venom gland a transamidaselike enzyme exchanges the glycine for an ammonia. This posttranslational modification has been reported for several other proteins. The prepromelittin is 70 amino acids long, 26 of which are melittin (Suchanek *et al.* 1978). Eighteen of the 21 amino acids of the pre part of prepromelittin are hydrophobic residues. The pro part of the polypeptide, 23 residues, contains all the acidic and most of the proline residues.

Kreil *et al.* (1980a,b) presented a theory relating the primary structure of prepromelittin to its conversion. The hydrophobic end of prepromelittin functions to transport the protein during translation through the endoplasmic reticulum for secretion. Located in the membrane or lumen of the endoplasmic reticulum is a protease termed the signal peptidase. The growing polypeptide chain is transported during synthesis through the membrane and cleaved. This would account for the lack of prepromelittin in intact translation systems. Cell-free systems have no endoplasmic reticulum associated with the ribosomes. Correct processing of prepromelittin takes place in the heterologous *Xenopus* system, since none was found. This type of processing is a feature of the biosynthesis of other proteins (Steiner *et al.*, 1980).

The promelittin is processed after the signal peptidase cleaves off the pre portion. A dipeptidylaminopeptidase was proposed and identified in the queen-bee venom gland. This enzyme cleaves dipeptides from the amino end of promelittin until melittin is formed. Inspection of the amino acid sequence reveals every second residue to be a proline or alanine. This is the first precursor molecule believed to be activated by the sequential removal of dipeptides. Considering the activity of melittin, this slower method is reasonable. A search for a specific endopeptidase to convert promelittin directly to melittin was unsuccessful. Further analysis of the processing of honey-bee melittin will add greatly to the detailed understanding of secretory protein processing *in vivo*.

Recombinant DNA technology has been used successfully with this melittin system (Vlasak *et al.*, 1983). The total mRNA from the venom gland of young queens was transcribed *in vitro* into cDNA using reverse transcriptase and DNA polymerase. Then this double-stranded cDNA was inserted into a circular plasmid after scission at a specific site by a restriction enzyme. This DNA was used to transform *Escherichia coli* X1776, and bacteria containing recombinant plasmids were isolated and characterized. One of the recombinants had an insert of 374 nucleotide base pairs in the plasmid DNA. This was sequenced by standard DNA sequencing techniques. The production of prepromelittin by the bacteria was not reported. Subsequently, the same researchers (Vlasak and Kreil, 1984) cloned and sequenced DNA coding for another major product of the queen honey-bee venom gland, preprosecapin. This protein, of 77 amino acids, is activated by a mechanism which is unknown, but different than that for melittin to produce the protein, secapin, of 25 amino acids.

IV. MEIOSIS

A. Spermatogenesis

Spermatogenesis occurs near the eighth day in the pupal stage (Hachinohe and Onishi, 1952) and spermiogenesis before emergence (Bishop, 1920). Spermatogonia can be located in the approximately 200 long, cylindrical tubules of larval drone testes. The upper ends of the tubules contain numerous round spermatocysts, masses of developing spermatogonia. Spermatocysts near the terminal end of the tubules are the most advanced in development. The spermatocysts pass down the sperm tubules, divide, and form a mass of primary spermatocytes, ready for meiosis. Cross sections of spermatocysts show that the meiotic divisions are synchronous. Hoage and Kessel (1968) believe that the synchrony is due to a continuous cytoplasm between the spermatocytes in each spermatocyst. They published light and electron micrographs showing cytoplasmic connections between the primary spermatocytes. MacKinnon and Basrur (1970) demonstrated that these intercellular bridges arise from incomplete cytokinesis in the mitoses preceding the formation of the primary spermatocytes. Intercellular bridges between germ cells have been noted previously in both invertebrates and vertebrates.

The drone honey-bee's germinal tissue is haploid. Thus, meiosis must be altered in drones for haploid gametes to be produced. Sharma *et al.* (1961) have aptly described honey-bee spermatogenesis as "abortive meiosis I and anomalous meiosis II": reductional division does not occur. Meves (1903, 1907), Mark and Copeland (1906), and Nachtsheim (1913) published descriptions of spermatogenesis in haploid drones based on light microscope observations. An irregular, extranuclear spindle forms in meiosis I, and it elongates and stretches the cell into an ellipsoid before degenerating. The nuclear membrane remains intact until meiosis II. A small, anucleate, cytoplasmic bud is formed rather than a secondary spermatocyte. A small "interzonal body" appears to pass into the cytoplasmic bud and remains there. Meiosis II is normal and forms two spermatids. However, the division of the cytoplasm is unequal; only the larger spermatid forms a functional sperm. Because of the special features of drone spermatogenesis, one gamete from the queen, the unfertilized egg, is faithfully multiplied into millions of gametes, or spermatozoa.

Several contemporary studies have revealed additional facts about spermatogenesis. Sharma *et al.* (1961) examined the process in *A. cerana*, which is identical to the process in *A. mellifera*. The nonnucleated cytoplasmic bud formed in meiosis I contains some spindle fibers from the extranuclear spindle and the "cytoplasmic body." This body is composed of protein and its origin and function are unknown. No interphase is observed between the clumped metaphase chromosomes of meiosis I and the metaphase of meiosis II. Also, the chromosomes form a metaphase plate in the abortive meiosis I. The spindle is extranuclear and irregular, and the chromosomes move haphazardly before returning to form an irregular mass of chromatin.

Hoage and Kessel (1968) studied spermatogenesis with the electron microscope. In meiosis I, they found that the extranuclear spindle is irregular and that the centrioles undergo additional replications before meiosis II. Up to 16 supernumerary centrioles were found in one dividing spermatocyte. These are eliminated from the cells in small cytoplasmic blebs about 8 μm in diameter before the onset of meiosis II. The nuclear envelope never breaks. During meiosis II a differentiated, or specialized, endoplasmic reticulum is present around the nucleus. The innermost layer of this nuclear membrane appears later to give rise to the spermatid nuclear membrane.

Spermatogenesis in the *A. mellifera* diploid drone is similar in all aspects to that of the haploid drone, including the abortive nature of the first meiotic division (Woyke and Skowronek, 1974). Chromosome reduction does not occur, and diploid spermatozoa are formed. Woyke and Skowronek hypothesize that homozygosity at the sex locus causes an abortive meiosis I. Measurements of DNA content, by Feulgen staining, in diploid drone testes show twice as much DNA in diploid spermatids and spermatozoa as in haploids (Woyke, 1975). Chaud-Netto (1980b,c) observed no allelic segregation in the spermatogenesis of heterozygous diploid drones. No chiasmata were found in the metaphase primary spermatocytes of *A. cerana* diploid drones (Hoshiba *et al.*, 1981). Therefore, not only is the reductional division missing even in the presence of homologous chromosomes, but also crossing over does not occur to any appreciable extent.

Other features of honey-bee spermiogenesis are similar to those described for other insects (Snodgrass, 1925, 1956; Hoage and Kessel, 1968; Sharma *et al.*, 1961).

B. Oogenesis

In the queen, the eggs have their origin in the upper reaches of the ovarioles (Snodgrass, 1925). The walls of the ovarioles are composed of small, flat epithelial cells. Inside is a multinucleate protoplasmic mass that is devoid of cell membranes. As these pass down the ovariole, they undergo differentiation and form primary oocytes. Farther down the ovarioles, distinct cells can be identified which ultimately give rise to the eggs after nourishment by nurse cells. At the time of oviposition the egg is usually in anaphase I (Nachtsheim, 1913). Meiosis forms four nuclei within the egg. These nuclei usually form a line perpendicular to the outside of the egg, and the innermost one becomes the female pronucleus (Snodgrass, 1925; DuPraw, 1967). Meiosis occurs in a cone-like thickening near the anterior pole of the egg, called the "Richtungsplasma" or maturation plasm (DuPraw, 1967). The three remaining polar nuclei fuse or divide several times before degenerating in the maturation plasm. Two of the polar nuclei may fuse and form the "Richtungskopulationskern," or polar copulation nucleus, which may divide several times before degenerating.

V. MITOSIS

A. Egg Cleavage

Honey-bee sperm are stored in the queen's spermatheca until use. Eggs pass down the lateral oviducts into the vagina, where they are fertilized before oviposition. Nelson (1915) could not find any pores in the obvious anteriorly located micropylar area of the egg in any transverse section. It is usually visible in other insects both externally and in sections. Even under the scanning electron microscope no micropylar pores are apparent (Bronskill and Salkeld, 1978). However, eggs taken from oviducts average in their micropylar area about 90 canals which at least penetrate deeply into the chorion (J. L. Williams, personal communication). These canals are not apparent in freshly laid eggs.

In most insects, including the honey bee, more than one sperm enter the egg. Nachtsheim (1913) reported from 3 to 10 spermatozoa in fertilized eggs. DuPraw (1967) concluded that all sperm entering an egg form male pronuclei. After meiosis II the female pronucleus migrates from the "Richtungsplasma" toward the dorsal side of the egg. Usually, it quickly encounters a male pronucleus and fuses with it; cleavage begins immediately thereafter. The accessory sperm degenerate (Nachtsheim, 1913). In a fertilized egg, meiosis and the first egg cleavage are completed in about the first 6 hr (Nelson, 1915). DuPraw (1967), with more advanced techniques, reports that after about 4.5 hr there are 8 cleavage nuclei. Petrunkewitsch (1903) noted the first mitosis 3–4 hr after an unfertilized egg is laid, although DuPraw believes mitosis is delayed in these eggs.

After fusion of the pronuclei, it it not technically correct to term the next few divisions egg cleavage (DuPraw, 1967). The egg is not itself cleaved; only nuclei are involved. No cell membranes are formed and the nuclei remain close together. The nuclei in the egg, before cell-membrane formation, undergo synchronous mitoses. This is undoubtedly due to their spatial relationship in the cytoplasmic syncytium. The zygotic nucleus migrates to or is near the so-called "cleavage center" for the early mitoses. This place is caudal from the cephalic pole of the egg about one-tenth of the egg's total length. During the last half of stage 1 of development, three mitoses produce 8 nuclei. These 8 nuclei migrate toward the caudal pole in stage 2 (about the next 3.6 hr) and undergo 4 more divisions. When the nuclei number 128, they spread out in the interior of the egg and migrate to the egg surface as three more mitoses occur. Blastoderm formation is complete about 2.5 hr later, when cell membranes are formed in the egg periplasm. Harbo and Bolten (1981) reported a time of 71.4 ± 1.2 hr for worker eggs to hatch. They also found that male eggs (haploid or diploid) required about 3 hr longer to develop.

B. Parthenogenesis

Parthenogenesis, an unusual occurrence in animals, is almost universal among Hymenoptera. The honey bee is no exception. The process was first suggested by Dzierzon (1845), although Nachtsheim (1912) points out that Aristotle wrote "bees produce drones without mating." Several statements of a descriptive nature can be made about honey-bee parthenogenesis (Suomalainen, 1950). It is a regular occurrence in the life cycle of the honey bee. Since the parthenogenetically produced individuals are haploid, at least in their germinal tissue, it is termed haploid or generative parthenogenesis. Unfertilized eggs develop into males, which is called arrhenotoky. Finally, parthenogenesis is facultative since eggs develop normally either with or without fertilization.

Cytological proof of Dzierzon's theory of parthenogenesis was first presented by Paulcke (1899), who found sperm only in female eggs. No sperm or traces of sperm were seen in numerous drone eggs. This was confirmed by Petrunkewitsch (1901) and Nachtsheim (1913).

DuPraw (1967) and others have noted that the female pronucleus migrates away from the maturation plasm after meiosis. If no sperm pronucleus encounters the egg pronucleus, it continues its migration until it reaches the dorsal surface of the egg. There it begins its first division. Consequently, male and female eggs initiate cleavage at different locations within the egg. DuPraw concluded not only that the first mitosis is delayed in drone eggs but also that there are fundamental differences between the development of male and female eggs.

In the vast majority of other animals, eggs are arrested in meiosis until fertilization. No development can or does occur until the egg is stimulated in some way by the sperm. Whatever this mechanism is, the honey bee seems to be unrestrained and can develop without extrinsic stimulation. A second mechanism which prevents egg development within the oviducts of queens apparently is operative. The nature of the stimulus to proceed with development after oviposition is unknown.

C. Endopolyploidy

Somatic polyploidy is virtually a universal phenomenon in the differentiated tissues of insects. Geitler (1938, 1939) first described high ploidy levels in the water strider *Gerris lateralis*. He suggested it occurred by chromosome duplication without karyokinesis (endomitosis) or cytokinesis. It appears that chromosome, and presumably DNA, replication occurs and the chromosomes begin to contract in prophase. Then mitosis is halted in late prophase without spindle formation, nuclear membrane breakage, karyokinesis, or cytokinesis. Cells with high ploidy levels probably have large numbers of separate chromosomes in multiples of the whole genome.

Painter (1945) first suggested that honey-bee somatic cells can attain high levels of polyploidy. By counting nucleoli, he estimated that some lateral pharyngeal gland cells were at least $16n$. Sanderson and Hall (1951) also noted high ploidy levels in follicle epithelial, hypodermal, and tracheal cells by direct chromosome counts. Risler (1954) estimated ploidy by a variety of techniques in tissue from epidermis, trachea, gut, brain, muscle, etc. He found high levels of ploidy throughout the course of development in both male and female individuals. Each tissue has its characteristic ploidy level and time of acquiring this level during development. Stekol'shchikov (1971) pioneered the use of radioactive label incorporated into DNA and photographic emulsion to detect polyploidy in honey-bee somatic tissues.

Merriam and Ris (1954) examined ploidy levels in tissues from drone, queen, and worker bees. They discovered that a single tissue may have more than one level of polyploidy. As an explanation of high ploidy, they suggest that there is a positive correlation between the amount of secretory activity of a cell and its ploidy level. Drone, worker, and queen tissues have roughly equivalent levels of ploidy. They reported no strict relationship between nuclear volume and DNA content. Mittwoch *et al.* (1966), examining differentiating cells in male and female larvae, concurred with this conclusion. Successive duplications of DNA during development were accompanied by increases in nuclear volume, but the relationship was not a simple one. Most dividing cells appeared to be either haploid in the male or diploid in the female. Nondividing cells were high in chromosome numbers. Stekol'shchikov (1971) has confirmed this in the spinning glands of bee larvae. Mitoses were never observed in gland cells which had increased their DNA content.

The finding of equivalent ploidy levels by Merriam and Ris (1954) in the drone, queen, and worker was questioned by Mello and Takahashi (1971). They examined larval silk glands and malpighian tubules and found a consistent relationship of worker < drone < queen in DNA content per cell. This relationship was suggested to reflect the difference between the biomass of the bees. They propose that an increased amount of DNA is required for the organ to function adequately in a larger insect.

Lerer and Dixon (1975) examined polyploidy levels in fat body nuclei to determine if higher ploidy levels are involved in the control of caste determination. No differences in ploidy levels were detected up to 84 hr after hatching in queen and worker larvae. They concluded that although differentiating tissues show higher ploidy levels, it is not involved in the early stages of honey-bee cast dimorphism.

Cells have a variety of ways to regulate the genome output. Endomitosis is just one method to vary the output of certain genes in a differentiated cell. Throughout the eukaryotic world, cells regulate themselves by polyteny, gene amplification, gene-dosage compensation, chromosome elimination, or chromosome inactivation.

VI. CYTOGENETIC ANOMALIES

A. Gynandromorphism

The first honey-bee gynandromorphs, or sex mosaics, were reported by Schirach (1766) and Laubender (1801). The gynandromorphs were called "Stacheldrohnen," or stinging drones, undoubtedly reflecting the mode of discovery. They were probably composed of a male head and a female abdomen with a sting. Later Menzel (1862) and von Siebold (1864) described more gynandromorphs. Von Siebold first discussed the intriguing question of their origin.

Gynandromorphs, ordinarily rare, seem to be present in larger numbers in certain colonies (Rothenbuhler *et al.*, 1949). Rothenbuhler *et al.* selected a gynandromorph-producing line of bees which reliably produces from 5 to over 40% of the worker brood as gynandromorphs. Rösch (1927) generated honey-bee gynandromorphs by cooling eggs shortly after oviposition. Stort and Soares (1978) found a gynandromorph after 600 rads of γ-rays from cobalt-60 on a colony.

Many descriptions of gynandromorphs exist in the literature in addition to those already mentioned. Eckert (1934, 1937) noted that gynandromorphs were worker-sized. Many have attempted to place gynandromorphs into morphological groups (e.g., Fyg, 1959) but they defy classification. This is because there are an infinite number of combinations of male and female tissue. However, it should be stressed that they are not a haphazard array of tissue (Milne, 1977; Fyg, 1959). Large areas of the adult are usually of the same sex, although even a single structure, such as the compound eye, can be mosaic. Stort and Soares (1978) reported that the antenna of a gynandromorph had a mixture of male and female olfactory plates. Gynandromorphs can range from a small proportion male to a small proportion female (Drescher, 1975). Internal tissue does not always correspond to the external cuticle. Buitkamp-Möbius (1975) identified female brain characteristics in gynandromorphs with drone heads. Gynandromorphs with gross body distortions were not found to live as long as other mosaics. Witherell (1971) believes that gynandromorphs do not live as long as normal workers and that many die within the hive. Milne (1977) observed that sex mosaics with one male and one female mandible often do not emerge. Drescher (1965) found that gynandromorphs require between 21 and 24 days for development from egg to adult. This period ranges from that required for a worker to that required for a drone. Fyg (1959) reported that a gynandromorph could be raised as a queen and cited four references to such an occurrence. No mention was made of their behavior. Rothenbuhler (1958b) found that large amounts of male tissue can survive the queen rearing process.

The mechanism by which a honey-bee gynandromorph is formed puzzled geneticists and cytologists for almost a century. Two major hypotheses were presented to account for honey-bee gynandromorphs, but neither had a significant amount of scientific support until the 1950s. Morgan (1905) proposed that the male haploid parts arose from the development of accessory sperm and that the diploid female tissue was derived from the zygotic nucleus. Boveri (1915) thought that the female pronucleus cleaved before fertilization and that only one of the two cleavage nuclei was fertilized. The two hypotheses can be distinguished by the origin of the male tissue. Rothenbuhler et al. (1949) suggested that the male parts are paternal, based upon the lineage of the male eye tissue. Further work by Rothenbuhler et al. (1952) on the gynandromorph-producing line demonstrated that the male tissue of gynandromorphs descends from accessory sperm. After examination of hundreds of genetically marked gynandromorphs, Rothenbuhler et al. concluded that all but a handful arise by androgenesis with zygogenesis. Drescher and Rothenbuhler (1963) used marked crosses of non-gynandromorph-producing lines to examine the origin further. Cooling eggs shortly after oviposition resulted in 82 gynandromorphs. Eighty had paternal male tissue. Only rarely were the male parts maternal and female parts biparental. Mackensen (1951) first described one such gynandromorph from a genetically marked queen.

Rothenbuhler (1955) and Rothenbuhler and Gowen (1955) examined the genetic basis of gynandromorph production in Rothenbuhler's selected line and concluded it is influenced by genes which primarily resided on the chromosomes. However, in the production of a gynandromorph the influence is found in the egg and not the sperm; sperm from gynandromorph-producing lines do not necessarily produce mosaics in non-gynandromorph-producing queens. Rothenbuhler et al. (1968) pointed out that there has been a selection in this gynandromorph-producing line to permit or facilitate the development of accessory sperm which normally degenerate. It is not known whether the mechanism normally leading to the degeneration is not able to handle the total number of sperm or is completely defective. Drescher (1965) believes the control of gynandromorph production is polygenic in nature.

Four investigations have addressed the interesting question of gynandromorph behavior: how do they behave and why? Sakagami and Takahashi (1956) examined the structure and behavior of 40 gynandromorphs and found worker-specific behavior, such as pollen collection and following dances, in gynandromorphs with an externally male head. They could not determine a correlation between behavior and the external sex characteristics of the head. Witherell (1971) could not confirm the collection of pollen by gynandromorphs. Jay and Jay (1966) observed a gynandromorph, which appeared to have a drone head, in an observation hive over a period of almost 4 weeks. Although not completely workerlike in its behavior, it inspected empty cells, visited brood cells, guarded the hive entrance, and foraged for nectar and pollen. Buitkamp-Möbius (1975) observed the behavior of 10 gynandromorphs in an observation hive before examining their brain structure. Worker behavior such as dancing and comb building was seen. Again, some mosaics with drone-looking heads possessed female brains and performed worker behavior. More complex worker behavior, such as foraging, was found only in those with a worker-like corpora pedunculata. A more complete understanding of the relationship between structure and behavior from gynandromorphs is possible by using the blastoderm fate mapping technique discussed later in this chapter.

There are two readily apparent applications of gynandromorphism. One is the use of the gynandromorph's patchwork-like construction to study behavior and identify structures responsible for behavior. Tucker and Laidlaw (1966) proposed another application to multiply sperm. Single-drone inseminations of queens are possible, but are less reliable and the queens rarely lay very long. They proposed inseminating a gynandromorph-producing queen only with semen from the drone of interest and then locating in her initial brood a number of gynandromorphs which might have drone reproductive organs and semen. Since the male parts are produced by androgenesis, the sperm from these gynandromorphs should be genetically identical to the original drone. A dozen or so gynandromorphs might yield 8 μl of semen for an adequate insemination. Kubasek *et al.* (1980) tested this hypothesis and found that genetically marked sperm can be recovered intact from gynandromorphs after two generations of androgenesis. No genetic influence was noted from the genetically marked gynandromorph-producing line.

B. Impaternate Females

Early in the twentieth century there were reports of exceptions to the rule of workers developing from fertilized eggs (Hewitt, 1892; Onions, 1912, 1914; Jack, 1917). These exceptions seemed to be found in the subspecies *A. m. capensis* of South Africa and the subspecies *A. m. intermissa* of North Africa. In each case, the exception was the finding of abundant female progeny developing from unfertilized eggs. Laying workers, believed to be unable to mate, laid eggs from which either males or females develop. A colony of *A. m. capensis* can remain queenless for months with no males seen in the brood before eventually requeening itself. Anderson (1963) revealed that the development of laying workers in *A. m. capensis* is different from that in *A. m. mellifera*. Anatomical laying workers were present in *A. m. capensis* colonies before dequeening. Physiological, or ovipositing, laying workers were found a few days after dequeening. *Apis m. mellifera* laying workers require much longer to develop. Brief internal fighting was observed in colonies of *A. m. capensis* after the queen was removed. Examination of *A. m. capensis* laying workers showed a large number of ovarioles per ovary, and these were typically asymmetrically distributed between the ovaries. These laying workers produce females parthenogenetically, which are, like haploid drones, without fathers (impaternate). Verma *et al.* (1982) confirmed that the laying workers were diploid, not tetraploid, in their germinal tissue.

Prior to Mackensen (1943), the production of females by parthenogenesis was unknown except in the Cape bee. Mackensen placed many clipped virgin queens of three stocks of chiefly European origin into colonies with queen excluders on the entrances. Twenty-one of the 50 queens which laid eggs produced some females parthenogenetically. He performed controls to insure that the eggs were from the virgin queens and not laying workers. The exact percentage of parthenogenetically produced females from unfertilized eggs could not be determined, since many of the drone eggs were not reared, but it was estimated at less than 1%. From 9 to 57% of the queens of each stock produced impaternate females. Mackensen believed that *A. m. capensis* queens were exported to the Western world during the nineteenth century and are the origin of this phenomenon in the stocks he studied. One colony was forced to rear a queen from an unfertilized egg. Then this virgin queen was forced to lay eggs, and again impaternate females were produced. This strongly indicates that the impaternate females are diploid.

In contrast to the parthenogenesis which produces haploid males, this one producing females is diploid and thelyotokous (Suomalainen, 1950). It is also tychoparthenogenetic, since it seems accidental and is probably not a part of the normal reproductive process even in *A. m. capensis*.

Tucker (1958) studied the cytological origin of these exceptional females. He did not look at the prolific production of the Cape bee, but at genetically marked lines of primarily European origin. Following the techniques of Mackensen (1943), Tucker forced heterozygous virgin queens to lay. Rare impaternate females were found to occur in every line Tucker investigated. He sought to determine whether the parthenogenetic females were the result of tetraploid eggs, automictic parthenogenesis, or apomictic parthenogenesis. Automictic parthenogenesis involves a meiotic division and chromosome reduction, but would require some mechanism such as polar body or cleavage nuclei fusion to restore diploidy. Apomictic parthenogeneis entails an omission of meiosis or at least the reductional division. Tucker inferred cytological events by the segregation of mutant genes in the impaternate females. His results were most consistent with the automixis with polar body fusion explanation, and he offered a hypothesis to account for the fusion. This states that the first meiotic division spindle is misoriented 90°. Instead of pointing into the interior of the egg, it is parallel to the egg surface. The second division is normal producing two nuclei toward the interior of the egg which are heterozygous for the sex locus. A diploid female is produced by the fusion of these two nuclei.

Verma *et al.* (1982) examined this microscopically in *A. m. capensis*. Both meiotic spindles are misoriented. This produces four haploid nuclei on a sagittal plane within the egg. The two central nuclei fuse to form the diploid female zygotic nucleus.

C. Others

A number of single-gene mutants have been discovered and are available for cytogenetic research. Also, instrumental insemination has been refined to a reliable procedure. Since it is easy to get hundreds of drones or tens of thousands of workers from a single queen, the isolation and study of rare exceptions (even 1 in 100,000) is not difficult. These three factors have enabled a number of other cytogenetic anomalies to be studied.

Mackensen (1951) isolated a gynandromorph which had male tissue of maternal origin and female tissue of biparental heritage. This could have arisen by the partial fertilization hypothesis of Boveri (1915). Alternately, it could have developed from a zygotic nucleus and a developing polar body. Genetic tests should be able to distinguish between these alternatives successfully.

Taber (1955) isolated 14 mosaic females from 470,000 workers. The queens producing these were homozygous for the recessive body color, cordovan (c), and were inseminated with semen from normal (+) and c drones. These mosaic females had patches of c and + tissue. The simplest explanation was that binucleate eggs were fertilized by two sperm, one + and one c. He estimated the frequency of binucleate eggs in the 35 queens producing the 14 mosaic females was 0.006%. The origin of these binucleate eggs is similar to that for the impaternate females.

Tucker (1958) examined thousands of bees from unmated queens. Ninety-seven percent of the exceptional progeny were workers, 2% were gynandromorphs, and 1% were mosaic drones. He assumed the same underlying mechanism that produced the impaternate females also produced the other exceptions. The gynandromorphs might have arisen by the mitotic cleavage of two haploid pronuclei followed by a fusion of two products, resulting in the formation of diploid female tissue. Male tissue would descend from the other haploid nuclei. Or, three pronuclei could be involved, two fusing to produce diploid female tissue ("Richtungskopulationskern"?) and the remainder producing male tissue. Mosaic males could be explained by the failure of two haploid, and genetically different, pronuclei to fuse.

Inbred matings designed to produce half diploid drones and half diploid workers in the fertilized progeny produce only workers (Mackensen, 1955). Half of the brood is eaten shortly after the egg hatches (Woyke, 1962), and no diploid drones have been observed in colonies of genetically marked lines designed to reveal them. Rothenbuhler (1957) isolated two gynandromorphs from the gynandromorph-producing line containing male tissue that could only be diploid. Thirty-eight unambiguous mosaic drones contained eye tissue that descended from a diploid zygotic nucleus homozygous at the sex locus. The origin of the female tissue was less clear. Two suggestions were offered: (1) binucleate eggs producing diploid female and diploid male tissue after fertilization, or (2) the union of 2 sperm resulting in diploid female tissue.

Drescher and Rothenbuhler (1964) examined similar crosses for exceptional mosaic males. Sixty-six mosaic males were found in 2477 progeny of one queen from the gynandromorph-producing line; 761 gynandromorphs were found, and only 2 contained male tissue of maternal origin. The 66 mosaic drones contained tissue of maternal and biparental origin. These appeared to have arisen from androgenesis producing haploid male tissue and zygogenesis producing diploid male tissue homozygous at the sex locus.

Laidlaw and Tucker (1964) looked for the fusion of accessory sperm to form diploid tissue. The gynandromorph-producing queens, homozygous for normal eye color $(+ +)$, were inseminated with sperm from two different allelic eye color mutants, snow (s) and tan (s^t). A fusion between a sperm containing s and a sperm containing s^t will produce a unique red eye color. Two of 2401 bees possessed these red female eyes and thus arose from the fusion of two sperm.

Finally, Harbo (1980) instrumentally inseminated genetically marked queens with semen from genetically marked drones. The semen had been stored in liquid nitrogen $(-196°C)$ before insemination. Mosaic males were isolated with tissue of paternal and maternal origin. These probably arose from haploid male and female pronuclei which failed to fuse but initiated cleavage. It was also disclosed that mosaic drones had mosaic testes and produced two types of genetically marked sperm.

VII. DEVELOPMENTAL CYTOGENETICS

A. Blastoderm Fate Mapping

The development of the insect blastoderm from the centrolecithal egg is unique and enables the perfection of a powerful analytical tool: blastoderm fate mapping. Early cleavage divisions are not accompanied by cell-membrane formation, and synchronous nuclear divisions take place in the interior of the egg. The nuclei migrate to the egg surface, where membranes are formed and make the cellular blastoderm. In insects there is a direct relationship between the location of a cell on the blastoderm surface and its fate in subsequent development. For example, a certain blastoderm cell will ultimately form part of the eye solely because of its location on the blastoderm. A gynandromorph develops from a mosaic blastoderm which has large patches of male and female cells. Cells on the blastoderm develop autonomously. A male cell in the center of female cells in the presumptive eye will ultimately develop male facets surrounded by female ones in the adult eye.

Sturtevant (1929) proposed the ideas supporting blastoderm fate mapping and constructed the first map. A blastoderm fate map is the demarcation on the blastoderm of prospective larval and adult structures. Fate maps are constructed from the data collected from gynandromorphs. The primary mapping postulate—which is quite similar to gene mapping—is that the closer two sites are for two structures on the blastoderm, the more often they will be of the same sex in a population of gynandromorphs. The percentage of gynandromorphs in which two structures differ in sex is equal to the distance, in units called sturts, on the blastoderm fate map separating them (Hotta and Benzer, 1972). Distances in sturts can be calculated between any pair of structures which can be reliably identified in the gynandromorphs. These distances between pairs of structures are used to construct a fate map by a process of sequential triangulation. The fate map is a diagram of one side of the blastoderm, showing areas which will develop into various structures. Blastoderm fate maps constructed for *Drosophila melanogaster* are consistent with early embryo maps derived by different methods (Janning, 1978). A more detailed account of blastoderm fate mapping can be found in Janning (1978), Benzer (1973), Janning *et al.* (1979), and Garcia-Bellido and Merriam (1969).

These maps not only tell us the developmental blueprint for the blastoderm but also provide a powerful analytical tool. Once a mutation is isolated, the identification of the primary tissue defect (focus) of the mutation is often difficult (Hotta and Benzer, 1972). However, once the fate map is developed, the identification is relatively straightforward. A population of gynandromorphs is produced in which either the male or the female tissue is genetically constructed to show the mutant phenotype. Then these gynandromorphs are scored for the mutation as well as structures on the fate map. The fate map is constructed and the mutation itself is located on the map. From the location of the mutation on the fate map, a focus can be reliably inferred.

The first blastoderm fate map constructed for any organism outside the genus *Drosophila* was for the honey bee (Milne, 1976). Data from 40 gynandromorphs isolated earlier (Drescher and Rothenbuhler, 1963) were used for this map of nine structures. Genetic markers were used to reliably identify the sex of some structures. Many honey-bee structures display strong sexual dimorphism and could be categorized easily in mosaics.

A more detailed blastoderm fate map was constructed by Milne and

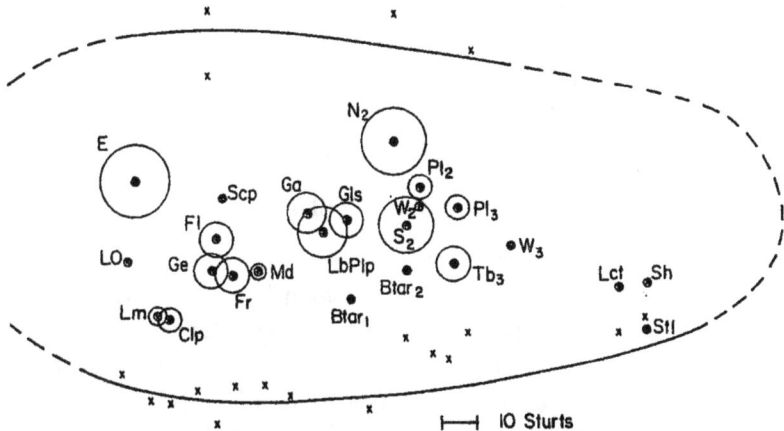

Honey-bee blastoderm fate map. [From Milne and Rothenbuhler (1983a), with
ı. Copyright 1983 by International Bee Research Association.] This map shows the
ɔn the side of the blastoderm of the primordial cells which will form 22 adult
Abbreviations used: $Btar_1$, prothoracic basitarsus; $Btar_2$, mesothoracic basitarsus;
ıs; E, compound eye; Fl, flagellum of antenna; Fr, frons; Ga, galea; Ge, gena; Gls,
'lp, labial palp; Lct, lancet of sting; Lm, labrum; LO, lateral ocellus; Md, mandible;
ın ocellus; N_2, mesonotum; Pl_2, mesopleuron; Pl_3, metapleuron; S_2, mesosternum;
of antenna; Sh, sheath of sting; Stl, stylet of sting; Tb_3, metathoracic tibia; W_2, fore
hind wing; x, distance to midline of blastoderm; — — —, tentative location of
. midline; MO, complex interacting focus located at LO, male bilaterally domineer-
ʒ, Fr, Ge, tentative locations; W_2, W_3, no diameters indicated.

ıhler (1983a) (Fig. 2). Twenty-five cuticular structures were scored
ʒynandromorphs isolated from eight gynandromorph-producing
ɛecessive genetic markers were used to identify the sex of structures
ing marked sexual dimorphism. Computer programs, written spe-
'or the honey-bee data, were used to calculate the distances be-
ırs of structures and construct a fate map from the distances (Milne,
ıe map from the chilled-egg gynandromorphs and the map from
ıdromorphs from the selected line are not different. This suggests
ıap is independent of the method used to produce the gynandro-
[he map is not influenced by the percentage of gynandromorphs a
)duces. Also, the map is independent of which queen produced the
ımorphs for the map. The detailed honey-bee map is consistent
ʲiously published maps for Drosophila and Habrobracon juglandis
l977).
ınd Rothenbuhler determined the sex of the ocelli by their location
ad surface. Examination of the data on the median ocellus indi-
ımplex interacting focus in which male tissue is bilaterally domi-
Iotta and Benzer, 1972). This single structure maps to both the left

and right lateral ocelli. The sex of these two structures alone determines the sex of the median ocellus. If either is male, the median ocellus is male. The median ocellus is female only if both lateral ocelli are female. This type of focus for a structural feature had not been reported previously.

B. Polarization of the Gynandromorphic Blastoderm

Drescher (undated) reports that sperm enter the micropylar area of the egg near its cephalic pole. Accessory sperm begin division there and form the male tissue of the gynandromorph. The zygotic nucleus is usually near the cleavage center when it initiates cleavage. Since the haploid nuclei appear anterior to the diploid nucleus, one would expect the gynandromorphic blastoderm to be polarized; male cells should tend to be localized in the anterior half and the female cells in the posterior half. Self-consistent blastoderm fate maps indicate little mixing of nuclei before blastoderm formation (Milne and Rothenbuhler, 1983a). The result of this polarized blastoderm should be polarized gynandromorphs.

Sakagami and Takahashi (1956) studied this in a group of 40 gynandromorphs. They concluded that the male tissue predominates in the head and female tissue in the abdomen. Drescher (1975), studying 50 structures in 185 gynandromorphs, concluded that only the compound eyes and ocelli differ significantly from an equal distribution of tissue between the sexes. These structures showed a higher tendency toward maleness in the gynandromorphs. Milne (1976), using the data from the 40 gynandromorphs of Drescher and Rothenbuhler (1963), noted a propensity for head structures to be male and abdominal structures to be female.

This polarization was examined in detail by Milne and Rothenbuhler (1983b) in a population of 1555 gynandromorphs. The percentage of male tissue was determined for 22 structures. In contrast to other studies, brood combs containing gynandromorphs were incubated and all gynandromorphs developing to the adult stage were found and examined, including those which did not emerge from brood cells. It is likely that the gynandromorphs studied elsewhere were picked from combs during hive inspections, both nonobvious and nonemerging gynandromorphs were missed, and a sampling bias existed. Milne and Rothenbuhler found no overall polarization when the structures were grouped by body region. Instead, a definite polarization was noted within the head region itself. Those structures nearer the cephalic pole on the fate map were more often male. The mouthparts, at the rear of the head region on the fate map, were more frequently female. The hypothesis presented to explain these data was that the eye and ocelli are near the location where the accessory sperm begin cleavage and the zygotic nucleus initiates cleavage near where the mouthparts are located on the blastoderm.

C. Estimation of Primordial Cell Numbers

One question central to developmental biology is how the genes control the developmental fate of cells. The success of blastoderm fate mapping indicates that until the blastoderm stage, cells have no developmental specificity. The exact time when determination of the fate of a cell occurs is apparently near or after the blastoderm stage. It would be useful to know not only when determination occurs, but also the number of cells restricted at this time toward the formation of a structure or group of structures.

Primordial cell numbers can be estimated from mosaic frequencies in gynandromorphs (Nissani and Lipow, 1977; Merriam, 1978). The formula

$$N = (n/2)(1 - \sqrt{1 - p^2})$$

estimates primordial cell number, where N is the number of primordial cells for a structure on the blastoderm, n is the total number of cells on the blastoderm, and p is the frequency of mosaic structures.

Milne (1985) used the data from 1555 gynandromorphs to estimate the number of primordial cells on the blastoderm giving rise to 22 adult cuticular structures. The percentage of mosaicism ranged from 0 to 46.67. An assumption of 3700 blastoderm cells was based on observations by Schnetter (1934) and DuPraw (1967). Estimates ranged from one blastoderm cell for the lateral ocellus, prothoracic basitarsus, antennal scape, and the stylet, lancet, and sheath of the sting, to 90 for the mesonotum and 214 for the compound eye. These estimates agree favorably with comparable estimates for *Drosophila*.

REFERENCES

Anderson, R. H. (1963). The laying worker in the Cape honeybee, *Apis mellifera capensis*. *J. Apic. Res.* 2, 85–92.

Armbruster, L. (1913). Über die Chromatinverhältnisse bei solitären Bienen und ihre Beziehung zur Frage der Geschlechtsbestimmung. *Ber. Naturforsch. Ges. Freiburg i. Br.* 20, 4–12.

Bachmayer, H., Kreil, G., and Suchanek, G. (1972). Synthesis of promelittin and melittin in the venom gland of queen and worker bees: pattern observed during maturation. *J. Insect Physiol.* 18, 1515–1521.

Benzer, S. (1973). Genetic dissection of behavior. *Sci. Am.* 229, 24–37.

Bishop, G. H. (1920). Fertilization in the honeybee, I. The male sexual organs: their histological structure and physiological functioning. *J. Exp. Zool.* 31, 225–226.

Boveri, T. (1915). Über die Entstehung der Eugsterschen Zwitterbienen. *Arch. Entwicklungsmech. Org.* 41, 264–311.

Brönskill, J. F., and Salkeld, E. H. (1978). The micropylar area of some hymenopterous eggs. *Can. Entomol.* 110, 663–665.

Buitkamp-Möbius, K. (1975). Strukturuntersuchungen an den Pilzkörpern in Oberschlund-ganglion von *Apis mellifica* gynandromorphen unter Berücksichtigung ihres Verhaltens. Ph.D. Dissertation, Rheinische Friedrich-Wilhelm Universität, Bonn, Federal Republic of Germany.

Chaud-Netto, J. (1980a). Estudos biológicos com rainhas triplóides de *Apis mellifera* 1. Produção de ovos abortivos por rainhas virgens. *Ciên. Cult. (São Paulo)* 32, 483–486.

Chaud-Netto, J. (1980b). Comprovação genética e citológica de triploidia em descendentes de cruzamentos controlados entre rainhas e zangões diplóides de *Apis mellifera*. (Hymenoptera, Apidae). *Ciên. Cult. (São Paulo)* 32, 351–355.

Chaud-Netto, J. (1980c). Estudos biológicos com abelhas triplóides de *Apis mellifera* (Hymenoptera, Apidae). *Ciên. Cult. (São Paulo)* 32, 611–615.

Crain, W. R., Davidson, E. H., and Britten, R. J. (1976). Contrasting patterns of DNA sequence arrangement in *Apis mellifera* (honeybee) and *Musca domestica* (housefly). *Chromosoma 59*, 1–12.

Davidson, E. H., Hough, B. R., Amenson, C. S., and Britten, R. J. (1973). General interspersion of repetitive with nonrepetitive sequence elements in the DNA of *Xenopus. J. Mol. Biol. 77*, 1–23.

De Lucca, F. L., Giorgini, J. F., and Calabrese, A. (1974). Effects of heat, urea and dimethylsulfoxide on ribosomal RNA of the honey bee (*Apis mellifera* L.). *Experientia 30*, 149–151.

Doncaster, L. (1906). Spermatogenesis of the hive bee (*Apis mellifica*). *Anat. Anz. 29*, 490–491.

Doncaster, L. (1907). Spermatogenesis of the honey bee (*Apis mellifica*). Correction. *Anat. Anz. 31*, 168–169.

Drescher, W. (undated). Genetish-embryologische Analyse der äusseren und inneren Faktoren bei der Erzeugung und Entwicklung von Gynandromorphen bei *Apis mellifica* L. Unpublished Habilitationsschrift.

Drescher, W. (1965). Der Einfluss von Umweltbedingungen auf die Bildung von Gynandromorphen bei der Honigbiene *Apis mellifica* L. *Insectes Soc. 12*, 201–218.

Drescher, W. (1975). Die räumliche und anteilmässig Verteilung männlichen und weiblichen Gewebes in der äusseren Morphologie gynandromorpher Individuen bei *Apis mellifica* L. *Insectes Soc. 22*, 13–26.

Drescher, W., and Rothenbuhler, W. C. (1963). Gynandromorph production by egg chilling. Cytological mechanisms in honey bees. *J. Hered. 54*, 195–201.

Drescher, W., and Rothenbuhler, W. C. (1964). Sex determination in the honey bee. *J. Hered. 55*, 91–96.

DuPraw, E. J. (1967). The honeybee embryo. *In* "Methods in Developmental Biology" (F. H. Wilt and N. K. Wessels, eds.), pp. 183–217. Crowell, New York.

Dzierzon, J. (1845). Gutachten über die von Hrn. Direktor Stöhr im ersten und zweiten Kapitel des General-Gutachtens aufgestellten Fragen. *Eichstädter. Bienenzeitung 1*, 109–113, 119–121.

Eckert, J. E. (1934). A gynandromorph honeybee. *J. Econ. Entomol. 28*, 1079–1082.

Eckert, J. E. (1937). Honeybee monstrosities. *Ann. Entomol. Soc. Am. 30*, 64–69.

Fahrenhorst, H. (1977a). Nachweis Übereinstimmender Chromosomen-Zahlen ($N = 16$) bei allen 4 *Apis*-Arten. *Apidologie 8*, 89–100.

Fahrenhorst, H. (1977b). Chromosome number in the tropical honeybee species *Apis dorsata* and *Apis florea. J. Apic. Res. 16*, 56–58.

Fyg, W. (1959). Normal and abnormal development in the honeybee. *Bee World 40*, 57–66, 85–96.

Garcia-Bellido, A., and Merriam, J. R. (1969). Cell lineage of the imaginal discs of *Drosophila* gynandromorphs. *J. Exp. Zool. 170*, 61–76.

Geitler, L. (1938). Die Entstehung der polyploiden somatischen Zellkerne bei Heteropteren durch wiederholte Chromosomenteilung ohne Spindelbildung und Kernteilung. *Naturwis-senschaften 26*, 722–723.

Geitler, L. (1939). Die Entstehung der polyploiden Somakern der Heteropteren, durch Chro-mosomenteilung ohne Kernteilung. *Chromosoma 1*, 1–22.

Giorgini, J. F. (1977). Low molecular weight RNA associated with 26S ribosomal RNA from *Apis mellifera* L. *Rev. Bras. Biol. 37*, 873–877.

Habermann, E., and Jentsch, J. (1967). Sequenzanalyse des Melittins aus den tryptischen und peptischen Spaltsluchen. *Hoppe-Seyler's Z. Physiol. Chem. 348*, 37–50.

Hachinohe, Y., and Onishi, N. (1952). On the meiosis of the drone honey bee *(Apis mellifica)*. *Bull. Natl. Inst. Agric. Sci. Ser. G 3*, 83–87.

Harbo, J. R. (1980). Mosaic male honey bees produced by queens inseminated with frozen spermatozoa. *J. Hered. 71*, 435–436.

Harbo, J. R., and Bolton, A. B. (1981). Development times of male and female eggs of the honey bee. *Ann. Entomol. Soc. Am. 74*, 504–506.

Hewitt, J. (1892). Fertile workers—their utility. *J. Hort. Soc. London 25*, 134.

Hoage, T. R., and Kessel, R. G. (1968). An electron microscope study of the process of differen-tiation during spermatogenesis in the drone honey bee *(Apis mellifera* L.) with special reference to centriole replication and elimination. *J. Ultrastruct. Res. 24*, 6–32.

Hoshiba, H. (1979). Chromosome of the diploid and haploid drone honeybee, *Apis mellifera. Proc. Inter. Apic. Cong. (Apimondia), 27*, 253–256.

Hoshiba, H. (1984). Karyotype and banding analyses on haploid males of the honey bee *(Apis mellifera). Proc. Jpn. Acad. Ser. B 60*, 122–124.

Hoshiba, H., and Kusanagi, A. (1978). Karyological study of honeybee. *J. Apic. Res. 17*, 105–109.

Hoshiba, H., Okada, I., and Kusanagi, A. (1981). The diploid drone of *Apis cerana japonica* and its chromosomes. *J. Apic. Res. 20*, 143–147.

Hotta, Y., and Benzer, S. (1972). Mapping of behavior in *Drosophila* mosaics. *Nature (Lond.) 240*, 527–535.

Jack, R. W. (1917). Parthenogenesis amongst the workers of the Cape honeybee: Mr. G. W. Onion's experiments. *Trans. R. Entomol. Soc. Lond. 64*, 396–403.

Janning, W. (1978). Gynandromorph fate maps in *Drosophila. In* "Results and Problems in Cell Differentiation, Vol. 9: Genetic Mosaics and Cell Differentiation" (W. J. Gehring, ed.), pp. 1–28. Springer-Verlag, Berlin-Heidelberg.

Janning, W., Pfreundt, J., and Tiemann, R. (1979). The distribution of anlagen in the early embryo of *Drosophila. In* "Cell Lineage, Stem Cells and Cell Differentiation. INSERM Symposium No. 10" (N. Le Doudrin, ed.), pp. 83–98. Elsevier, North-Holland Biomedical Press, Amsterdam.

Jay, S. C., and Jay, D. H. (1966). A behavioral study of a gynandromorphic honey bee. *Can. Entomol. 98*, 170–174.

Jegen, G. (1920). Zur Geschlechtsbestimmung bei *Apis mellifica. In* "Festschrift für Zschokke" No. 38, Basel.

Jordan, R. A., and Brosemer, R. W. (1974). Characterization of DNA from three bee species. *J. Insect Physiol. 20*, 2513–2520.

Kerr, W. E., and Laidlaw, H. H., Jr. (1956). General genetics of bees. *Adv. Genet. 8*, 109–153.

Kindås-Mügge, I., Lane, C. D., and Kreil, G. (1974). Insect protein synthesis in frog cells: the translation of honey bee promelittin messenger RNA in *Xenopus* oocytes. *J. Mol. Biol. 87*, 451–462.

Kindås-Mügge, I., Frasel, L., and Digglemann, H. (1976). Characterization of promelittin messenger RNA from the venom gland of young queen bees. *J. Mol. Biol. 105*, 177–181.

Kreil, G. (1973a). Structure of melittin isolated from two species of honey bees. *FEBS Lett. 33*, 241–244.

Kreil, G. (1973b). Biosynthesis of melittin, a toxic peptide from bee venom. Amino-acid sequence of the precursor. *Eur. J. Biochem. 33*, 558–566.

Kreil, G. (1975). The structure of *Apis dorsata* melittin: phylogenetic relationships between
honeybees as deduced from sequence data. *FEBS Lett. 54*, 100–102.

Kreil, G., and Bachmayer, H. (1971). Biosynthesis of melittin, a toxic peptide from bee venom.
Detection of a possible precursor. *Eur. J. Biochem. 20*, 344–350.

Kreil, G., Haiml, L., Suchanek, G. (1980a). Stepwise cleavage of the pro part of promelittin by
dipeptidylpeptidase IV. *Eur. J. Biochem. 111*, 49–58.

Kreil, G., Mollay, G., Kaschnitz, R., Haiml, L., and Vilas, V. (1980b). Prepromelittin: specific
cleavage of the pre and the propeptide *in vitro*. In "Precursor Processing in the Biosynthesis
of Proteins" (M. Zimmerman, R. A. Mumford, and D. F. Steiner, eds.). *Ann. N.Y. Acad. Sci.
343*, 338–346.

Kubasek, K. J., Rinderer, T. E., and Lee, W. R. (1980). Isogenic sperm line maintenance in the
honey bee. *J. Hered. 71*, 278–280.

Laidlaw, H. H., and Tucker, K. W. (1964). Diploid tissue derived from accessory sperm in the
honey bee. *Genetics 50*, 1439–1442.

Laubender, B. (1801). Einige Bemerkungen uber die von Herrn Schulmeister Lukas neu ent-
deckten Stacheldrohnen. *Okonomische 17*, 429.

Lee, W. R. (1958). The dosage response curve for radiation-induced dominant lethal mutations
in the honeybee. *Genetics 43*, 480–492.

Lerer, H., and Dixon, S. E. (1975). Polyploidy in fat-body of honeybees during larval growth
and development. *J. Apic. Res. 14*, 105–111.

Mackensen, O. (1943). The occurrence of parthenogenetic females in some strains of honeybees. *J. Econ. Entomol. 36*, 465–467.

Mackensen, O. (1951). Viability and sex determination in the honeybee (*Apis mellifera* L.).
Genetics 36, 500–509.

Mackensen, O. (1955). Further studies on a lethal series in the honeybee. *J. Hered. 46*, 72–74.

MacKinnon, E. A., and Basrur, P. K. (1970). Cytokinesis in the gonocysts of the drone honey
bee (*Apis mellifera* L.). *Can. J. Zool. 48*, 1163–1166.

Manning, J. E., Schmid, C. W., and Davidson, N. (1975). Interspersion of repetitive and
non-repetitive DNA sequences in the *Drosophila melanogaster* genome. *Cell 4*, 141–155.

Mark, E. L., and Copeland, M. (1906). Some stages in the spermatogenesis of the honey bee.
Proc. Am. Acad. Arts Sci. 42, 103–111.

Mello, M. L. S., and Takahashi, C. S. (1971). DNA content and nuclear volume in larval organs
of *Apis mellifera. J. Apic. Res. 10*, 125–132.

Menzel, A. (1862). Über Zwitterbienen. *Nördlinger Bienenztg. 18*, 167–168, 186–187.

Merriam, J. R. (1978). Estimating primordial cell numbers in *Drosophila* imaginal discs and
histoblasts. In "Results and Problems in Cell Differentiation, Vol. 9: Genetic Mosaics and
Cell Differentiation" (W. J. Gehring, ed.), pp. 71–96. Springer-Verlag, Berlin–Heidelberg.

Merriam, R. W., and Ris, H. (1954). Size and DNA content of nuclei in various tissues of male,
female and worker honeybees. *Chromosoma 6*, 522–538.

Meves, F. (1903). Über Richtungskörperbildung im Hoden von Hymenopteren. *Anat. Anz. 24*,
29–32.

Meves, F. (1907). Die Spermatocytenteilungen bei der Honigbiene (*Apis mellifica* L.), nebst
Bemerkungen über Chromatinreduktion. *Arch. Mikrosk. Anat. Entwicklungsmech. 70*,
414–491.

Milne, C. P., Jr. (1976). Morphogenetic fate map of prospective adult structures of the honey
bee. *Develop. Biol. 48*, 473–476.

Milne, C. P., Jr. (1977). Blastoderm fate map of the honey bee (*Apis mellifera* L.). Ph.D.
dissertation, The Ohio State University, University Microfilms International, Ann Arbor,
Mich.

Milne, C. P., Jr. (1985). Estimating primordial cell numbers giving rise to honeybee adult structures. *J. Apic Res. 24*, 7–12.

Milne, C. P., Jr., and Rothenbuhler, W. C. (1983a). The honeybee blastoderm fate map. *J. Apic. Res. 22*, 69–78.

Milne, C. P., Jr., and Rothenbuhler, W. C. (1983b). Polarization of the honey bee gynandromorphic blastoderm. *Can. J. Genet. Cytol. 25*, 561–566.

Mittwoch, V., Kalmus, H., and Webster, W. S. (1966). Deoxyribonucleic acid values in dividing and non-dividing cells of male and female larvae of the honey bee. *Nature (Lond.) 210*, 264–266.

Morgan, T. H. (1905). An alternative interpretation of the origin of gynandromorphous insects. *Science 21*, 632–634.

Nachtsheim, H. (1912). Parthenogenese, Eireifung und Geschlechtsbestimmung bei der Honigbiene. *Sitzungsber. Ges. Morph. Physiol. München 28*, 22–29.

Nachtsheim, H. (1913). Cytologische Studien über die Geschlechtsbestimmung bei der Honigbiene (*Apis mellifica* L.). *Arch. Zellforsch. 11*, 169–241.

Nelson, J. A. (1915). "The Embryology of the Honey Bee." Princeton University Press, Princeton, N.J.

Nissani, M., and Lipow, C. (1977). A method for estimating the number of blastoderm cells which give rise to *Drosophila* imaginal discs. *Theor. Appl. Genet. 49*, 3–8.

Onions, G. W. (1912). South African "fertile worker bees." *Agric. J. Union S. Afr. 3*, 720–728.

Onions, G. W. (1914). South African "fertile" worker bees. *Agric. J. Union S. Afr. 7*, 44–46.

Painter, T. S. (1945). Nuclear phenomena associated with secretion in certain gland cells with especial reference to the origin of cytoplasmic nucleic acid. *J. Exp. Zool. 100*, 523–541.

Paulcke, W. (1899). Zur Frage der parthenogenetischen Entstehung der Drohnen (*Apis mellifera ♂*). *Anat. Anz. 16*, 474–476.

Pene, J. J., Knight, E., Jr., and Darnell, J. E., Jr. (1968). Characterization of a new low molecular weight RNA in HeLa cell ribosomes. *J. Mol. Biol. 33*, 609–623.

Petrunkewitsch, A. (1901). Die Richtungskörper und ihr Schicksal im befruchteten und unbefruchteten Bienenei. *Zool. Jahrb., Abt. Anat. Ontog. Tiere 14*, 573–608.

Petrunkewitsch, A. (1903). Das Schicksal der Richtungskörper im Drohnenei. Ein Beitrag zur Kenntniss der natürlichen Parthenogenese. *Zool. Jahrb., Abt. Anat. Ontog. Tiere 17*, 481–516.

Petters, R. M. (1977). A morphogenetic fate map constructed from *Habrobracon juglandis* gynandromorphs. *Genetics 85*, 279–287.

Risler, H. (1954). Die somatische Polyploidie in der Entwicklung der Honigbiene (*Apis mellifica* L.) und die Wiederherstellung der Diploidie bei den Drohnen. *Z. Zellforsch. Mikrosk. Anat. 41*, 1–78.

Rösch, G. A. (1927). Über einen Weg, Zwitter der Honigbiene (*Apis mellifica* L.) im Experiment zu erzeugen. *Sitzungsber. Ges. Morph. Physiol. München 37*, 71–81.

Rothenbuhler, W. C. (1955). Hereditary aspects of gynandromorph occurrence in honey bees (*Apis mellifera* L.). *Iowa State Coll. J. Sci. 29*, 487–488.

Rothenbuhler, W. C. (1957). Diploid male tissue as new evidence on sex determination in honey bees. *J. Hered. 48*, 160–168.

Rothenbuhler, W. C. (1958a). Genetics and breeding of the honey bee. *Annu. Rev. Entomol. 3*, 61–80.

Rothenbuhler, W. C. (1958b). Progress and problems in the analyses of gynandromorphic honey bees. *Proc. Int. Congr. Entomol. 10th*, 867–873.

Rothenbuhler, W. C. (1975). The honey bee, *Apis mellifera, Handb. Genet. 3*, 165–172.

Rothenbuhler, W. C., and Gowen, J. W. (1955). Chromosomal localization of the hereditary basis for gynandromorph production in honey bees (*Apis mellifera* L.). *Genetics 40*, 592–593.

Rothenbuhler, W. C., Polhemus, M. S., Gowen, J. W., and Park, O. W. (1949). Gynandro-morphic honey bees. *J. Hered.* 40, 308–311.

Rothenbuhler, W. C., Gowen, J. W., and Park, O. W. (1952). Androgenesis with zygogenesis in gynandromorphic honeybees (*Apis mellifera* L.). *Science* 115, 637–638.

Rothenbuhler, W. C., Kulinčević, J. M., and Kerr, W. E. (1968). Bee genetics. *Annu. Rev. Genet.* 2, 413–438.

Ruttner, F., and Mackensen, O. (1952). The genetics of the honeybee. *Bee World* 33, 53–62, 71–79.

Sakagami, S. F., and Takahashi, H. (1956). Beobachtungen über die gynandromorphen Honig-biene, mit besonderer Berücksichtigung ihrer Handlungen innerhalb des Volkes. *Insectes Soc.* 3, 513–529.

Sanderson, A. R., and Hall, D. W. (1948). The cytology of the honey bee, *Apis mellifica,* L. *Nature (Lond.)* 162, 34–35.

Sanderson, A. R., and Hall, D. W. (1951). Sex in the honey-bee. *Endeavour* 10, 33–39.

Schirach, A. G. (1766). "Sachsischer Bienenvater, oder des Herrn Palteau." A. J. Spiekermann, Leipzig and Zitlau.

Schnetter, M. (1934). Morphologische Untersuchungen über das Differenzierungszentrum in der Embryonalentwicklung der Honigbiene. *Z. Morph. Oekol. Tiere,* 29, 114.

Sharma, G. P., Gupta, B. L., and Kumbkarni, C. G. (1961). Cytology of spermatogenesis in the honey-bee *Apis indica* (F.). *J. R. Microsc. Soc.* 79, 337–351.

Siebold, C. T. von. (1864). Über Zwitterbiene. *Zeit. Wiss. Zool. Abt. A.* 14, 73–80.

Snodgrass, R. E. (1925). "Anatomy and Physiology of the Honeybee." McGraw-Hill, New York.

Snodgrass, R. E. (1956). "Anatomy of the Honey Bee." Cornell University Press, Ithaca, N.Y.

Steiner, D. F., Quinn, P. S., Chan, S. J., Marsh, J., and Tager, H. S. (1980). Processing mecha-nisms in the biosynthesis of proteins. *In* "Precursor Processing in the Biosynthesis of Pro-teins" (M. Zimmerman, R. A. Mumford, and D. F. Steiner, eds.). *Ann. N.Y. Acad. Sci.* 343, 1–16.

Stekol'shchikov, M. G. (1971). Soderzhanie DNK v yadrakh kletok pautinnykh zhelez lichinok pchel. *Tsitol. Genet.* 5, 536–540.

Stort, A. C., and Soares, A. E. (1978). Occurrence of a new type of mosaicism in *Apis mellifera. Experientia* 34, 31–32.

Sturtevant, A. H. (1929). The claret mutant type of *Drosophila simulans,* a study of chromosome elimination and cell-lineage. *Zeit. Wiss. Zool. Abt. A.* 135, 323–356.

Suchanek, G., and Kreil, G. (1977). Translation of melittin messenger RNA *in vitro* yields a product terminating with glutaminylglycine rather than with glutaminamide. *Proc. Natl. Acad. Sci. U.S.A.* 74, 975–978.

Suchanek, G., Kindås-Mügge, I., Kreil, G., and Schreier, M. H. (1975). Translation of honeybee promelittin messenger RNA. Formation of a larger product in a mammalian cell-free system. *Eur. J. Biochem.* 60, 309–315.

Suchanek, G., Kreil, G., and Hermodson, M. A. (1978). Amino acid sequence of honeybee prepromelittin synthesized *in vitro. Proc. Natl. Acad. Sci. U.S.A.* 75, 701–704.

Suomalainen, E. (1950). Parthenogenesis in animals. *Adv. Genet.* 3, 193–253.

Swanson, C. P., Merz, T., and Young, W. J. (1981). "Cytogenetics." Prentice-Hall, Inc., Engle-wood Cliffs, N.J.

Taber, S., III. (1955). Evidence of binucleate eggs in the honey bee. *J. Hered.* 46, 156.

Tucker, K. W. (1958). Automictic parthenogenesis in the honey bee. *Genetics* 43, 299–316.

Tucker, K. W., and Laidlaw, H. H. (1966). The potential for multiplying a clone of honey bee sperm by androgenesis. *J. Hered.* 57, 213–214.

Verma, L. R. (1972). Cytochemical analysis of nuclear histone in honeybee (*Apis mellifera* L.) spermatozoa. *Can. J. Zool.* 50, 1241–1242.

Verma, S., Ruttner, F., and Verma, L. R. (1982). Cytological mechanism of syngamy in the automatic laying workers of *Apis mellifera capensis*. *In* "The Biology of Social Insects" (M. D. Breed, C. D. Michener, and H. E. Evans, eds.), pp. 412–413. *Proc. 9th Cong. IUSSI*, Westview Press, Boulder, Col.

Vlasak, R., and Kreil, G. (1984). Nucleotide sequence of cloned cDNAs coding for preprosecapin, a major product of queen bee venom glands. *Eur. J. Biochem.* 145, 279–282.

Vlasak, R., Unger-Ullmann, C., Kreil, G., and Frischauf, A. (1983). Nucleotide sequence of cloned cDNA coding for honeybee prepromelittin. *Eur. J. Biochem.* 135, 123–126.

Witherell, P. C. (1971). Note on behavior of gynandromorphic honey bees, *Apis mellifera. Ann. Entomol. Soc. Am.* 64, 951.

Woyke, J. (1962). The hatchability of 'lethal' eggs in a two sex-allele fraternity of honeybees. *J. Apic. Res.* 1, 6–13.

Woyke, J. (1975). DNA content of spermatids and spermatozoa of haploid and diploid drone honeybees. *J. Apic. Res.* 14, 3–8.

Woyke, J., and Skowronek, W. (1974). Spermatogenesis in diploid drones of the honeybee. *J. Apic. Res.* 13, 183–190.

Woyke, J., Knytel, A., and Bergandy, K. (1966). Cytological proof of the origin of drones from inseminated eggs of the honey bee. *Bull. Acad. Pol. Sci. Ser. Sci. Biol.* 14, 65–67.

CHAPTER 9

Population Genetics

I. INTRODUCTION

The first studies of the population genetics of honey bees began in the 1950s (Crow and Roberts, 1950; Kerr, 1951), but developments in this field (involving mathematical modeling and field observations of evolutionary mechanisms) did not interfere with apidologists' investigations until the mid 1970s. The main exceptions concerned specialized topics such as genetic load, sex-limited genes, sex determination, and the evolution of haplo–diploidy by Kerr and his associates in Brazil (Kerr, 1951, 1967, 1976).

During the last 10 years, studies of the population genetics of honey bees have become more numerous. In Europe, a lasting interest in subspecific taxonomy has opened the way to works on the differentiation and the evolution of *Apis mellifera* (L.). As a consequence of the Africanized bee problem, Americans recently have begun to share this preoccupation, introducing new methods for assessing genetic variability and for discriminating between populations. Also, studies have been undertaken on the effects of inbreeding and on the mechanism of the sex locus aimed specifically at the genetic improvement of the honey bee.

II. GENETIC VARIABILITY

A. Morphological Variability

So far, no true morphological polymorphisms [rare variants due to mutations are not included (Ford, 1940)] have been found in honey-bee populations. On the other hand, many morphological studies have been carried

TABLE 1. Estimated Heritabilities of Some
Biometrical Characters

Character	Estimate of Roberts (1961)	Estimate of Gonçalves (1970)
Fore-wing length	High	High
Fore-wing width	High	Low
Number of hamuli	Low	High
Tongue length	High	Not studied
Cubital index	High	Not studied

out, mainly in order to characterize geographical subspecies. Within the species, these characters show continuous variability that excludes simple genetic determinism. However, even if environmental influences on the phenotype may not be excluded, there must be some genetic determinism, likely polygenic, so that the expressed phenotype variation is a rather good indicator of genetic variation. This statement is supported by several arguments.

The morphological characters used in taxonomic studies are usually selected for their ability to characterize populations independently of environmental conditions and are thus presumably genetically determined.

Second, studies have shown (Cornuet et al., 1975) that F_1 hybrids exhibit intermediate values with respect to the subspecies from which they are bred. One may conclude that the dominance effects of the genes involved are weak in comparison to their additive effects. Consequently, the heritabilities of these characters should be high.

Roberts (1961), having done a true genetic study of seven characters, found a high heritability for six of them. However, his estimates must have included variance of dominance effects since he also found significant heterosis in these characters. His results agree moderately with those of Gonçalves (1970), who studied the genetic determinism of 11 characters (four are common to both authors). For all characters used in those biometrical studies, the heritability was found to be high by at least one author (see Table 1 for examples).

Since the early works of Alpatov and Goetze [cited in Ruttner (1968)], who introduced biometrical techniques into honey-bee taxonomy, quantitative studies of morphological variability have increased greatly at three levels—population sampling, characters variety, and statistical analysis methods.

Besides the more limited studies of Dupraw (1965), Fresnaye (1965), Tomassone and Fresnaye (1971), Louis and Lefevre (1971), Cornuet et al. (1975), Daly (1975), and Gadbin et al. (1979), there is the considerable work of Ruttner (Ruttner *et al.*, 1978; Ruttner, 1981, 1982), and it may be assumed that no major subspecies has been ignored by honey-bee biometricians.

On the whole, 41 characters have been subject to measurements by different authors. They are all listed and described fully in Ruttner *et al.* (1978). They concern pilosity (3 characters), size (15), pigmentation (7), wing venation (15), and number of wing hooks.

The most important feature of honey-bee morphology appears to be its intense and world-wide variability. As an example, Table 2 gives the lowest and highest mean value of colonies drawn from Ruttner *et al.* (1978) for four different characters, the last one actually being a ratio.

B. Ecoethological Characters

Ecoethological characters encompass the behavior expressed by the colony to face ecological conditions. Since honey bees have evolved in a wide variety of conditions that support flowering plant life, from equatorial forest to cold Bachkirian lands, the species as a whole must face many different ecological conditions. This, in combination with geographical isolation, has led to the evolution of numerous subspecies with genomes adapted to their own specific environments. Many observations and experiments have compared ecologically related behavior between subspecies. Generally, the subspecies are considered the best adapted to their own local conditions. However, the evidence is not always clear, especially when the flora has changed drastically, e.g., in Israel (Ruttner, 1982). A striking example of differences in ecological adaptation between subspecies is shown by the way a few Africanized swarms have invaded South America where the climate and flora are more similar to Africa than to Europe.

A subspecies such as *A. m. mellifera* is distributed over large areas and therefore must tolerate a variety of environments. So the question arises of how closely a population is adapted to its own environment. An experiment by Louveaux (1966) provides some insight into this point. Hives placed in

TABLE 2. Variation of Morphological
Characters among *Apis mellifera* Colonies[a]

Character	Lowest value (mm)	Largest value (mm)
Fore-wing length	7.98	9.69
Proboscis length	5.31	7.19
Hair length	0.158	0.477
Cubital index	1.58	3.62

[a] After Ruttner *et al.* (1978).

three distinct French locations were monitored periodically for brood surface, weight, and pollen preference. In each place, the annual brood production curve was characteristic and particularly adapted to local food resources. This means that there was a peak in brood rearing before the
beginning of the main nectar production season, which brought colonies to
a peak in adult population just at the time of main nectar production. Then
the colonies were exchanged between locations, and it was observed that
they maintained their developmental rhythm and choice of plants, leading
to poor efficiency. Generally, they underwent queen replacement during
the following year. For the next generation, in which queens were mated
with local drones, hive behavior was intermediate between local and imported colonies. This result confirms the genetic basis of environmentally
adapted behavior and shows the ecological genetic differentiation within a
subspecies which led Louveaux to bestow the term "ecotypes" on these
populations.

C. Allozymes

Allozyme variation, detected by electrophoresis and histochemical staining, is considered the best way to assess genetic variability in natural populations because of the direct relationship between phenotype and genotype.
The description and discussion of gene-enzyme variation are presented in
another chapter in this volume (Sylvester, Chapter 7). The main conclusion,
however, is that honey bees have low variability for this set of traits and
little differentiation between subspecies.

III. GENETIC STRUCTURE

The previous section has shown the large amount of morphological and
ecoethological variability that exists in the cosmopolitan *Apis mellifera* species. Such variation, largely space structured, has led to the use of an in-
fraspecific taxonomic division: the geographical subspecies (Buttel-Reepen, 1906). This concept, which has proved to be highly effective in
describing the genetic structure of the species, has been defined by Ruttner
(1968) as the "lowest taxonomic unity which can be morphologically characterized." This definition emphasizes the morphological aspect of variation and the primary role played by biometrics in characterizing subspecies.
Thanks to the recent introduction of computerized multivariate analysis and
classification methods (Dupraw, 1965; Tomassone and Fresnaye, 1971;
Cornuet, 1982a; Daly *et al.*, 1982), our knowledge of the genetic structure of
the species has been improved greatly.

A few attempts have been made to substitute allozymic characters for biometrical ones (Mestriner and Contel, 1972; Sylvester, 1976; Martins *et al.*, 1977; Gartside, 1980). Most of these have been made on American populations in order to distinguish "European and Africanized" bees, which can not be considered as well-defined taxonomic entities (Sylvester, 1982). Nevertheless, although enzymic polymorphisms are very few, they provide additional information on genetic differentiation and gene flow in *Apis mellifera*.

A. Geographical Differentiation in the Old World

1. *Present Status*

Europe, Africa, and Western Asia encompass the area in which honey bees naturally evolved. According to Ruttner (1982), 24 subspecies have been discriminated and characterized. They are classified in four groups (Rothenbuhler *et al.*, 1968):

Northwest group:	*sahariensis, intermissa, major, iberica, mellifera*
Southwest European group:	*cecropia, carnica, ligustica, sicula, adami*
Middle Eastern group:	*cypria, syriaca, anatolica, persica, armenica, caucasica*
African group:	*lamarckii, adansonii, yemenetica, scutellata, littorea, monticola, capensis, unicolor*

Most of the time, a subspecies is not a panmictic unity, but it is composed of populations distinguishable not only by their ecological behavior (Louveaux, 1966) but also by biometric analysis (Cornuet *et al.*, 1975). A study of a Cevenol (Fr.) population (Cornuet *et al.*, 1978) shows the degree of differentiation of the *mellifera* subspecies: two apiaries, 8 km apart but separated by a 1300-m-high ridge, are partly separated by discriminant analysis. This subspecies is presumably composed of small populations exchanging genes with surrounding ones by mating and swarming (there seems to be no large-scale migration in *A. m. mellifera*). Mating flights rarely exceed 10 km and swarming 3 km, so that gene flow between populations decreases rapidly as distance between them increases allowing sharp differentiation. However, not all subspecies behave the same way. For instance, samples of *adansonii* taken more than 100 km apart could not be discriminated (Gabdin *et al.*, 1979). This has been interpreted as a possible consequence of the well known migratory behavior of African subspecies (Smith, 1961). Actually, this behavior and the lack of topographic barriers have been considered as an explanation for taxonomic difficulties in South Africa. Furthermore, ecological differences more than geographical isolation seem responsible for subspecies differentiation in Africa (Ruttner, 1975).

2. Hypothetical Evolutionary History

Apis mellifera (as well as *A. cerana, A. florea,* and *A. dorsata*), probably originated in Southeastern Asia. According to Ruttner (1982), it spread through the Middle East, Near East, Africa, and Europe about 30,000 years ago. The late glaciations (*ca.* 7000 B.C.) forced European populations into the southern peninsula of Spain, Italy, and Greece. This effective isolation accelerated the formation of local subspecies (Ruttner, 1952).

Finding a characteristic "Y-shaped" pattern of subspecies on the first plane of a principal components analysis, Ruttner *et al.* (1978) suggested the following hypothesis about the spread of *Apis mellifera*. From a dispersal center situated in northeast Africa – Near East, the species evolved in three main branches (A, M, and C, Fig. 1).

The Southern A (for African) branch gave rise to the African group. The Western M (for *mellifera*) branch is composed of the sequence *sahariensis – intermissa – iberica – mellifera*. The Eastern C (for *carnica*) branch contains *cecropia, sicula, ligustica,* and *carnica*.

The pattern of variation in one allozyme locus, malic dehydrogenase (MDH) (Fig. 1), as drawn from data presented by Nunamaker and Wilson (1981) and Cornuet (1982b), enhances the probability of the assumption. This pattern shows clear evidence of a progressive transition from *intermissa* to *mellifera* through *iberica*. So far, the B allele seems characteristic of the western branch, as is the C allele in the eastern branch.

B. Recent Evolution in the New World

Although New World countries are among the best honey producers, there were no *Apis mellifera* in this part of the world before the sixteenth century. Honey-bee introduction was accomplished by European settlers, so that present populations originated mainly from *mellifera, iberica, ligustica, carnica,* and *caucasica* subspecies (Rothenbuhler, 1979). No new subspecies has emerged in this small lapse of time and, with so many successive introductions from Europe, the so-called "Creolian" bee is simply a genetically mixed population.

So far, the only biometrical study performed on American populations (Daly and Balling, 1978) does not give any insight into the structure of populations on this continent because all samples of European or African ancestry were pooled. The MDH locus analysis in some South and North

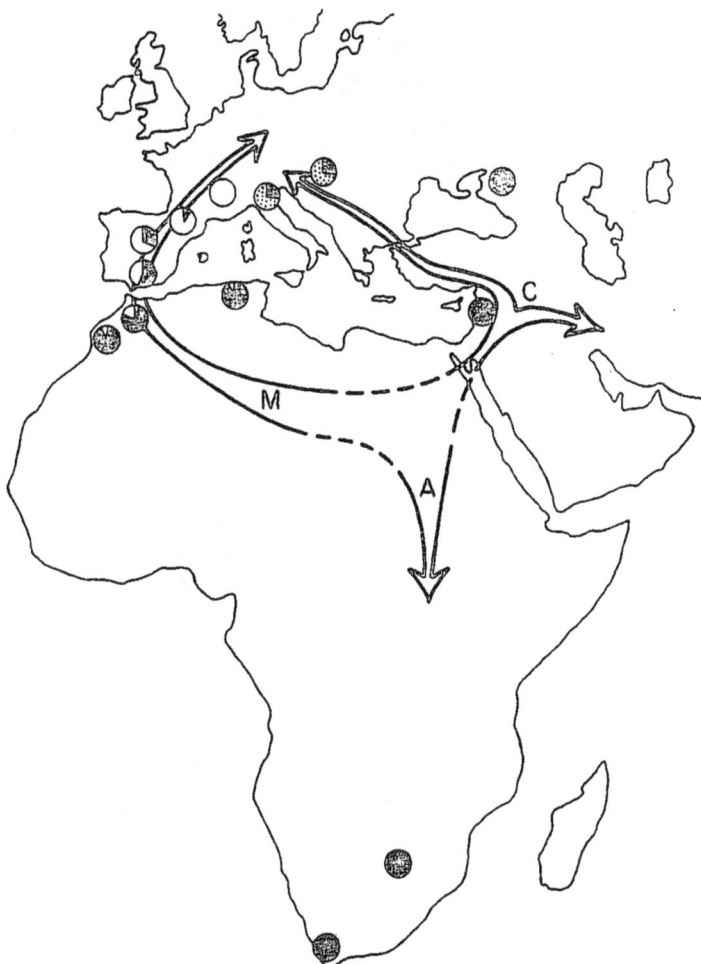

Fig. 1. Pattern of MDH locus variation (black, slow allele; white, medium allele; dots, fast allele) and presumable evolution of A. *mellifera* subspecies. Because of reduced samples, the MDH pattern for *carnica* and *syriaca* might not be representative.

American stocks enhances the above assumption of the generality of mixed population; all three alleles always have been found.

The most drastic event in the evolution of American honey bees has been the spreading of the Africanized bees. A starting population of 26 swarms escaped from a Brazilian apiary in 1957 and invaded the whole South American continent in some 25 years. A striking fact is that the founding African genome has not been diluted in the large South American gene pool, but has remained rather pure. Actually, some samples from Colombia studied biometrically looked like pure *adansonii* (J. Fresnaye, personal communication). This was confirmed by Nunamaker and Wilson's (1981) finding of a strong shift of MDH gene pool toward African genotypes in Brazil. Besides the fact that African genes have been selected because of their higher fitness in equatorial and tropical environments, there must have been a degree of sexual isolation as well (Kerr and Bueno, 1970).

IV. INBREEDING

By increasing the probability that alleles will occur in the homozygous state, inbreeding allows the expression of normally "hidden" recessive alleles. As some of them are deleterious, the result is a decrease of the fitness of inbred individuals. However, one explanation for the existence of genetic variability in a population is the selective advantage of heterozygotes. In haplo–diploid species such as *A. mellifera* this mechanism can be effective only in the diploid sex and traits important to males must not rely on heterosis (Brücker, 1978).

Consequently, in haplo–diploid systems the gene pool should contain fewer deleterious alleles and individuals should be less sensitive to inbreeding depression. However, honey bees have developed a mating behavior that tends to prevent excessive inbreeding: queens mate outside the hive and are inseminated by several drones. Second, heterosis is known in honey bees for many kinds of characters, e.g., morphology, behavior, and economic value. Third, the mechanism of sex alleles does induce an inbreeding depression, since, with inbreeding, the probability of the homozygous state for the sex locus and therefore of gaps in the brood is higher.

A. Effect of Inbreeding

1. Morphology

Morphological characters are subject to change under inbreeding (Mackensen, 1956). For example, inbreeding has led to a clear disruptive selection pattern for a wing vein (A of cubital index) (Cornuet *et al.*, 1975): one group of inbred lines showed values significantly above the highest noninbred line and the other group showed values significantly below the lowest noninbred line. Also, variances are different between inbred and noninbred lines. Brückner (1976a) found that in inbred workers the intracolony variance is reduced and the intraindividual variance (wing characters are measured once per wing and such a variance is related positively to dissymmetry) is increased with respect to noninbred workers. The same results have been found for the number of wing hooks. These findings tend to prove the existence of homeostasis in morphological traits.

2. Physiology

Laboratory experiments on groups of workers show that thermoregulatory ability was lowered significantly in inbred bees compared to outbred ones (Brückner, 1976b).

3. Behavior

Plass (1953) observed that inbred colonies had brood-rearing deficiencies and a decrease in nest-cleaning behavior. This last characteristic may explain why inbred colonies are more sensitive to parasitic diseases.

The recruiting ability is also affected by inbreeding (Brückner, 1978): non-inbred workers recruited more than inbred ones.

4. Economic Characters

The negative effect of inbreeding upon the preceding characters demonstrates why economic characters like honey yield are so badly affected by inbreeding. A practical consequence is that care must be taken to avoid consanguinity in breeding.

B. Theoretical Studies

1. Definition of Inbreeding Coefficients

Since there are different inbreeding coefficients with different meanings, it is necessary to define the coefficients that will be used in the following paragraph.

The consanguinity coefficient, denoted F, of a diploid individual gives the probability that at a given locus both alleles are identical by descent (they are copies of a same ancestral allele) (Malécot, 1948). In honey bees, this coefficient is irrelevant for haploid males.

The kinship coefficient, denoted Φ, between two individuals is the probability that at a given locus, one allele taken at random in one individual is identical by descent with one allele taken at random in the other individual. In honey bees, because males are haploid, we have the following relationship:

$$\Phi_{ab} = \Phi_{mb} \tag{1}$$

where m is the mother of the drone a.

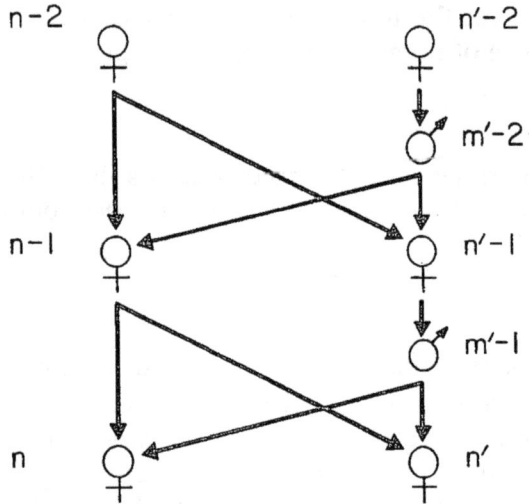

Fig. 2. Aunt–nephew inbreeding scheme.

A relationship links these two coefficients:

$$F_d = \Phi_{mf} \tag{2}$$

where d is the daughter of m and f.

2. Systems of Inbreeding

Crow and Robert (1950) have given formulas for the increase of inbreeding with several breeding systems like brother–sister mating and aunt–nephew mating. As an example of how they may be obtained, let us take the aunt–nephew mating (Fig. 2). Let F_n be the consanguinity coefficient of queen n. According to Eqs. (1) and (2),

$$F_n = \Phi_{n-1, n'-1} \tag{3}$$

In the computation of $\Phi_{n-1, n'-1}$, there are three cases to consider.

a. Both genes taken at random in $(n - 1)$ and $(n' - 1)$ come from the mother $(n - 2)$.

b. Both genes taken at random in $(n - 1)$ and $(n' - 1)$ come from the father $(m' - 2)$.

c. One gene comes from the mother and the other from the father.

The probability of these three events are $(A) = \frac{1}{4}$, $(B) = \frac{1}{4}$, and $(C) = \frac{1}{2}$.

In case A, there are still two eventualities: (1) both genes are copies of the same gene [probability of case $(a) = \frac{1}{2}$]; (2) both genes are copies of both mother genes [probability of case $(b) = \frac{1}{2}$]. In the latter case, they may be identical with the probability F_{n-2}. In case B, the genes are identical necessarily. In case C, the two genes are identitical with the probability F_{n-1} according to Eq. (2). Combining all these probabilities,

$$\tfrac{1}{4}(1 + F_{n-2})/2 + \tfrac{1}{4} + F_{n-1}/2 = \tfrac{3}{8} + F_{n-1}/2 + F_{n-2}/8 \qquad (4)$$

This theoretical example shows the way in which consanguinity coefficients are computed in more normal and generally complicated breeding schemes. As in other domestic species, programs now are available to compute the inbreeding coefficient.

3. Change with Time of Consanguinity

Since it has been shown that inbreeding produces negative effects upon characters of economical importance, bee breeders must avoid matings between close relatives. However, selection—reducing the number of reproductives—leads inevitably to the increase of the mean consanguinity. Hence, it is of some interest to have a good idea of its value. Chevalet and Cornuet (1982b) have computed the change with time of sanguinity in a particular case of selection, but the method they adopted allows the extension of their results to various situations.

Their computations are fitted to the following breeding scheme. At each generation, a fixed number of queens (R), are selected (according to a selection index which takes into account the value of colonies and sister colonies). Each queen gives birth to S daughters. All the RS daughters are naturally mated by males of two distinct origins: males that are sons of P-selected queens (with a probability β), and males coming from queens outside of the breeding scheme (probability $= 1 - \beta$). The RS resulting colonies are then tested and among them will be chosen the R- and P-selected queens (the R- and P-queen sets overlap generally). This means that the selected queens may produce daughter queens and sons. The rationale, too long to be reiterated here, leads to the recurrence formulas:

$$F_g = \beta[F_{g-1}/2 + (2R)^{-1}(1 + F_{g-2})/2] + (1 - R^{-1})\Phi_{g-2}/2 \qquad (5)$$

$$\Phi_g = F_g/2 + (\beta/2)^2[1 - (Rf)^{-1}][(1 + F_{g-2})/2P + (1 - P^{-1})\Phi_{g-2}]$$
$$+ (4Rf)^{-1} + [(1 + F_{g-1})/2R + (1 - R^{-1})\Phi_{g-1}]/4 \qquad (6)$$

with F_g = mean consanguinity coefficient at the gth generation;

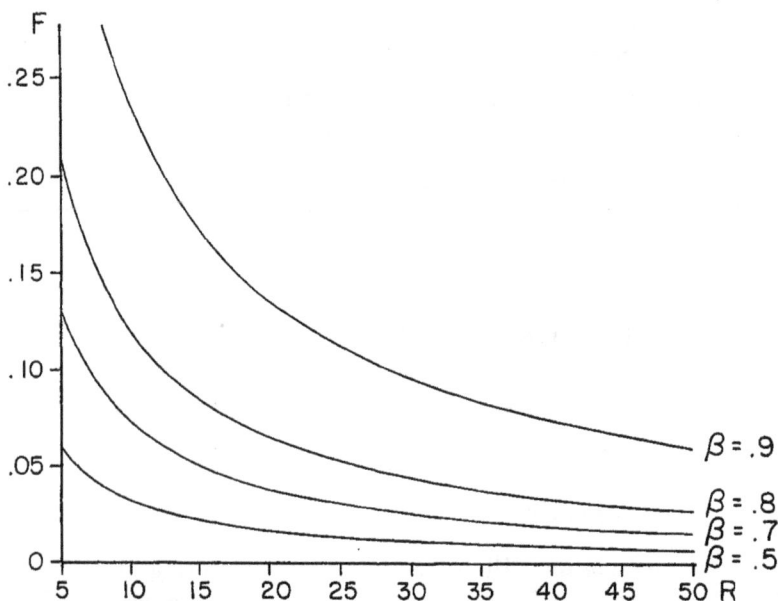

Fig. 3. Representation of the limit consanguinity coefficient (F_∞) as a function of the number of hives (R) and the isolation coefficient (β).

Φ_g = mean kinship coefficient between two queens of the gth generation; and

f = the inverse of the probability that two sisters have the same father ("effective number" of males inseminating a queen).

As F and Φ are increasing functions of g, the number of generations, the limit values F_∞ and Φ_∞ are also upper limits for these coefficients.

Figure 3 shows the variation of F_∞ as a function of R with the following assumption: $P = R$ and $f = 6$. It appears that when β is lower than 0.7 it is sufficient to select 10 queens at each generation to keep the consanguinity coefficient lower than 0.1. When β is higher than 0.7, it is necessary to keep more selected queens.

In a closed population ($\beta = 1$), the reciprocal of the effective population size, which characterizes the speed of convergence of F_g and Φ_g toward 1, is found to be

$$1/N_e \simeq 1/9P + (3f + 2)/9Rf \tag{7}$$

When no selection is assumed, Eq. (7) is equivalent to the one established by Wright (1933) for sex-linked loci.

V. POPULATION GENETICS OF SEX ALLELES

Since honey-bee sex alleles have been discovered (see Woyke, Chapter 4), several studies have been conducted to estimate their number in actual populations. Besides having theoretical interest, there is a practical benefit that comes from the close relationship between the number of sex alleles and the mean brood survival (Woyke, 1976). In order to understand the rationale on which most estimation formulas are based, I shall first develop some theoretical aspects.

A. Theoretical Aspects

The most thorough study in this field was by Yokoyama and Nei (1979), and the following is a summary of their findings.

1. Sex Allele Frequencies

Let $x_{i,t}^F$ and $x_{i,t}^M$ be the frequencies of allele $X_i (i \in [1, n])$ in queens and drones at the tth generation in the population under study. With the assumptions of random mating, equal variability and fertility among heterozygotes (females) and hemizygotes (males), respectively, and no mutation at the sex locus, we have in the next generation:

$$x_{i, t+1}^F = \frac{x_{i, t}^M(1 - x_{i, t}^F) + x_{i, t}^F(1 - x_{i, t}^M)}{2[1 - \sum_{j=1}^n x_{j, t}^M x_{j, t}^F]} \tag{8}$$

and

$$x_{i, t+1}^M = X_{i, t}^F \tag{9}$$

The equilibrium frequency is obviously $x_i^F = x_i^M = 1/n$. Yokoyama and Nei (1979) show that this equilibrium is stable and is approached at an approximate rate of $2/(3n)$ per generation. They then derived two approximate formulas for allele frequency distribution based on two kinds of assumptions.

As an example, we give the first one (the other one being only slightly different and giving negligible differences for practical purposes):

$$\Phi(x) = 4N_e v x^{-1} \exp\{8N_e x/3(1 - J)$$
$$+ [4N_e(2/3 + v) - 1]\ln(1 - x)\} \tag{10}$$

The parameters are

N_e = effective population size
 v = mutation rate per generation
 J = expected homozygosity

Then $\Phi(x)$ is completely defined by two additional conditions:

$$\int_0^1 x\Phi(x)\,dx = 1 \tag{11}$$

$$\int_0^1 x^2\Phi(x)\,dx = J \tag{12}$$

where $\Phi(x)\,dx$ is the expected number of alleles whose frequencies are from x to $x + dx$.

Figure 4 shows the allele frequency distributions for three values of N_e

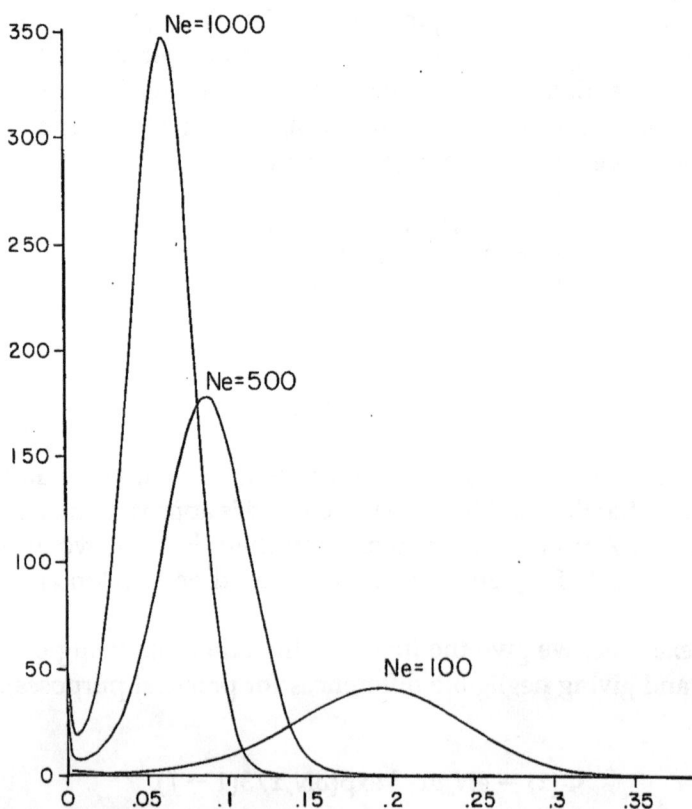

Fig. 4. Sex allele frequency distribution as a function of the effective population size (N_e) with a fixed mutation rate ($v = 10^{-5}$).

(100, 500, 1000) and one value of v (10^{-5}). It can be considered that allele frequencies fluctuate largely around the equilibrium value, especially in small populations.

2. Number of Sex Alleles in Theoretical Populations

If we assume that all allele frequencies are equal, then the number of sex alleles becomes equal to the reciprocal of the expected homozygosity since

$$J = \Sigma_i x_i^2 = n(1/n^2) = 1/n$$

This number has been defined as the "effective number of alleles" (n_e) by Kimura and Crow (1964), who gave the formula

$$n_e = \left(\frac{-4N_e}{3 \ln(v \sqrt{37.5N_e})} \right)^{\frac{1}{4}} \qquad (13)$$

Yokoyama and Nei (1979) introduced another coefficient, namely the "actual number of alleles" (n_a), which takes into account the fluctuation of frequencies. In this case, n_a is computed from the allele frequency distribution function $\Phi(x)$.

$$n_a = \int_{1/2N}^{1} \Phi(x) \, dx \qquad (14)$$

Figure 5 shows the variation of n_a and n_e as functions of the effective population size for the fixed value of 10^{-5} of the mutation rate.

For small populations, n_a and n_e are close. When N_e increases, both numbers increase, n_a always being larger than n_e, as expected. Assuming equal sex allele frequencies leads to underestimating their actual number.

As far as the accuracy of Eqs. (13) and (14) is concerned, computer simulations have shown that, on the average, the allele number is in good agreement with the corresponding theoretical value of n_a but that individual values fluctuate widely.

These numbers are valid only for populations at equilibrium. Page and Marks (1982) have simulated the effect of breeding small closed populations starting with a common value of the sex allele number ($n = 12$). Their results show that equilibrium values are approached quite slowly: e.g., with 50 colonies it takes an average of more than 30 generations to loose only two alleles out of 12 (the corresponding equilibrium value of n_a being 5.5). In that case, the loss of brood variability is negligible.

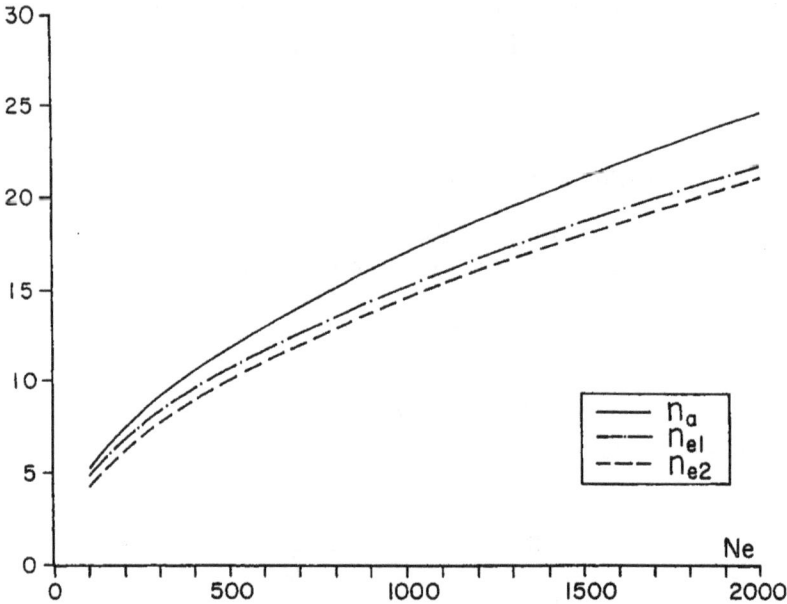

Fig. 5. Expected number of sex alleles in closed panmictic populations as a function of the effective population size (N_e) with a fixed mutation rate ($v = 10^{-5}$); n_a is computed from Eq. (9), n_{e1} is equal to $1/J$ (Eq. 7), and n_{e2} is computed from approximated Eq. (8).

B. Estimations in Actual Populations

Different methods have been applied to actual populations to estimate the number of sex alleles. Basically two kinds of observations have been used: (1) measuring brood viability and (2) counting diploid larval drones.

Mackensen (1951) conducted a diallele cross from inbred lines among which only two had the same geographical origin. Only one cross out of 15 exhibited poor brood viability, and it could be concluded that out of 12 tested sex alleles, 11 were different. This was only a proof of the existence of at least 11 sex alleles in *Apis mellifera*.

The first population study was performed by Laidlaw *et al.* (1956) in an apiary at Piracicaba, Brazil. They inseminated virgin queens from 61 different hives with sperm of drones of a 62nd hive (one drone per queen). Assuming equal allele frequencies and random mating, their experiment led to an estimate of 12.4 ± 3.6 different sex alleles. This method gives an estimate of the effective number of alleles. With a mutation rate of 10^{-5} and population size of 266 hives, the above formulas give theoretical values of $n_e = 11.4$ and $n_a = 12.7$, very close to their estimate.

Woyke's (1976) estimate on Kangaroo Island (Australia) rests on the simple relationship between the number of sex alleles and the average brood viability, when alleles frequencies are supposed to be equal:

$$n = 1/(1 - V) \qquad (15)$$

where n is number of sex alleles and V is mean brood viability.

This direct method looks very attractive. However, it becomes inaccurate for large numbers of sex alleles and requires special precautions in order to separate the loss of viability only due to sex allele homozygosity from other sources of mortality. Woyke found a somewhat low value of 6. A possible explanation is a bottleneck effect experienced by the Kangaroo Island population during its introduction, since only six colonies were imported one century ago.

Adams *et al.* (1977), sexing samples of 1-day-old larvae from 90 colonies of hybrids between A. *mellifera* and A. *m. ligustica,* found a proportion of diploid males of 0.0503 ± 0.0309. They estimated the number of sex alleles to be 18.9 (between 16 and 22 at the 90% level). Considering the population size of 500 colonies and a mutation rate of 10^{-5}, we get theoretical values of $n_e = 16.0$ and $n_a = 17.9$, once more close to the estimate. In the latter case, since the population under study was the result of a cross between two subspecies, it is not surprising that the estimated value of n_e is larger than the theoretical one which assumes equilibrium. Moreover, one must keep in mind that all these estimates are based on the assumption of equal sex allele frequencies. Actual numbers are therefore larger.

VI. CONCLUSION

The current genetic structure of the young *Apis mellifera* species is the consequence of its success in spreading over any environment supporting flowering plant life. The generally low effective population sizes and dispersal capabilities have promoted rapid, strong differentiation into subspecies and, in some subspecies, into ecotypes. The only exceptions concern a few African races that have developed a highly dispersive behavior. Since, like many haplodiploid species, honey bees express a very low level of allozyme variation, we do not find the same differentiation in enzyme genes. Yet, the MDH locus has some discriminating power.

Kerr (1976) estimated that a substantial part of the genome (around 14%) is only expressed in the female sex. These sex-limited genes, which are not submitted to selection in the haploid state, could therefore present a wider variability than those coding for proteins necessary to both sexes, such as basic enzymes revealed by electrophoresis.

Of more practical interest are the recent studies on closed or semiclosed populations with regard to inbreeding and sex alleles. Their combined results provide a valuable guide for devising breeding schemes for genetic improvement or isolation. With other theoretical studies in quantitative genetics (e.g., Chevalet and Cornuet, 1982a), honey-bee breeding has been given new tools.

Among all the troubles it brought, the Africanized bee provided the occasion to throw light upon the poor knowledge of American populations and to promote a new interest in the population genetics of this species in the Americas. Unfortunately, there is little time left to study relic populations such as "Creolian" bees from a taxonomic and evolutionary standpoint.

In other respects, as old American populations originated from several quite distinct European races, it is a little simplistic to characterize "European" versus Africanized bees.

Besides the old but still valuable biometrical techniques, the electrophoretic one, which has not yet provided an alternative for discriminating populations, can still be improved. Likewise, other techniques are still to be explored.

REFERENCES

Adams, J., Rothman, E. D., Kerr, W. E., and Paulino, Z. L. (1977). Estimation of the number of sex alleles and queen mating from diploid males frequencies in a population of *Apis mellifera*. *Genetics 86*, 583–596.

Badino, G., Celebrano, G., and Manino, A. (1982). Genetic variability of *Apis mellifera ligustica* Spin. in a marginal area of its geographical distribution. *Experientia 38*, 540–541.

Brückner, D. (1976a). The influence of genetic variability on wing symmetry in honey bees (*Apis mellifera*). *Evolution 30*, 100–108.

Brückner, D. (1976b). Nachweis von Heterosis für das Merkmal Temperaturregulierung bei der Honigbiene (*Apis mellifera*). *Apidologie 7*, 243–249.

Brückner, D. (1978). Why are there inbreeding effects in haplo–diploid systems. *Evolution 32*, 456–458.

Buttel-Reepen, H. V. (1906). Apistica, Beiträge zur Systematik, Biologie, sowie zur geschichtlichen und geographischen Verbreitung der Honigbiene (*Apis mellifica* L.), ihrer Varietäten und der übrigen Apisarten. *Mitt. Zool. Museum Berlin, Bd. 3, Heft 2*, 119–201.

Chevalet, C., and Cornuet, J. M. (1982a). Etude théorique sur la sélection du caractère production de miel chez l'Abeille. 1. Modèle génétique et statistique. *Apidologie 13*, 39–65.

Chevalet, C., and Cornuet, J. M. (1982b). Evolution de la consanguinité dans une population d'abeilles soumises a selection. *Apidologie 13*, 157–168.

Cornuet, J. M. (1982a). Représentation graphique de populations multinormales par des ellipses de confiance. *Apidologie 13*, 15–20.

Cornuet, J. M. (1982b). The MDH polymorphism in some west Mediterranean honey bee populations. *Proc. Cong. IUSSI 9th (Boulder) Suppl. 3*.

Cornuet, J. M., Fresnaye, J., and Tassencourt, L. (1975). Discrimination et classification de populations d'abeilles à partir de caractères biométriques. *Apidologie 6*, 145–187.

Cornuet, J. M., Fresnaye, J., and Lavie, P. (1978). Etude biométrique de deux population d'abeilles cevénoles. *Apidologie 9*, 41–55.

Crow, J. F., and Roberts, W. C. (1950). Inbreeding and homozygosis in bee. *Genetics 35*, 612–621.

Daly, H. D. (1975). Identification des abeilles africanisées par la technique de la morphologie multivariationnelle. *Proc. Inter. Apiculture Cong. (Apimondia) 25*, 379–381.

Daly, M. V., and Balling, S. S. (1978). Identification of Africanized honey bees in the Western Hemisphere by discriminant analysis. *J. Kansas Entomol. Soc. 51*, 857–869.

Daly, M. V., Hoelmer, K., Norman, P., and Allen, T. (1982). Computer-assisted measurements and identification of honey bees (Hymenoptera: Apidae). *Ann. Entomol. Soc. Am. 75*, 591–594.

Dupraw, E. J. (1965). The recognition and handling of honey bee specimens in non-Linnean taxonomy. *J. Apic. Res. 4*, 71–84.

Ford, E. B. (1940). Polymorphism and taxonomy. *In* "The New Systematics" (J. Huxley, ed.) pp. 493–513. Clarendon Press, Oxford.

Fresnaye, J. (1965). Etude biométrique de quelques caractères morphologiques de l'abeille noire francaise *(Apis mellifica mellifica)*. *Ann. Abeille 8*, 271–283.

Gadbin, C., Cornuet, J. M., and Fresnaye, J. (1979). Approche biométrique de la variété locale d'*Apis mellifera* L. dans le sud tchadien. *Apidologie 10*, 137–148.

Gartside, D. F. (1980). Similar allozyme polymorphism in honeybees *(Apis mellifera* L.) from different continents. *Experienta 36*, 649–650.

Gonçalves, L. (1970). Analise genetica do cruzamento entre *Apis mellifera ligustica* e *Apis mellifera adansonii*. Escolha e analise genetica de caracters morfologicos de cabeça e do torax. Ph.D. Thesis, Faculdade de Medecina de Ribeiro Preto USP.

Kerr, W. E. (1951). Bases para o estudo da genetics dos Hymenoptera em geral e dos Apinae sociais em particular. *Anais da Esc. Sup. de Aq. Luiz de Queiroz 8*, 220–235.

Kerr, W. E. (1967). Multiple alleles and genetic load in bees. *J. Apic. Res. 6*, 61–64.

Kerr, W. E. (1976). Population genetic studies in bees. 2. Sex limited genes. *Evolution 30*, 94–99.

Kerr, W. E., and Bueno, D. (1970). Natural crossing between *Apis mellifera adansonii* and *Apis mellifera ligustica*. *Evolution 24*, 145–148.

Kimura, M., and Crow, J. F. (1964). The number of alleles that can be maintained in a finite population. *Genetics 49*, 725–738.

Laidlaw, H. H., Gomes, F. P., and Kerr, W. E. (1956). Estimation of the number of lethal alleles in a panmictic population of *Apis mellifera* L. *Genetics 41*, 179–188.

Louis, J., and Lefevre, J. (1971). Les races d'abeilles *(Apis mellifera* L.) 1. Détermination par l'analyse canonique. *Biométrie-Praximétrie XII*, 1–41.

Louveaux, J. (1966). Les modalités de l'adaptation des abeilles *(Apis mellifera* L.) au milieu naturel. *Ann. Abeille 9*, 323–350.

Mackensen, O. (1951). Viability and sex determination in the honey bee *(Apis mellifera* L.). *Genetics 36*, 500–509.

Mackensen, O. (1956). Some effects of inbreeding in the honeybees. *Apic. Abstr.* 247/56.

Malécot, G. (1948). "Les Mathématiques de l'Hérédite." Masson et Cie, Paris, France.

Martins, E., Mestriner, M. A., and Contel, E. P. B. (1977). Alcohol dehydrogenase polymorphism in *Apis mellifera*. *Biochem. Gen. 15*, 357–366.

Mestriner, M. A., and Contel, E. P. B. (1972). The P-3 and EST loci in the honey bee *Apis mellifera*. *Genetics 72*, 733–738.

Nunamaker, R. A., and Wilson, W. T. (1981). Comparison of MDH allozyme patterns in the African honey bee *(Apis mellifera adansonii* L.) and the Africanized populations of Brazil. *J. Kansas Entomol. Soc. 54*, 705–710.

Page, R. E., and Marks, R. W. (1982). The population genetics of sex determination in honeybees: random mating in closed populations. *Heredity 48*, 263–270.

Plass, F. (1953). Inzuchwirkung und Heterosiseffekt bei der Honigbiene. Bad Godesberg, Landwirtschaftl. *Informationsdienst 66*, 49–68.

Roberts, W. C. (1961). Heterosis in the honey bee as shown by morphological characters in inbred and hybrid bees. *Ann. Entomol. Soc. Am. 54*, 878–882.

Rothenbuhler, W. C. (1979). Semidomesticated insects: honeybee breeding. *In* "Genetics and Beneficial Organisms" (M. A. Moy and J. J. McKelvey, eds.). Rockfeller Foundation, Bellagio, Italy.

Rothenbuhler, W. C., Kulincĕvić, J. M., and Kerr, W. E. (1968). Bee genetics. *Annu. Rev. Genetics 2*, 413–438.

Ruttner, F. (1952). Alter and Herlsunftder Bienenrassen Europe. *Oest. Irulser 2*, 8–12.

Ruttner, F. (1968). L'amélioration du cheptel. *In* "Traite de Biologie de l'Abeille (R. Chauvin, ed.), pp. 181–236. Masson et Cie, Paris, France.

Ruttner, F. (1975). African races of honey bees. *Proc. Inter. Apic. Cong. (Apimondia) 25*, 325–344.

Ruttner, F. (1981). Taxonomie des abeilles mellifères de l'Afrique tropicale. *Proc. Inter. Apic. Cong. (Apimondia) 28*, 275–281.

Ruttner, F. (1982). Taxonomie et évolution des races d'abeilles du bassin méditerranéen considéré d'un point de vue économique. *Bull. Tech. Apic. 9*, 61–68.

Ruttner, F., Tassencourt, L., and Louveaux, J. (1978). Biometrical statistical analysis of the geographic variability of *Apis mellifera* L. I. Material and methods. *Apidologie 9*, 363–382.

Smith, F. G. (1961). The races of honeybees in Africa. *Bee World 42*, 255–260.

Sylvester, H. A. (1976). Allozyme variation in honeybees (*Apis mellifera* L.). Ph.D. dissertation, University of California, Davis, Calif.

Sylvester, H. A. (1982). Electrophoretic identification of Africanized honeybees. *J. Apic. Res. 21*, 93–97.

Tomassone, R., and Fresnaye, J. (1971). Etude d'une méthode biométrique et statistique permettant la discrimination et la classification de populations d'abeilles (*Apis mellifera* L.) *Apidologie 2*, 49–65.

Woyke, J. (1976). Population genetics studies on sex alleles in the honey bee using the example of the Kangaroo Island bee sanctuary. *J. Apic. Res. 15*, 105–123.

Wright, S. (1933). Inbreeding and homozygosis. *Proc. Natl. Acad. Sci. U.S.A. 19*, 411–420.

Yokoyama, S., and Nei, M. (1979). Population dynamics of sex-determining alleles in honey bees and self-incompatibility alleles in plants. *Genetics 91*, 609–626.

CHAPTER 10

Reproduction and Mating Behavior

GUDRUN KOENIGER

I. INTRODUCTION

Honey bees are eusocial insects. As such, they have "a reproductive division of labor in the colony, that is, a worker caste cares for the young of the reproductive caste." Also, "there is an overlap in generations so that offspring assist parents" (Wilson, 1972). Thus, individuals in a honey-bee colony are replaced in succession without dissolving the colony. Furthermore, the honey-bee queen is not able to raise brood without workers.

Honey-bee reproduction must be considered as two separate processes: (1) *replacement and multiplication of the members in the colony* when transmission of genes to a new reproductive generation only occurs through mating drones or when the old queen is replaced, and (2) *multiplications of the colony through swarming* when the old and 1 to n young queens each found new colonies with a part of the workers of the old colony. The remaining workers stay with the last young queen at the old nesting site. In the second case genes are transmitted to a new reproductive generation through both queens and drones. (Fig. 1).

This chapter is limited to mating behavior and reproduction within the colony. Most of the experiments described here were carried out with the European subspecies *Apis mellifera carnica* and *A. m. ligustica*. There are many large differences between the subspecies of honey bees (Ruttner,

replacement and
  multiplication within
  the colony

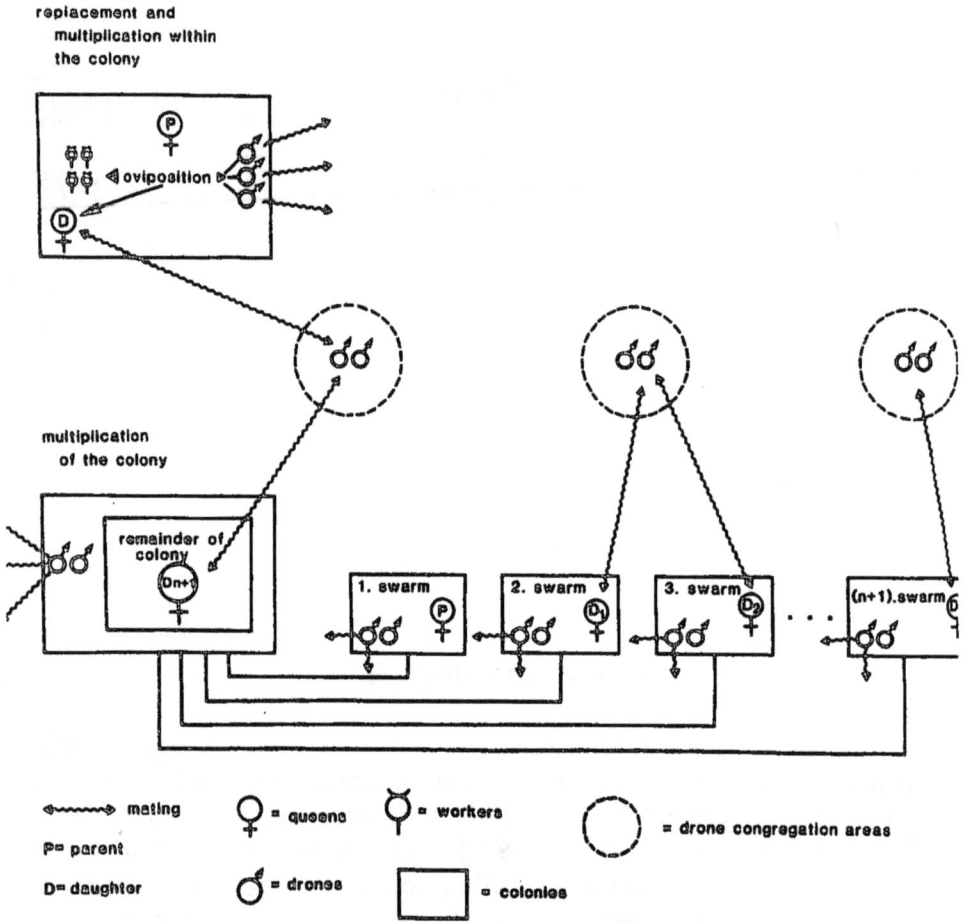

multiplication
of the colony

Fig. 1. Reproduction of *Apis mellifera*. [Drawing by Sandra Kleinpeter.]

Chapter 2), but the central features of reproduction and mating behavior seem to be very similar for all subspecies.

II. THE PREMATING PERIOD

A. Queens

Shortly after emergence the queen can be observed moving about the brood area of the nest. She will destroy sealed queen cells she encounters if the colony is not preparing to swarm. Worker bees apparently do not show reactions to newly emerged queens during the first few hours.

However, from noon of the first day, some workers start to follow her. They lick and touch her with their antennae and front legs, and three to four times a day they feed her. At the same time, the first antagonistic behavior is seen; the workers push and bite the virgin queen, pull her legs, or even "ball" her. "Balling" involves enough bees biting and holding a queen that only a ball of bees can be seen. The queen runs from these antagonistic activities, and when she is caught she starts to produce a peculiar sound by vibration of the thoracic muscles — "she pipes" (Simpson, 1964). Piping results in a "stop reaction" of the workers (Little, 1962). The aggressive treatment of the queen by the workers always has a peak in the early afternoon hours (Hammann, 1956). Since this is when queens normally take mating flights, one gets the impression that the workers' aggressive activities result in chasing the virgin queen out for a mating flight (Huber, 1792). There is experimental evidence for this hypothesis. A queen kept with workers not older than 3 days was not attacked but also did not leave the hive (Hammann, 1956).

Within 3–5 days after emergence the queen displays a very specific activity. Simultaneously, she spreads her sting chamber and vibrates her abdomen. Normally, this occurs when the queen is quiet on the comb. The activity stops the aggressiveness of workers and they form a court around the young queen. The spreading activity increases daily until a mating flight takes place (Alber et al., 1955; Hammann, 1956).

As these behavioral changes take place, the glands of the young queen develop. The size of the mandibular glands increases and they begin producing queen substances (Butler, 1960; Butler et al., 1962; Pain, 1961), which are composed of several fatty acids. The relative concentrations of the acids are age-dependent. For example, the amount of 9-oxodecenoic acid (9-ODA) goes up from 7.2 μg when queens are 1–2 days old to 132.5 μg when they are 5–10 days old (Butler and Paton, 1962). During the same period the concentration of 10-hydroxy-2-decenoic acid (10-HDA), a "worker-like" signal, decreases (Crewe, 1982). Also, the size and activity of the tergite glands both increase (Renner and Baumann, 1964). The odor of the queen becomes strongly aromatic and is clearly different from that of an egg-laying queen (Hammann, 1956). This strong odor seems to be produced "at least partly" by the tergite glands, whose maximum activity occurs at mating time (Renner and Baumann, 1964).

The development of the ovary in many insects is controlled by the median neurosecretory cells of the brain (located in the pars intercerebralis) and by the corpora allata (CA), which produce juvenile hormone (JH) (Engelmann, 1970). In the maturing queen, too, the neurosecretory production is increasing and reaches its maximum on day 6 (Fig. 2a). The neurosecretion is transported through the axons to the corpora cardiaca (CC) and to the CA, where it seems to be stored. The activity of the CA decreases during this time (Herrmann, 1969). This decrease of CA activity in queens differs from both other insects and also from worker bees, where JH, produced in the CA,

a

b

Days

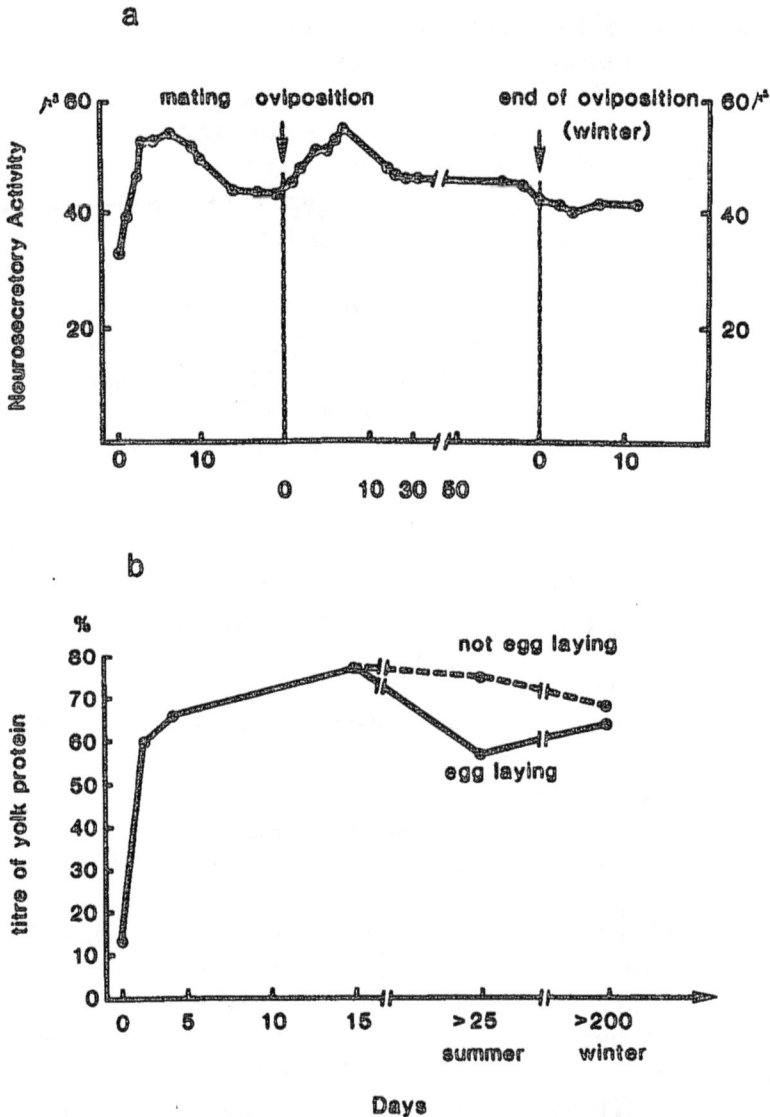

Fig. 2. Activity of (*a*) the neurosecretory cells and (*b*) the blood yolk protein titre in relation to the age of the queen. [Figure 2*a* is redrawn after Herrmann (1969), with permission. Copyright 1969 by Deutscher Imkerbund eV. Villip b. Bad Godesberg. Figure 2*b* is redrawn after Engels and Fahrenhorst (1974), with permission. Copyright 1974 by Springer Verlag/Heidelberg.]

stimulates the synthesis of yolk protein (vitellogenin) (Rutz *et al.*, 1976). The blood titre of yolk protein in the queen honey bee increases during the first 2 weeks (Fig. 2*b*) (Engels and Fahrenhorst, 1974), but no correlation is evident between the amount of vitellogenin and the JH titre (Engels, 1981; Fluri *et al.*, 1981). Maturing of eggs starts only after mating.

The first flight usually occurs on day 5 or 6 when the activity of the glands and neurosecretory cells as well as the vitellogenin titre indicate that the queen is sexually matured. Then the queen runs over the combs while whirring her wings. She becomes attracted to the light, runs out the hive entrance, and performs an orientation flight which takes 2–5 min. Shortly after, she will frequently start again for mating. The workers do not follow her, but many gather around the hive entrance exposing their scent glands.

B. Drones

For the first 3 days after emergence, drones are fed by workers. When they are only 1 day old they get less food than older drones; probably they are still unable to beg for food effectively (Free, 1956). Until they are 5 days old, they stay within the brood nest (Örösi-Pal, 1959). They spend 70–80% of their time inactively, only occasionally cleaning themselves or wandering over the comb and soliciting food. Mainly, they are fed by nurse bees though the food originates from the honey sac rather than the pharyngial glands (Free, 1957; Mindt, 1962). As the drones grow older, they move from the brood nest to the honey combs and start to feed themselves from the cells. Workers take more care of young drones. More workers clustered on cages with drones 6–7 days old than on cages with drones 3 weeks old (Jaycox, 1961). Also, workers still feed young drones when the older ones are already driven out of the colony in the fall.

The genetical aspects of spermiogenesis are completed by the time drones emerge from the cell. In adult drones 3–8 days old, spermatozoa migrate from the testes into the vesicula seminalis, where they undergo several days of final maturation. They are positioned in the vesicula with their heads against the glandular epithelial walls (Kurennoi, 1953; Mindt, 1962). These cells secrete by terminal constrictions. In this way, the epithelium shrinks while the cavity of the vesicula seminalis enlarges (Bishop, 1920). Eventually the contents of this organ consist of a small amount of lymphlike fluid, with about $11 \cdot 10^6$ spermatozoa. Mucus production in accessory glands starts only after the emergence of the male and is finished at the age of 6–7 days. Sperm and mucus then stay in the *vesicula seminalis* and in the mucous glands, respectively, until mating.

When young drones were kept isolated in cages, the migration of spermatozoa proved to be independent of food but was influenced by temperature (Mindt, 1962). However, drones in small colonies kept in cold ambient temperatures showed a normal amount of maturing spermatozoa (Effinowicz, 1978). For normal production of mucus, drones had to be kept with nurse bees (Mindt, 1962).

The majority of drones take their first flight at the age of 6–8 days, and the reported range spans from 4 to 14 days (Howell and Usinger, 1933;

Kurennoi, 1954; Oertel, 1956). These reports also indicate that the daily flight time, normally between 1 p.m. and 4 p.m. standard time, also varies widely. Taber (1964) showed that daily flight time depends on photoperiod. Differences in daily flight time also result from variations in weather, climate, and subspecies.

Before flying, drones feed and, paying special attention to the antennae and eyes, clean themselves. The first flights last between 6 and 16 min and serve for orientation. Mating flights last from 25 to 57 min and are made by drones more than 12 days old (Witherell, 1971). Only then have the spermatozoa matured.

III. MATING BEHAVIOR

Until the beginning of the eighteenth century, most people thought that queens did not mate because nobody ever observed mating. Rather, it was thought that eggs were fertilized by the drones within the cells. Others claimed that mating occurred in the hive. Janscha (1775) described the orientation flight of the queen, her mating with drones high in the air, and her return with something white in her sting chamber. The first experiments concerning mating probably were done by Huber (1792). He found that queens produced workers only when they could fly out of the hive. He described the "mating sign" as a part of the drone's copulatory apparatus stuck in the open sting chamber of the returning queen.

Observations of natural mating are still rare because mating takes place high in the air. Nevertheless, we now have substantive evidence on this complex and important aspect of bee biology.

A. Environmental Factors

Queens and drones always leave their hives for mating. This increases the risk that a colony may loose their only queen. It is not amazing then that queens are very sensitive to weather conditions and only fly on fine days during the warmest period of the day, which is usually in the early afternoon. During experiment with 64 queens on a Mediterranean island, no successful matings were observed when the maximum temperature was below 20°C. The same was true when there was a strong wind or many clouds, even though the temperature was sometimes above 20°C (Fig. 3, April 15 and 16). Some queens tried to fly but returned after a few minutes. Even wind velocities over 23 km/hr decreased the number of successful matings (Alber et al., 1955). Drones show similar behavior, but do fly under less favorable conditions.

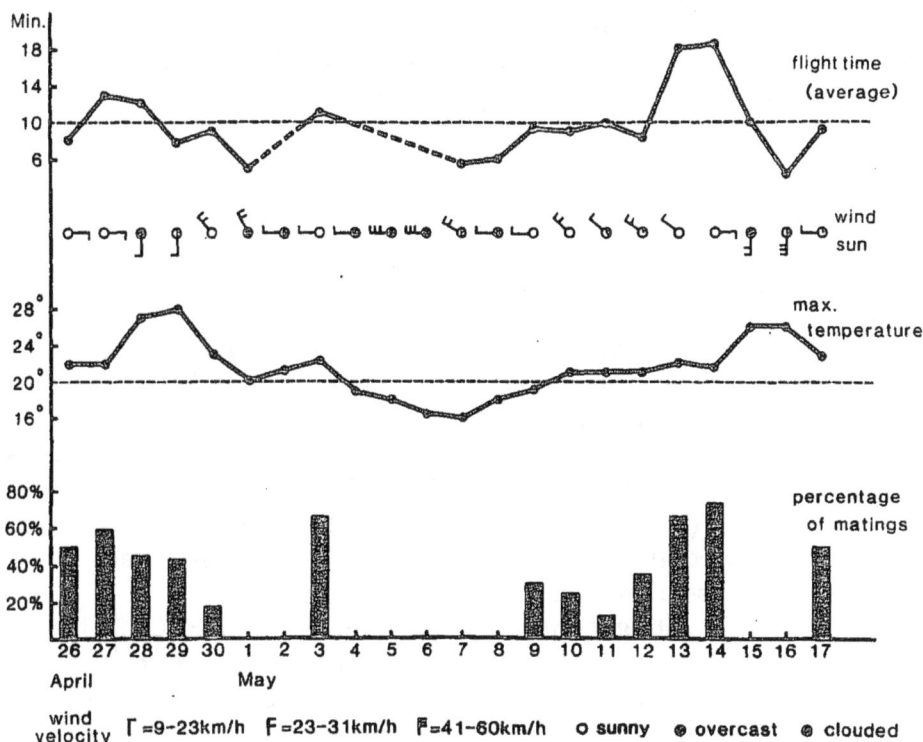

Fig. 3. Frequency of matings and weather conditions during the mating of 64 queens (*Apis mellifera carnica*). [Redrawn after Ruttner (1955), with permission. Copyright 1955 by Österreich. Imkerbund.]

B. Flight Range

The flight range of queens and drones cannot be observed directly but can be inferred from certain experiments. Ruttner and Ruttner (1972) distributed 250 genetically marked virgin queens and three groups of genetically marked drones through a bee-free alpine valley. The comparison of the worker progeny of these queens with the composition of the drone populations showed that only a few matings occurred with drones from the vicinity of the queen's colony. The average distance between apiary and mating place was more than 2 km, and the maximum distance was 5 km. These results are similar to those of Woyke (1960), who found queens had better mating success when drones were placed 2.5 km away rather than in the same apiary. The flight range of drones is about 6 km, with flights of 5 km being common. Thus, drones can range over an area of roughly 78 km^2 (Ruttner and Ruttner, 1972). In an area containing only experimental bees, the greatest mating distance was more than 16 km (Peer, 1957), but only three of 12 queens from droneless nuclei mated at this distance and only after 4 weeks. Distances of 5.7 and 12.8 km resulted in 89 and 67% of mated queens, respectively.

C. Duration and Number of Matings

Mating flights of queens normally last between 5 and 30 min (Oertel, 1940) and average 14 (Taber, 1954), 18 (Roberts, 1944), or 25 min (Woyke, 1960). The number of matings was estimated by (1) counting spermatozoa in the oviducts of queens returning from mating flights and comparing the count with the number of spermatozoa produced by one drone (Trjasko, 1956; Woyke, 1955, 1960), (2) using genetically marked queens and drones and evaluating the progeny (Alber et al., 1955; Peer, 1956; Roberts, 1944; Taber, 1954), and (3) determining the distribution of the frequency of diploid drones in a population (Adams et al., 1977). The results range from seven to 17 successful matings before oviposition.

The number of mating flights is usually from one to three. Of 303 once-mated queens, 63% performed further flights and 38% returned with the same average volume of semen (seven to eight drones) as from the first. An additional 6% mated successfully a third time (Woyke, 1960). If queens do not meet any drones, they fly frequently for at least 3 weeks. According to Woyke (1966), the concentration of spermatozoa in the spermatheca rather than the absolute number of spermatozoa explains the behavior of queens during the mating period. With an average concentration of $4.6 \times 10^6/$ mm^3, queens start for a second flight, whereas with a concentration of $5.9 \times 10^6/$mm^3 they do not fly again.

D. Drone Congregation Areas

The probability of a queen meeting drones within a specific time is not entirely dependent on the general concentration of drones. Considering the normal flight range and the duration of flight, mating with only one drone is a much rarer occurrence than would be expected statistically. Normally, queens returning from flights are either unmated or mated by several drones (Ruttner, 1966).

Experiments were carried out to find more about the distribution of drones. The method of Gary (1963) which uses a balloon with tethered or caged queens is common to these studies. Drones begin to visit queens when they are raised 6–10 m above the ground. The maximum number of drones appears at 15–25 m in Austria. In warmer climates, drones often follow queens readily when they are only 2–3 m above the ground. The daily height of the flight is greatly dependent on meteorological conditions, but in specific conditions drones stay at a rather constant altitude and pay no attention to queens offered at a different height (Butler and Fairey, 1964; Gary, 1963; Ruttner and Ruttner, 1963). Thus, the vertical distribution of drones is not random.

The horizontal distribution of drones is not random either. Abundant drones are attracted by queens in some areas but few or none at all in others. This phenomenon occurs in various countries and continents [Africa: personal reports; Austria: Ruttner and Ruttner (1963); France: Jean-Prost (1957); United States: Zmarlicki and Morse (1963); Gary (1963); Venezuela: Otis and Taylor (1982)]. Areas of drone congregation remain the same for years. Some have been known to be constant for more than 15 years. The drones congregate regardless of the presence of a queen. The limits of the congregation areas fluctuate daily to some extent. The usual size lies between 30 and 200 m in diameter (Ruttner, 1966). Little is known concerning the origins of these "drone congregation areas." Light distribution might be an influence (Praagh et al., 1976). The flight direction of drones starting from the apiary is influenced by structural characteristics of the near and far horizon in a hilly district. They prefer the direction of depressions in the horizon (Ruttner and Ruttner, 1966). Drones caught in congregation areas always come from both different colonies and different apiaries (Ruttner and Ruttner, 1966). By this behavior, assuming that queens use the same information for finding a congregation area, several matings with drones of different origin can occur in a short time.

In summary, the immensely complex mating-flight behavior can be understood to result in queens mating with several unrelated drones. Such matings must occur away from the queen's hive, which increases flight times, survival risks, and energy expenditures. Congregation areas reduce these costs and risks. However, they are still substantial and must be counterbalanced by fitness increasing benefits. Inferences from present knowledge indicate that these benefits are those that result from reduced chances of inbreeding.

E. Drone Pursuit

Drones attracted by a fixed queen form a comet-like swarm numbering between 50 and 300 under good conditions. The swarm instantly forms and suddenly disintegrates as if one drone was being chased by the others. The swarm always orients from downwind of the queen. Also, the drones fly below the queen's level, fixing her with the dorsal–frontal part of the eye in an angle between 0 and 40° (Gary, 1963; Praagh et al., 1980). Thus, the pursuit seems to be influenced by visual as well as olfactory cues.

The primary source of the pheromones which attract drones is the oily secretion of the mandibular glands (Gary, 1962; Pain and Ruttner, 1963), which are the "queen substances" described by Butler (1954). The most effective compound of the pheromone is 9-oxodec-2-enoic acid (9-ODA; Gary, 1962). A secondary component, 9-hydroxodec-2-enoic acid (9-HDA), does not show a synergistic effect (Boch et al., 1975). The range of active attraction is more than 50 m (Butler and Fairey, 1964). The secretion of the abdominal tergite glands also is effective, but only in addition to 9-ODA and within short distances (less than 30 cm). It increases the duration of pursuit and copulatory activity (Renner and Vierling, 1977). In good weather, drones in a congregation area pursue butterflies, dragonflies, or stones thrown high in the air (Gary, 1963). Visual stimuli from moving objects with high contrast against the sky apparently induce an initial drone pursuit (Gerig, 1971). However, drones follow these objects for only a moment unless the pheromones are added.

Drones must chase and grasp a *flying* queen, in contrast to workers, which normally look for flowers. The eyes of drones are larger than those of the workers. Drones have an average of 9973 ± 56 eye facets ($n = 4$), while workers have only 4752 ± 35 facets ($n = 4$). (Praagh et al., 1980). The facets of drones are also generally larger; the largest is found in the dorsofrontal region that is used to fixate the flying queen. Because of the largeness of the eye, the angle of divergence is still very small; the smallest horizontal angle ($2\Delta\psi^h = 1.0°$) is in the dorsal region. There, the binocular field covers 26° starting at a distance of 0.1 mm from the head in the vertical plane (Seidl, 1980).

The size of the eye, its subdivision into specialized regions, and the small angles of divergence indicate that drones have good vision. This, together with the observation that drones usually fix the queen from a distance of 2–5 cm from an angle between 0 and 40° below the queen, supports the view that visual as well as olfactory guidance plays an important role in the mating behavior of the drone.

F. Copulation

Actual mating has only been observed with queens glued on a support in front of a camera installed about 10 m high in a drone congregation area. For mating to occur, the sting chamber of the queen must be held open artificially (G. Koeniger et al., 1979). When a drone is following a glued queen, his hind legs are hanging and spread slightly to the side and back. As the drone approaches the queen, these legs are bowed, but before he touches her with them, he seizes her dorsally with the distal parts of the tarsi of the first and second pairs of legs (Fig. 4). Within a split second, all six legs

Fig. 4. The drone seizes the queen first with the tarsi of the first and second pairs of legs. [After G. Koeniger et al., (1979), with permission. Copyright 1979 by International Bee Research Association.]

seize the queen in a dorsoventral clasp with the first pair of legs usually touching the back of the third abdominal segment, the second pair touching the fifth, and the hind legs always grasping the tergites of the fourth and fifth segments where they overlap the sternites (Fig. 5). Hind leg grasping is done with metatarsi that are covered with special adhering hairs (Ruttner, 1975). Palpation has never been observed. After mounting, the abdomen of the drone is curled under and mating occurs at once when the sting chamber is open.

In experiments with wooden queen models, sting-chamber openings with diameters of 2, 3, and 4 mm were most successful in causing matings (Gary and Marston, 1971). Drones possess a "genital sense field" (Fig. 6), which receives tactile stimuli (Ruttner, 1962; Schlegel, 1967). Only 300 msec after contacting the open sting chamber, a drone becomes paralyzed when the membranous endophallus is everted. This eversion results from the pressure of hemolymph, which is pushed back by contractions of strong abdominal muscles. The drone swings back and, while completing the eversion, injects the semen into the oviducts of the queen. After this, the mating sign, which consists of mucus and the chitinized plates of the endophallus, is deposited. A few seconds after swinging back, the drone separates from the queen with an explosive snap (Gary, 1963) and leaves the mating sign closing the sting chamber. The next drone is able to insert his

Fig. 5. After having seized the queen in a dorsoventral clasp, the drone everts his endophallus into the queen and becomes paralyzed. [After G. Koeniger *et al.* (1979), with permission. Copyright 1979 by International Bee Research Association.]

endophallus into the queen ventral of the mating sign. The mating sign gets tangled at the hairy basal part of the next drone's endophallus, and completion of the eversion pulls out the old mating sign (Fig. 7) and leaves the new one. In this way, several drones can mate a queen one after another (Trjasko, 1957; G. Koeniger, 1981).

During copulation, some active cooperation of the queen is necessary. The sperm can only be injected into the oviducts when the queen folds back the valvula vaginales (Fyg, 1966; Laidlaw, 1944). Also, drones seldom separate from queens whose sting chambers are held open artificially: such eversions stop half way. Some active movement of the queen's abdomen must aid the eversion.

G. Control of Natural Mating

There are still too many facts missing to design a plan for controlled natural mating. The readiness of queens and drones to mate is determined by their treatment from workers, hormonal conditions, and many external conditions. Mating behavior rarely occurs in confinements, even though some tests have been made in huge cages (Harbo, 1971). Also, queens prepared for photography only rarely opened their sting chambers and allowed drones to copulate even though all other conditions seemed to be optimum. At present, natural mating can only be controlled at isolated mating stations (Laidlaw and Page, Chapter 13; Ruttner, 1983).

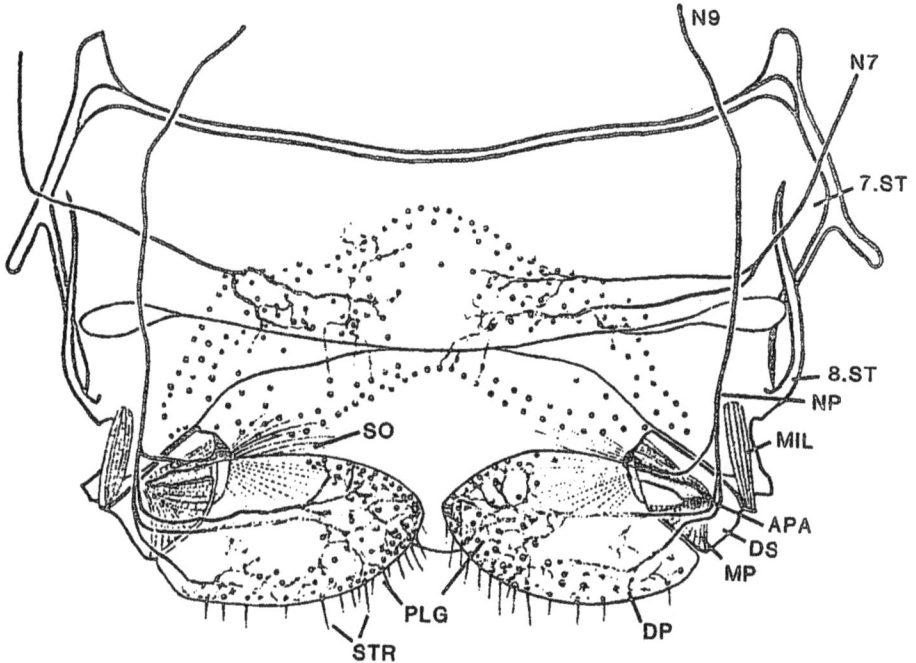

Fig. 6. Genital sense field of the drone. APA, basal apodeme of aedeagus; DP, penis valves; DS, parameralis plate; MP, parameralis muscle; MIL, inter segmental muscle (sternite 8); N7–N9, nerves of segments 7 and 9; NP, parameralis nerve; PLG, plexus of penis valves; SO, sensory hairs of the parameral sensory organ; STR, sensilla trichoidea; ST, sternit. [From Ruttner, (1962), with permission. Copyright by Institut National de la Recherche Agronomique.]

IV. SPERM MOVEMENT IN THE QUEEN

A. Sperm Transfer

After one mating flight the oviducts contain an average of $87 \cdot 10^6$ spermatozoa. The oviducts are empty 24 hr later. After the start of oviposition an average of $5.3 \cdot 10^6$ spermatozoa are found in the spermatheca (Woyke, 1960).

The transfer of the spermatozoa into the spermatheca is a complex process, involving the muscles of the queen, the fluid of the spermatheca and its gland, and individual movements of spermatozoa. Spermatozoa orient to the wall of the oviducts. By contraction of the oviduct and the abdominal muscles, some spermatozoa are transported to the orifice of the spermaduct although the majority are discarded (Ruttner, 1956). When Ruttner and Koeniger (1971) paralyzed the muscles of queens, only a few spermatozoa reached the spermatheca and the oviducts were still full with sperm. Spermatozoa are able to actively migrate through the spermaduct from its ori-

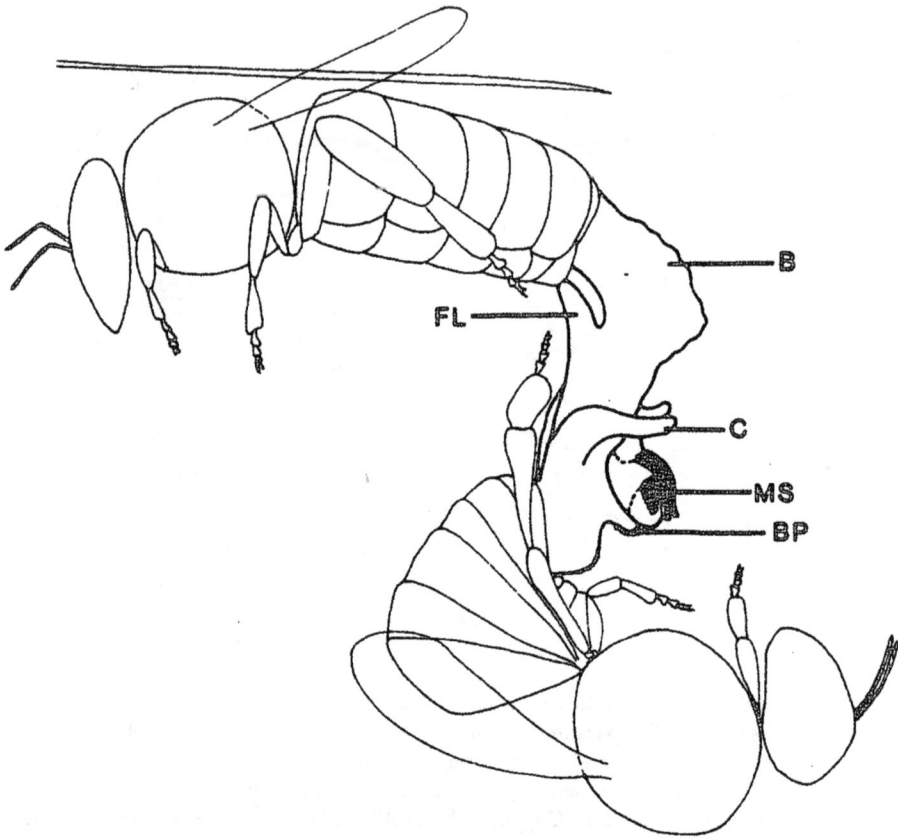

Fig. 7. Position of the mating sign from the preceding drone on the everted endophallus. B, bulb; BP, hairy basal plate; C, cornua; FL, fimbriate lobe; MS, mating sign with chitinized plates. [From G. Koeniger (1983), with permission. Copyright by Delta-Verlag M. Buske GmbH.]

fice. Speed of migration is fastest in ducts less than 35 μm in diameter (the diameter of the spermaduct ranges between 10 and 25μm) (Gessner and Ruttner, 1977). The sperm pump (Bresslau, 1905) supports migration but the mechanism is not fully understood. After introducing plastic granulae (diameter = 7 μm) into the median oviduct of a normal young queen, 0.3% of the number predicted from an assumption that all movement is passive reached the spermatheca (Gessner and Ruttner, 1977). Frozen (nonfunctional) semen does not enter the spermatheca (Harbo, 1976).

The spermathecal gland also must have some influence on the migration of spermatozoa. Queens, whose glands were extirpated at the age of 1–4 days and who were mated naturally about 2 weeks later, stored less than 10% of the normal number of spermatozoa (G. Koeniger, 1968). Also, the liquid (composition not known) within the duct must be important. A normal number of spermatozoa reached the spermatheca when inseminations

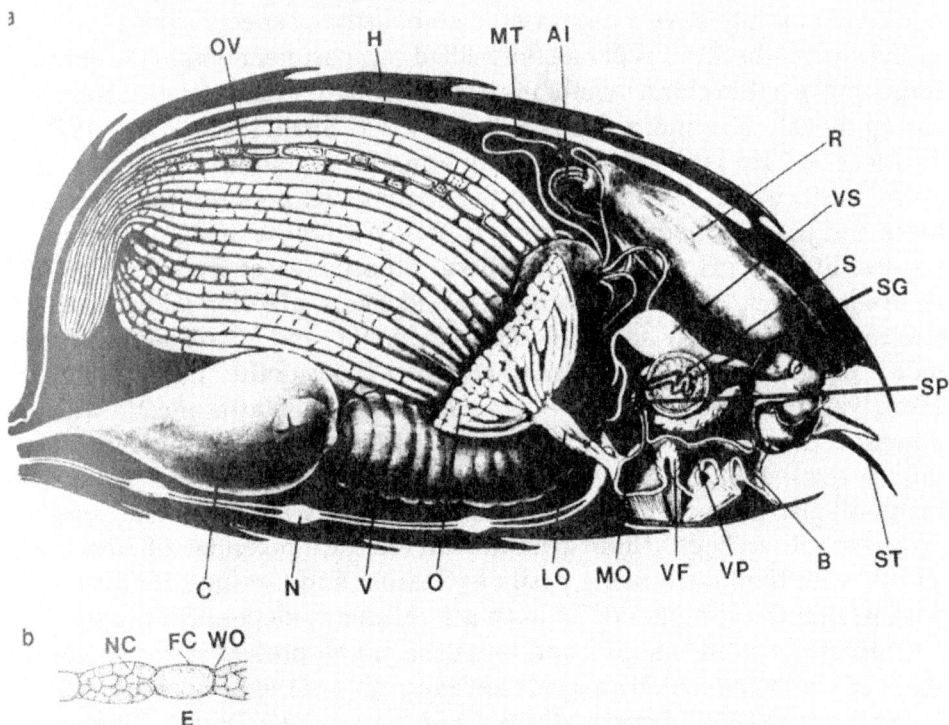

Fig. 8. (a) The reproductive systems and other internal organs of the queen honey bee. (b) The detail of an ovariole. (a) AI, anterior intestine; B, Bursa copulatrix; C, crop; H, heart; LO, lateral oviduct; MO, median oviduct; MT, malpighian tubules; N, nerve cord; O, ovary; OV, ovariole; R, rectum; S, spermatheca with tracheal net; SG, spermathecal gland; SP, spermathecal pump; ST, sting; V, ventriculus; VF, valve fold; VP, vaginal passage; VS, venom sack. (b) E, egg; FC, follicle cells; NC, nurse cells; WO, wall of ovariole. [After Ruttner, (1960), with permission. Copyright by Ehrenwirth Verlag München.]

were done immediately following the mechanical puncturing of the spermatheca. Less than 1% of the normal number did so when the spermathecal fluid was removed following puncturing and the spermatheca was refilled with the hemolymph (Gessner and Ruttner, 1977). The number of spermatozoa reaching the spermatheca is also influenced by external conditions, such as temperature and number of worker bees attending the queen (Foti, 1973; Woyke and Jasinski, 1973, 1982).

B. Sperm Storage

The queen is able to store spermatozoa in her spermatheca for several years. Their fertilizing capacity is preserved even though they are not the

queen's own body cells. They are kept under special physiological conditions and possibly have a depressed metabolism for energy saving.

The spermatheca is a separate fluid-filled compartment within the queen
consisting of a thin chitin membrane surrounded by a single layer of colum-
nar epithelial cells and a dense tracheal net (Dallai, 1975; Poole, 1970;
Ruttner et al., 1971). A two-branched spermathecal gland is attached to the
dorsal surface and discharges into the spermaduct between the spermatheca and the spermathecal pump (Fig. 8a). Within the first 3 days of the
queens' life the pH of the spermathecal fluid increases from 7.4 to 8.6,
where it remains. The spermathecal gland has a pH of 7 (Gessner and
Gessner, 1976; Lensky and Schindler, 1967). Na$^+$ and K$^+$ are significantly
more concentrated in the spermatheca than in any other tissue or fluid.
Moreover, fresh semen becomes reversibly immobile after mixing with a
diluent of the same Na$^+$ and K$^+$ concentration or with diluted spermathecal
fluid (Verma, 1973). When spermathecal fluid is mixed with crushed spermathecal glands, sperm motility is induced (Lensky and Schindler, 1967).
The spermathecal epithelium generates an electrical potential difference of
21 mV with the lumen being positive (Gessner and Gessner, 1976). They
suggest that the secretion of K$^+$ is an active, energy-dependent process.

Glucose, fructose, sucrose, and trehalose are all present in the spermatheca of virgin and mated queens. However, most of these sugars occur at a
lower concentration in spermathecal fluid than in hemolymph. There is a
high concentration of trehalase in the spermatheca. This may provide an
additional or alternative source of energy for spermatozoa, especially when
they are activated for fertilization (Alumot et al., 1969). Histochemical analysis shows that succinate dehydrogenase and acid phosphatase are also in
the spermatheca (Molodyak and Belyaeva, 1977).

When the spermathecal glands of mated queens are removed, the turgor
of the spermatheca is diminished and the queens lay unfertilized eggs.
Spermatozoa remain motile at least for 77 days after spermathecal gland
removal. However, they must be changed in some way since when they are
reinseminated into virgin queens less than 2% of the normal number reach
the spermatheca. The removal of 50–70% of the tracheal net results in a
degeneration of the epithelial cells and the immobility of spermatozoa in
diluents. The turgor does not change (G. Koeniger, 1970; Poole, 1972).
There are no data on the metabolic rate of the spermatozoa in the spermatheca, but many from studies of spermatozoa in vitro. Reviews are given by
Taber (1975) and Verma (1974).

The distribution of semen in the spermatheca was determined by using
genetic markers. The progeny are not homogenous. Rather, the eggs are
fertilized by each semen type at rather constant frequencies. This is only
possible when spermatozoa are mixed while in the spermatheca (H. H.

Laidlaw, Jr., personal communication, 1983; Moritz, 1983). In contrast, Taber (1955) and Woyke (1964b) found evidence that semen of the same type must cluster within the spermatheca.

V. EGG MATURATION

When a queen returns from her mating flight, she receives much more physical contact from workers than ever before. Sometimes the queen is attacked when she enters the hive and is even balled. However, later the workers lick her, touch her with their antennae, and form a court. They often feed her with protein-rich food which is important for the development of eggs (Allen, 1955; Rutz and Lüscher, 1974). The queen often bends her abdomen, which aids the filling of the spermatheca (Ruttner, 1956). During these movements the mating sign is removed; often this is done with the help of workers. If the queen needs more mating flights she normally starts again the same or the following day. The first eggs are deposited 1–3 days after the last successful mating.

A. Regulating Factors

After having sufficient matings the activity of the neurosecretory cells in the pars intercerebralis increases, as does the delivery of the stored neurosecretion (Fig. 2a). This process releases egg production (Biedermann, 1964). The neurosecretion can be induced artificially by two or three CO_2 narcoses (Herrmann, 1969; Mackensen, 1947).

The factors which induce the intensified neurosecretion normally must occur during the mating, since unmated queens do not start to lay eggs before they are between 40 and 60 days old (Koeniger, 1976; Mackensen, 1947). The mechanism involved in the stimulation is not well understood. In insects generally, a wide range of stimuli are known. They can be purely mechanical or chemical, or both; they can be caused by the spermatozoa, the male accessory glands, the copulatory organ, or several other agents. For the queen bee only some factors have been experimentally examined, such as the influence of active flight, the mating sign, the insemination, and different contacts between drones and queens.

The active flight of the queen has little influence on the start of oviposition. Queens which do not meet drones during their flights fly out frequently. Yet, their ovaries remain undeveloped at least to the age of 23 days. When the active flight is omitted from the mating process by fixing queens to a wire most of them start to lay eggs a few days later (G. Koeniger, 1981). Also, queens permitted to fly before being instrumentally inseminated (with

CO_2 narcoses) do not lay eggs earlier than queens without flights (Ebadi and Gary, 1980).

The transfer of spermatozoa and the mating sign are not very important oviposition stimuli. During the mating of free-flying drones with queens fixed to wires, the eversion of the endophallus is mostly incomplete so only few or no spermatozoa are injected into the queen and no mating sign occurs. However, the large majority of queens mated in this way lay eggs before they are 23 days old (G. Koeniger, 1981). Instrumental insemination without CO_2 narcosis does not trigger oviposition (Mackensen, 1947).

The perception of the stimuli leading to neurosecretion must mainly occur in the sting chamber and bursa copulatrix. Only when the endophallus of active drones is inserted into them will a high percentage of queens begin to lay eggs. External contacts during mating do show a slight effect (18% of the queens stimulated in this way started oviposition).

There are some factors which when tested in isolation do not show any effect but which seem to influence oviposition when they occur in addition to the main factors. Ebadi and Gary (1980) found a significant negative correlation ($r = -0.99$) between the number of spermatozoa and the delay in oviposition onset after instrumental insemination (one application of 75% CO_2). Kaftanoglu and Peng (1982) suggest that "the presence of spermatozoa and/or seminal fluid . . . accelerate[d] the onset of oviposition," even though they did not find a significant difference between oviposition onset of virgin queens treated three times with CO_2 and queens instrumentally inseminated with normal or washed sperm that were also treated three times with CO_2.

B. The Maturation Process

As soon as the queen starts oviposition, 85% of the protein synthesized by her fat body is vitellogenin. This is three times higher than that of nonfemale specific proteins. No vitellogenin is synthesized in the ovary. The vitellogenins, released by the fat body into the hemolymph, are taken up pinocytically by the egg cell. The synthesis of other proteins in the ovary is controlled by the nurse cells. The period of vitellogenesis for a single follicle is estimated to be 2 days. Another 2 days are required for glycogen synthesis and chorion formation. Toward the end of oogenesis, the contents of nurse cells is readsorbed into the egg cell (Fig. 8b) (Engels, 1965, 1968, 1972). Before the egg enters the oviduct it is surrounded by the chorion.

During the height of brood-rearing season, a good queen in a strong colony can lay up to 1500 eggs per day (Gerig and Wille, 1975; Nolan, 1925). Thus, the queen is able to produce more egg substance than her own body weight during a single day. Queens have about 360 ovarioles, and each ovariole has 20–30 egg follicles growing simultaneously. Special physiological conditions of taking up extra oocytic components into the oocyte enable the rapid growth of the egg cells. In addition, proteins are transmitted from workers to the queen and probably are not completely digested in the midgut but rather are transferred directly into the oocytes (Lacomb, 1977; Rutz and Lüscher, 1974).

VI. OVIPOSITION

A. Behavior and Seasonality

After the development of eggs in the ovaries, the queen starts oviposition. The deposition of eggs does not occur regularly: there are periods of egglaying and long periods of rest. During these rest periods the workers form a court and examine, clean, and feed the queen.

When laying, the queen seems to move in a random manner on a comb area suitable for egg laying. She periodically stops and inserts her head and forelegs into a cell. After about 3 sec she withdraws. When the "cell inspection" was prevented, the queen deposited several eggs into one cell (N. Koeniger, 1969). On average, every second inspection is followed by an oviposition. To do this the queen lifts her body from the comb, curves her abdomen, and pushes it into the inspected cell. Ten to 25 sec later she removes her abdomen and the egg is attached to the middle of the cell bottom by one end.

In temperate climates the egg-laying rate of the queen undergoes an annual cycle. Brood rearing starts in January on a small but increasing scale, until in about May it reaches its maximum, after which it again declines (Fig. 9). Brood rearing usually stops in November and December for only a few weeks.

The annual cycle of oviposition is correlated with physiological changes. Between November and January the queen has smaller and less developed ovaries. A gradual build-up of the queen's fat body stores begins in November and continues through March. The beginning of intensive egg laying coincides with a large drop in the queen's lipid and protein reserves (Shehata et al., 1981).

The way in which environmental factors initiate the physiological and behavioral annual changes is not yet clear. The photoperiod is the most likely external regulator (Avitable, 1978; Kefuss, 1978; Morse, 1975).

Other external conditions are less important than seasonal changes, but they play a role in regulation of egg laying within a specific season. For

Gudrun Koeniger

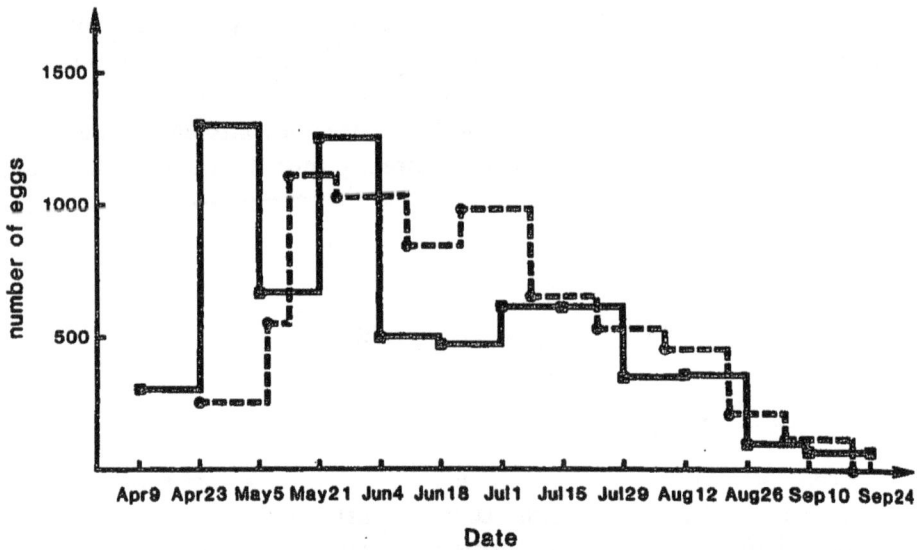

Fig. 9. Average daily egg-laying rate of queen bees measured at 14-day intervals from April to September in Liebefeld, Switzerland. Solid line, 1968 ($n = 6$); dotted line, 1973 ($n = 4$). [Redrawn after Gerig and Wille (1975), with permission. Copyright 1975 by *Schweiz. Entomol. Ges.*]

example, oviposition is highly correlated with pollen-collecting activities (Cale, 1967). Nectar coming into the hive only shows a great influence when bees are near starvation (Chauvin, 1956).

Internal factors, such as the filling of the spermatheca, also have an influence. Queens inseminated with live semen laid 3.0–9.6 times more eggs than did queens that were not inseminated or inseminated with frozen sperm (Harbo, 1976).

B. Oviposition of Fertilized Eggs

During oviposition into worker cells the discharged egg is fertilized. The mechanism of the release of spermatozoa from the spermatheca seems to be as complex as the process involved in filling it. The queen must dispense spermatozoa gradually because the supply is sufficient for several years. Between one and 10 spermatozoa enter an egg (Nachtsheim, 1914; Woyke et al., 1966). By counting the spermatozoa remaining in the spermatheca after a known number of eggs had been fertilized, Harbo (1979) found that the rate of sperm depletion is nonlinear. Queens seem to release a specific volume of fluid containing spermatozoa. The volume is estimated to be 1/153,000 of the total spermathecal volume. The number of spermatozoa in this volume is between 30 and 35 for the first eggs. After deposition of

63,000 fertilized eggs the queen releases only about 20 spermatozoa for one egg.

When a queen becomes old or when the colony multiplies by swarming, new queens are produced. They develop from fertilized eggs which are deposited into special "queen cells." The hatched larvae are nourished with a special food, "royal jelly," which induces development of these larvae into queens instead of into workers. By feeding "royal jelly," workers are able to rear queens from 1- to 3-day-old worker larvae should the old queen accidentally become lost.

C. Oviposition of Unfertilized Eggs

The queen lays not only worker eggs, but also drone eggs. Dzierzon (1849) discovered that drones develop from unfertilized eggs. This discovery raised the question of how the queen controls fertilization. Prevention of the fertilization of the egg depends primarily on the season and colony conditions. When queens of a strong colony are confined to drone combs in spring and summer they produce only drone eggs. In autumn they produce worker brood (Berlepsch, 1873; N. Koeniger, 1970).

During the "drone season" the specific shape of drone cells prevents fertilization. Drone cells are wider than worker cells. The queen recognizes the size of the cell during the cell inspection with her forelegs. Prevention of cell inspection or amputation of the queen's forelegs results in her laying fertilized eggs into drone cells (N. Koeniger, 1970). This seems to indicate that the "normal" process is the addition of spermatozoa to an egg which is prevented by certain stimuli from the drone cell. Queens oviposit fertilized eggs into queen cells which, although they are large, differ totally from drone cells.

Normally, fertilization of the egg determines its sex. However, inbred line drones can be raised from fertilized eggs (Woyke, 1963), and workers of *Apis mellifera capensis* can raise queens and workers from their own unfertilized eggs (Onions, 1912; Ruttner, 1977; Woyke, Chapter 4).

REFERENCES

Adams, J., Rothmann, E. D., Kerr, W. E., and Paulino, Z. L. (1977). Estimation of sex alleles and queen matings from diploid male frequencies in a population of *Apis mellifera. Genetics 86,* 583–596.

Alber, M., Jordan, R., Ruttner, F., and Ruttner, H. (1955). Von der Paarung der Honigbiene. Z. *Bienenforsch. 3,* 1–28.

Allen, D. (1955). Observations on honeybees attending their queens. *Br. J. Anim. Behav. 3,* 66–69.

Alumot, E., Lensky, Y., and Holstein, P. (1969). Sugars and trehalase in the reproductive organs and hemolymph of the queen and drone honeybees. *Comp. Biochem. Physiol. 28,* 1419–1425.

Avitable, A. (1978). Brood rearing in honeybee colonies from late autumn to early spring. *J. Apic. Res. 17,* 69–73.

Berlepsch, A. (1873). "Die Biene und ihre Zucht mit beweglichen Waben in Gegenden ohne Spätsommertracht." J. Schneider, Mannheim.

Biedermann, M. (1964). Neurosekretion bei Arbeiterinnen und Königinnen von *Apis mellifera* L. unter natürlichen und experimentellen Bedingungen. *Z. Wiss. Zoologie 170,* 256–308.

Bishop, G. H. (1920). Fertilization in the honeybee. *J. Exp. Zool. 31,* 225–286.

Boch, R., Shearer, D. A., and Young, J. C. (1975). Honeybee pheromones: field tests of natural and artificial queen substance. *J. Chem. Ecol. 1,* 133–148.

Bresslau, E. (1905). Der Samenblasengang der Bienenkönigin. *Zool. Anz. 29,* 299–325.

Butler, C. G. (1954). The method and importance of the recognition by a colony of honeybees (*A. mellifera*) of the presence of its queen. *Trans. R. Entomol. Soc. (London) 105,* 11–19.

Butler, C. G. (1960). Queen recognition by worker honeybees (*Apis mellifera* L.). *Experientia 16,* 424–426.

Butler, C. G., and Fairey, E. M. (1964). Pheromones of the honeybee: Biological studies of the mandibular gland secretion of the queen. *J. Apic. Res. 3,* 65–76.

Butler, C. G., and Paton, P. N. (1962). Inhibition of queen rearing by queen honeybees (*Apis mellifera* L.) of different ages. *Proc. R. Ent. Soc. (London) A 37,* 114–116.

Butler, C. G., Callow, R. K., and Johnston, N. C. (1962). The isolation and synthesis of queen substance 9-oxodec-*trans*-2-enoic acid, a honeybee pheromone. *Proc. R. Soc. Ser. B 155,* 417–432.

Cale, G. H. (1967). Pollen gathering relationship to honey collection and egg laying in honeybees. *Proc. Inter. Apic. Cong. (Apimondia) 21,* 230–232.

Chauvin, R. (1956). La ponte chez la reine des abeilles. *Insectes Soc. 3,* 499–504.

Crewe, R. M. (1982). Compositional variability: The key to the social signals produced by the honeybee mandibular gland. *Proc. Congr. IUSSI 9th (Boulder),* 318–322.

Dallai, R. (1975). Fine structure of the spermatheca of *Apis mellifera. J. Insect Physiol. 21,* 89–109.

Dzierzon, J. (1849). "Neue verbesserte Bienenzucht des Pfarrers Dzierson." Selbstverl. d. Herausgebers, Bruckisch 3. Aufl.

Ebadi, R., and Gary, N. E. (1980). Factors affecting survival, migration of spermatozoa and onset of oviposition in instrumentally inseminated queen honeybees. *J. Apic. Res. 19,* 96–104.

Effinowicz, G. (1978). Einfluss der Temperatur auf die sexuelle Reifung der Drohnen. Diplomarbeit Universität Frankfurt.

Engelmann, F. (1970). "The Physiology of Insect Reproduction." Pergamon Press, Oxford.

Engels, W. (1965). Der zeitliche Ablauf von Protein- und Kohlehydratsynthesen in der Oogenese bei *Apis mellifera* L. *Verhandl. Dt. Zoolog. Ges. Jena.* 43–51.

Engels, W. (1968). Extraoocytäre Komponenten des Eiwachstums bei *Apis mellifica. Insectes Soc. 15,* 271–288.

Engels, W. (1972). Quantitative Untersuchungen zum Dotterprotein-Haushalt der Honigbiene (*Apis mellifica*). *Wilhelm Roux, Arch. Entwickelungsmech. Org. 171,* 55–86.

Engels, W. (1981). Control of fertility in bees. *Proc. Int. Conf. Regulation of Insect Develop. and Behav. Karpasz, Poland.* 655–660.

Engels, W., and Fahrenhorst, H. (1974). Alters- und Kastenspezifische Veränderungen der Haemolymph-Protein-Spektren bei *Apis mellifica. Wilh. Roux, Arch. Entwicklungsmech. Org. 174,* 285–296.

Fluri, P., Sabatini, A. G., Vecchi, M. A., and Wille, H. (1981). Blood juvenile hormone, protein and vitellogenin titres in laying and non-laying queen honeybees. *J. Apic. Res. 20,* 221–225.

Foti, N. (1973). Kontrollierte Spermatozoenmigration bei Bienenköniginnen (Apis mellifica L.). Proc. Inter. Apic. Cong. (Apimondia) 24, 345–347.

Free, J. B. (1956). A study of the stimuli which release the food begging and offering responses of worker honeybees. Br. J. Anim. Behav. 34, 94–101.

Free, J. B. (1957). The food of adult drone honeybees (Apis mellifera). Anim. Behav. 5, 7–11.

Fyg, W. (1966). Über den Bau und die Funktion der Valvula vaginalis der Bienenkönigin (Apis mellifica L.). Z. Bienenforsch. 8, 256–266.

Gary, N. E. (1962). Chemical mating attractants in the queen honeybee. Science 136, 773–774.

Gary, N. E. (1963). Observations of mating behaviour in the honeybee. J. Apic. Res. 2, 3–13.

Gary, N. E., and Marston, J. (1971). Mating behaviour of drone honey bees with queen models (Apis mellifera L.). Anim. Behav. 19, 299–304.

Gerig, L. (1971). Wie Drohnen auf Königinnen-Attrappen reagieren. Schweiz. Bienenz. 94, 558–562.

Gerig, L., and Wille, H. (1975). Periodizität in der Eiablage der Bienenkönigin (Apis mellifica L.). Mitt. Schweiz. Entomol. Ges. 48, 91–97.

Gessner, B., and Gessner, K. (1976). Inorganic ions in the spermathecal fluid and their transport across the spermathecal membrane of the queen bee, Apis mellifera. J. Insect Physiol. 22, 1469–1474.

Gessner, B., and Ruttner, F. (1977). Transfer der Spermatozoen in die Spermatheka der Bienenkönigin. Apidologie 8, 1–18.

Hammann, E. (1956). Wer hat die Initiative bei den Ausflügen der Königin, die Königin oder die Arbeitsbienen? Ph. D. Diss. Zool. Inst. d. Freien Univ., Berlin.

Harbo, J. R. (1971). "Annotated Bibliography on Attempts at Mating Honeybees in Confinement." Intern. Bee Res. Association Bibliogr. No. 12, Gerrards Cross, England.

Harbo, J. R. (1976). The effect of insemination on the egg laying behavior of honey bees. Ann. Entomol. Soc. Am. 69, 1036–1038.

Harbo, J. R. (1979). The rate of depletion of spermatozoa in the queen honeybee spermatheca. J. Apic. Res. 18, 204–207.

Herrmann, H. (1969). Die neurohormonale Kontrolle der Paarungsflüge und der Eilegetätigkeit bei der Bienenkönigin. Z. Bienenforsch. 9, 509–544.

Howell, D. E., and Usinger, R. L. (1933). Observation on the flight and length of life of drone bees. Ann. Entomol. Soc. Am. 26, 239–246.

Huber, F. (1792). "Nouvelles Observations sur les Abeilles." Translation of Kleine, G. (1856). Ehlers, Einbeck.

Janscha, A. (1775). "Vollständige Lehre von der Bienenzucht," G. Münzberg, Wien.

Jaycox, E. R. (1961). The effects of various foods and temperatures on sexual maturity of the drone honey bee (Apis mellifera). Ann. Entomol. Soc. Am. 54, 519–523.

Jean-Prost, P. (1957). Observation sur le vol nuptial des reines d'abeilles. C. R. Acad. Sci. 245, 2107–2110.

Kaftanoglu, O., and Peng, Y. S. (1982). Effects of insemination on the initiation of oviposition in the queen honeybee. J. Apic. Res. 21, 3–6.

Kefuss, J. A. (1978). Influence of photoperiod on the behaviour and brood rearing activities of honeybees in a flight room. J. Apic. Res. 17, 137–151.

Koeniger, G. (1968). Experimenteller Beitrag zur Physiologie der Spermatheka der Bienenkönigin (Apis mellifica L.). Ph.D. Diss. Universität Frankfurt.

Koeniger, G. (1970). Bedeutung der Tracheenhülle und der Anhangsdrüse der Spermatheka für die Befruchtungsfähigkeit der Spermatozoen in der Bienenkönigin (Apis mellifica L.). Apidologie 1, 55–71.

Koeniger, G. (1976). Einfluss der Kopulation auf den Beginn der Eiablage bei der Bienenkönigin. Apidologie 7, 343–355.

Koeniger, G. (1981). Entfernung des Begattungszeichens durch den sich paarenden Drohn. Proc. Inter. Apic. Cong. (Apimondia) 28, 235–237.

Koeniger, G. (1983). Die Entfernung des Begattungszeichens bei der Mehrfachpaarung d. Bienenkönigin. *Allg. dt. Imkerztg. 17*, 244–245.

Koeniger, G., Koeniger, N., and Fabritius, M. (1979). Some detailed observations of mating i the honeybee. *Bee World 60*, 53–57.

Koeniger, N. (1969). Über die Eiablage bei der Honigbiene. *Proc. Congr. IUSSI 6th Bern*, p 115–120.

Koeniger, N. (1970). Über die Fähigkeit der Bienenkönigin (*Apis mellifica* L.) zwischen Arbe terinnen und Drohnenzellen zu unterscheiden. *Apidologie 1*, 115–142.

Kurennoi, N. M. (1953). When are drones sexually mature? *Pchelovodstvo 30*, 28–32, (I Russian). *Bee Res. Assoc. Translation* E182.

Kurennoi, N. M. (1954). Flight activity and sexual maturity of drones *Pchelovodstvo 31*, 24–2((In Russian). *Apic. Abstr.* 189/56.

Lacomb, M. (1977). Feinstruktur des Mitteldarmes bei Bienen und Proteinernährung der Kön gin. Ph.D. Diss. Universität Münster.

Laidlaw, H. H., Jr. (1944). Artificial insemination of queen bees. *J. Morph. 74*, 429–465.

Lensky, Y., and Schindler, H. (1967). Motility and reversible inactivation of honeybee sperma tozoa in vivo and in vitro. *Ann. Abeille 10*, 5–16.

Little, H. F. (1962). Reactions of the honey bee, *Apis mellifera* L. to artificial sounds an vibrations of known frequencies. *Ann. Entomol. Soc. Am. 55*, 82–98.

Mackensen, O. (1947). Effect of carbon dioxide on initial oviposition of artificially inseminate and virgin queen bees. *J. Econ. Entomol. 40*, 344–349.

Mindt, B. (1962). Untersuchungen über das Leben der Drohnen, insbesondere Ernährung un Geschlechtsreife. *Z. Bienenforsch. 6*, 9–33.

Molodyak, A. V., and Belyaeva, E. N. (1977). Activity of enzymes in the queen's spermatheca *Pchelovodstvo 54*, 20–22, (in Russian). *Apic. Abstr.* 919/79.

Moritz, R. F. A. (1983). Homogeneous mixing of honeybee semen by centrifugation. *J. Api Res. 22*, 249–255.

Morse, R. A. (1975). "Bees and Beekeeping." Cornell University Press, Ithaca, N.Y.

Nachtsheim, H. (1914). Cytologische Studien über die Geschlechtsbestimmung bei der Honig biene (*Apis mellifica* L.). *Arch. Zellforsch. 11*, 169–241.

Nolan, J. W. (1925). "The Broodrearing Cycle of the Honeybee." U.S.D.A. Bull. No. 1349.

Oertel, E. (1940). Mating flights of the queen bees. *Glean. Bee Cult. 68*, 292–293.

Oertel, E. (1956). Observations on the flight of drone honey bees. *Ann. Entomol. Soc. Am. 4* 497–500.

Onions, G. W. (1912). South African "fertile worker bees." *Agr. J. Union S. Africa 3*, 720.

Örösi-Pal, Z. (1959). The behavior and nutrition of drones. *Bee World 40*, 141–146.

Otis, G. W., and Taylor, O. R. (1982). Comparative flight behavior of European and Afr canized drone honeybees. *Proc. Congr. IUSSI 9th (Boulder)*. IUSSI Congress Supplemet Part II.

Pain, J. (1961). Sur la phéromone des reines d'abeilles et ses effects physiologiques. *Ann. Abeil, 4*, 73–152.

Pain, J., and Ruttner, F. (1963). Les extraits des glandes manidublaires des reines d'abeille attirent les mâles, lors du vol nuptial. *C. R. Acad. Sci. 256*, 512–515.

Peer, D. F. (1956). Multiple mating of queen honey bees. *J. Econ. Entomol. 49*, 741–743.

Peer, D. F. (1957). Further studies on the mating range of the honeybee, *Apis mellifera. Cat Entomol. 89*, 108–110.

Poole, H. K. (1970). The wall structure of the honey bee spermatheca with comments about i function. *Ann. Entomol. Soc. Am. 63*, 1625–1628.

Poole, H. K. (1972). The effect of tracheal interruption on the spermathecal wall of the quee honeybee. *Proc. Soc. Exp. Biol. Med. 139*, 701–703.

Praagh, J. P. van, Arendse, M. C., and Ruttner, F. (1976). Measurement and characterization (

light- or radiance fields at drone congregation areas (*Apis mellifera* L.). *Verh. Dtsch. Zool. Ges.* 273.

Praagh, J. P. van, Ribi, W., Wehrhahn, C., and Wittmann, D. (1980). Drone bees fixate the queen with the dorsal-frontal part of their compound eyes. *J. Comp. Physiol. 136*, 263–266.

Renner, M., and Baumann, M. (1964). Über Komplexe von subepidermalen Drüsenzellen (Duftdrüsen?) der Bienenkönigin. *Naturwissenschaften 51*, 68–69.

Renner, M., and Vierling, G. (1977). Die Rolle des Taschendrüsenpheromons beim Hochzeitsflug der Bienenkönigin. *Behav. Ecol. Sociobiol. 2*, 329–338.

Roberts, W. C. (1944). Multiple mating of queen bees proved by progeny and flight tests. *Glean. Bee Cult. 72*, 281–283.

Ruttner, F. (1955). Einfache und mehrfache Paarung der Königinerwiesen aus der Nachkommenschaft. *Bienenvater 76*, 1–19.

Ruttner, F. (1956). Zur Frage der Spermaübertragung bei der Bienenkönigin. *Insectes Soc. 3*, 351–359.

Ruttner, F. (1960). Fortpflanzung und Vererbung. in "Biene und Bienenzucht" (Büdel, A., and Herold, E., eds.). Ehrenwirth Verlag, München.

Ruttner, F. (1962). Bau und Funktion des peripheren Nervensystems an den Fortpflanzungsorganen der Drohnen. *Ann. Abeille 5*, 5–58.

Ruttner, F. (1966). The life and flight activity of drones. *Bee World 47*, 93–100.

Ruttner, F. (1975). Ein metatarsaler Haftapparat bei den Drohnen der Gattung Apis (Hymenoptera: Apidae). *Ent. Germ. 2*, 22–29.

Ruttner, F. (1977). The problem of the cape bee (*Apis mellifera capensis* Escholtz): parthenogenesis—size of population—evolution. *Apidologie 8*, 281–294.

Ruttner, F. (1983). "Zuchttechnik und Zuchtauslese bei der Biene: Anleitungen zur Aufzucht von Königinnen und zur Kör- und Belegstellenpraxis." Ehrenwirth Verlag, München.

Ruttner, F., and Koeniger, G. (1971). Die Füllung der Spermatheka der Bienenkönigin. Aktive Wanderung oder passiver Transport der Spermatozoen? *Z. Vergl. Physiol. 72*, 411–422.

Ruttner, H., and Ruttner, F. (1963). Untersuchungen über die Flugaktivität und das Paarungsverhalten der Drohnen. *Bienenvater 84*, 297–301.

Ruttner, H., and Ruttner, F. (1966). Untersuchungen über die Flugaktivität und das Paarungsverhalten der Drohnen. 3. Flugweite und Flugrichtung der Drohnen. *Z. Bienenforsch. 8*, 332–354.

Ruttner, H., and Ruttner, F. (1972). Untersuchungen über die Flugaktivität und das Paarungsverhalten der Drohnen. 5. Drohnensammelplätze und Paarungsdistanz. *Apilologie 3*, 203–232.

Ruttner, F., Enbergs, H., and Kriesten, K. (1971). Die Feinstruktur der Spermatheka der Bienenkönigin. *Apidologie 2*, 67–97.

Rutz, W., and Lüscher, M. (1974). The occurrence of vitellogenin in workers and queens of *Apis mellifica* and the possibility of its transmission to the queen. *J. Insect Physiol. 20*, 897–909.

Rutz, W., Gerig, L., Wille, H., and Lüscher, M. (1976). The function of juvenile hormone in adult worker honeybees *Apis mellifera*. *J. Insect Physiol. 22*, 1485–1491.

Schlegel, P. (1967). Elektrophysiologische Beobachtungen an den Borstenfeld-Sensillen des äusseren Geschlechtsapparates der Drohnen (*Apis mellifica ♂*). *Naturwissenschaften 54*, 26.

Seidl, R. (1980). The visual fields and interommatidial angles in the three castes of honeybee. *Verh. Dtsch. Zool. Ges.* 367.

Shehata, S. M., Townsend, G. F., and Shuel, R. W. (1981). Seasonal physiological changes in queen and worker honeybees. *J. Apic. Res. 20*, 69–78.

Simpson, J. (1964). The mechanism of honeybee queen piping. *Z. Vergl. Physiol. 48*, 277–282.

Taber, S., III. (1954). The frequency of multiple mating of queen honey bees. *J. Econ. Entomol. 47*, 995–998.

Taber, S., III. (1955). Sperm distribution in the spermatheca of multiple-mated queen hone bees. *J. Econ. Entomol.* 48, 522–525.

Taber, S., III. (1964). Factors influencing the circadian flight rhythm of drone honey bees. *Ann Entomol. Soc. Am.* 57, 769–775.

Taber, S., III. (1975). Semen of *Apis mellifera*: fertility, chemical and physical characteristic *Adv. Invertebr. Repro. ISIR Calicut India* 1:219–251.

Trjasko, V. V. (1956). Multiple matings of queen bees. *Pchelovodstvo* 33, 50–54 (in Russian)

Trjasko, V. V. (1957). Über Drohnen welche sich mit Königinnen paaren. *Pchelovodstvo* 34 29–31 (in Russian).

Verma, L. R. (1973). An ionic basis for a possible mechanism of sperm survival in the sper matheca of the queen honeybee (*Apis mellifera* L.). *Comp. Biochem. Physiol. ser. A,* 44 1325–1331.

Verma, L. R. (1974). Honeybee spermatozoa and their survival in the queen's spermatheca. *Bee World* 55, 53–61.

Wilson, E. O. (1972). "The Insect Societies." Belknap Press, Harvard University, Cambridge Mass.

Witherell, P. C. (1971). Duration of flight and of interflight time of drone honey bees, *Apis mellifera. Ann. Entomol. Soc. Am.* 64, 609–612.

Woyke, J. (1955). Multiple mating of the honeybee queen (*Apis mellifica* L.) in one nuptia flight. *Bull. Acad. Polon. Sci. II* 3, 175–180.

Woyke, J. (1956). Anatomo-physiological changes in queen bees returning from mating fligh and the process of multiple mating. *Bull. Acad. Polon. Sci. II* 4, 81–87.

Woyke, J. (1960). Natural and artificial insemination of queen honeybees. *Pszczelinicze Zeszyty Nankowe* 4, 183–275.

Woyke, J. (1963). Drone larvae from fertilized eggs of the honeybee. *J. Apic. Res.* 2, 19–24.

Woyke, J. (1964a). Causes of repeated mating flights by queen honeybees. *J. Apic. Res.* 3, 17–23.

Woyke, J. (1964b). Die Wirkung aufeinanderfolgender Drohnen auf die Besamung der Köni- gin. *Proc. Inter. Apic. Cong. (Apimondia)* 19, 550–552.

Woyke, J. (1966). Wovon hängt die Zahl der Spermien in der Samenblase der auf natürlicher Wege begatteten Königinnen ab? *Z. Bienenforsch.* 8, 236–248.

Woyke, J., and Jasinski, Z. (1973). Influence of external conditions on the number of spermato- zoa entering the spermatheca of instrumentally inseminated honeybee queens. *J. Apic. Res.* 12, 145–149.

Woyke, J., and Jasinski, Z. (1982). Influence of the number of attendant workers on the number of spermatozoa entering the spermatheca of instrumentally inseminated queens outdoors in mating nuclei. *J. Apic. Res.* 21, 129–133.

Woyke, J., Kuytel, A., and Bergandy, K. (1966). The presence of spermatozoa in eggs as proof that drones can develop from inseminated eggs of the honeybee. *J. Apic. Res.* 5, 71–78.

Zmarlicki, C., and Morse, A. (1963). Drone congregation areas. *J. Apic. Res.* 2, 64–66.

Part II

Breeding

CHAPTER 11

Quantitative Genetics

ANITA M. COLLINS

I. INTRODUCTION

Quantitative genetics of honey bees has received less attention than other areas of bee genetics. The majority of work has dealt with geographic variation and classification (Ruttner, Chapter 2), population genetics, especially in relation to sex alleles (Cornuet, Chapter 9), and, of course, visible mutations (Tucker, Chapter 3). The study of visible mutants has been a major topic because bee biologists, like Mendel with his peas, used mutants to test the fundamentals of inheritance in honey bees. Crosses between individuals differing in discrete characters produced offspring which had uniform phenotypes. Relationships between offspring and parents were easily seen and described. Results of crosses where the trait examined was not easily divided into clear groups were difficult to interpret. The methodology for analyzing such variation and describing the inheritance of the characters involved was developed in part because of this difficulty. Provine (1971) presented an interesting history of a conflict between the Mendelians and the biometricians early in this century which stimulated the development of statistics as a science. This conflict parented the study of continuously variable characters, called quantitative genetics, which now complements Mendelian genetics.

There is no longer any argument that the biological mechanics of the inheritance of quantitative traits differ in any fundamental way from those that determine discrete traits. The major difference between them is the number of genes having significant control on their expression. A discrete trait has distinct visible differences that are attributable to a single locus (gene) with several possible variants called alleles. A quantitative trait is controlled by a number of genetic loci, each of which may have many alleles, and possibly significant modifiers at other loci. This multifactorial nature of the inheritance of traits such as tongue length, wing length, weight, honey production, and defensive behavior requires the study of populations of individuals and descriptions of the properties of these traits at the population level. Important measures include the population mean value, the variation, and the relationships between different continuous characters that may be genetically or phenotypically interactive. Such descriptions are more complex than the simple classical Mendelian descriptions of the inheritance of discrete traits.

Quantitative genetic theory describes the biological bases for the changes observed in animal and plant breeding programs. This theory deals with many aspects of populations and their quantitative traits including the structure of hypothetical randomly breeding populations, results of crosses between inbred lines, and mechanisms of selection. For more extensive explanations of this basic theoretical work, the reader is referred to Falconer (1981), Hill (1984a,b), Lush (1945), and Mather (1949). This chapter will present the basic ideas of theoretical quantitative genetics and their application to honey bees.

II. APPROACHING QUANTITATIVE CHARACTERS

A. Measurement

Quantitative traits usually involve many genes, each contributing a small effect. In some cases there are modifier genes at other loci which have indirect, or pleiotropic, effects on the genes directly affecting the character. There is no difference in basic chromosomal mechanics for these two types of genes, although in some cases the genes involved in controlling a continuous character may be rather closely associated on a chromosome (Falconer, 1981). Such groups of many genes affecting a single character are called polygenes.

Many of the characters of honey bees that have economic importance are quantitative. To aid in more clearly defining the underlying genetic complexity, the visible expression of characters, the phenotype, must be clearly described. Some traits, such as morphological ones, are easily measured. Physiological traits, such as hormone or pheromone levels and disease resistance, may require more complex assessment. Behavioral traits, such as pollen collection and honey production, may require the development of a measurement system that divides the complex behavior into smaller more easily studied parts.

Measurement of quantitative traits has three major requirements. First, the measurement techniques must be appropriate for the characteristic. Second, the scale must be appropriate to show existing variation between individuals. Third, the distribution of the measurements must be normal, falling symmetrically around the population mean, to meet assumptions made in the basic theory. Sometimes this is achieved by measuring in one scale and then recasting the values using a logarithmic or other transformation.

Some traits may be measurable only on certain individuals or at certain times. For example, egg-laying capacity is expressed only by reproductive females, and levels of chemicals such as alarm pheromones vary with age and must be measured at specific times in the development of an individual. Other traits may require an organism's death and dissection to measure (ovariole development in queens or laying workers) or may be evaluated by measurement of close relatives—progeny or siblings. Also, the behavior of social insects is often group behavior rather than individual behavior. Here, measurements must be made on a group of related individuals, either a subsample or the entire colony.

B. Objectives

One of the basic objectives of quantitative genetics is predicting the outcome of a selection or breeding program based on observations of existing populations. Measurements are made on groups of relatives which are used to predict how future offspring will express a character. This process uses estimates of population parameters such as means, variances, and covariances.

A second objective of quantitative genetics is to discover how an organism's phenotype is influenced by its genotype and the environment in which it develops and exists. This can be done by comparing individuals or groups having the same or similar genotypes (genetic constitution) in different environmental conditions. Alternately, individuals or groups with different genotypes can be compared in similar environmental conditions.

The basic relationship underlying the value of such comparisons is expressed in Eq. (1). The phenotype (P) equals the sum of the effect of the genotype (G) and the environment (E).

$$P = G + E \tag{1}$$

If we could reduce variation due to environmental influences to zero, then the phenotype would directly reflect the underlying genotype. For some discrete traits this may be true, but it is rarely true for quantitative characters. Theoretically, it is possible to measure a population having a single genotype over all ranges of its normal environment and, by calculating the mean value, have a measure of the actual genotypic value. However, this is not practical, and rough estimates of genotypic value are based on phenotypes expressed in one or only a few environments.

III. RESEMBLANCE—WHY SISTERS ARE ALIKE

A. Average Effect and Breeding Value

There are a number of concepts that are important to understanding genotypic value in the context of a population. First, the *average effect* of a gene is the mean deviation from the population mean of individuals which receive that gene from one parent. Thus it is the mean deviation caused by replacing a single gene in an array of genotypes with one allele of the gene in question. The inclusion of this single allele would change the phenotypes in several specific ways. The average of these changes is the average effect of this gene. Expressed another way, it is the average effect a gene will have against the whole background of possible genotypes in the population.

Not only are we interested in the average effect of a single gene, but we are interested in the complex of genes that an individual can pass on to its offspring. The value associated with this collection of genes carried by an individual and potentially transmitted to its offspring is referred to as the individual's *breeding value*. This is frequently estimated by making measurements on the progeny and assigning the mean value of these measurements as the breeding value of the parent. This is the sum of the average effects of all the genes that this parent carries.

This concept considers only the additive effects of genes. It does not include any interaction between alleles of the same locus or between genes at different loci. In the development of a genetic theory, Mendel looked at single genes or loci occuring in diploid organisms having two alleles, although a locus might have many possible alleles or forms. Quantitative genetics looks at many genes, each of which may have two alleles present at any one time in the standard diploid organism. In honey bees, we must remember that while queens and workers are diploid (arising from fertilized eggs) and will have two alleles for each locus, drones are haploid (arising from unfertilized eggs) and have only one allele present per locus. This must be taken into account when applying normal diploid quantiative genetics to honey bees.

B. Dominance and Epistasis

There are two possible types of interaction between loci that also influence expression of the phenotype. The interaction between alleles of the same gene when it occurs in a diploid organism is referred to as *dominance*. Dominance can only be expressed in diploid organisms. The parameter "degree of dominance" is the deviation of the heterozygote from the midpoint value for the homozygous parents. If we assign a value of $-A$ to a parent who is homozygous for an allele and the value of $+A$ to the other parent who is homozygous for the other allele, then the value of the F_1 heterozygote will indicate the degree of dominance. If there is no dominance, the value of the heterozygote will be the zero point midway between the two homozygotes. However, if there is dominance, the value of the heterozygote will deviate to one side or the other of the midpoint. With incomplete dominance, the heterozygote value will fall somewhere between the midpoint and one of the parents. With complete dominance, the value of the heterozygote will coincide with the value of one of the homozygotes. Overdominance occurs when the heterozygote value lies outside the range of either of the two parents.

If dominance is a factor in the expression of the genotype of a diploid individual, it will have a bearing on the average effect of the gene. The average effect of a gene will be dependent on its dominance relationship to its allelic pair. For example, an allele that is recessive and paired with a dominant allele will have no effect on the phenotype of the heterozygote. However, if that same allele is inserted against the background of an equivalent recessive allele, the phenotype does change. Quantitative genetics assesses the average affect of a gene across a population of genotypic backgrounds. Therefore, the average effect of a gene having dominance will be considerably influenced by the genetic composition of the population. The relative allelic frequencies in the population will change the average effect.

The other interaction occurring in polygenic systems is the interaction between genes at different loci. This is also referred to as the epistatic deviation, or *epistasis*. These interactions may be very complex and are not readily amenable to dissection.

The G in Eq. (1) has now been separated into its component parts. Genotypic value is equal to the additive effects (A), or breeding value, plus the effect of dominance (D) and interactions between multiple genes (I):

$$G = A + D + I \tag{2}$$

IV. VARIATION—WHY COLONIES DIFFER

A phenotype is the product of the genotype and the environment in which it is expressed. But something more is necessary before a genetic study of a metric character is possible. That is the variation present in a population. As the phenotype of an individual is the sum of effects of both genotype and environment, a population's variation can also be partitioned as

$$V_P = V_G + V_E$$
$$= V_A + V_D + V_I + V_E \tag{3}$$

The phenotypic variation, V_P, is equal to the genotypic variation, V_G, plus the environmental variation, V_E. The genotypic variation can be further subdivided into an additive portion (V_A), a deviation due to dominance interactions (V_D), and an interaction (epistatic) deviation (V_I).

In reality, when measurements are made on individuals and average values are calculated for populations of individuals, they are estimates of these component parts. Earlier it was proposed that if one could eliminate the effects of environment, then the phenotype would directly reflect the value of the genotype. While the variation due to environment cannot be completely removed, it can be reduced in controlled experiments. If the variation due to the environment is reduced, then the variation due to genotype can more easily be seen. Environmental influences that are controlled during an experiment might include those due to nutrition, climate, errors of measurement, or maternal effects. For the honey bee, the effect of a common colony environment might be thought of as equivalent to maternal effects in mammals. Also, there are a number of intangible, and possibly uncontrollable, influences in both developmental and immediate environments.

It is possible to estimate the effects due to environment by controlling all the variation due to genotype. This can be done by using a group of identical genotypes (i.e., highly inbred individuals or F_1 hybrids of inbred lines). This is routinely done with a number of mammals where highly inbred lines have been maintained for many generations. However, this is difficult with honey bees since inbred lines very quickly lose their fitness value and often cannot be maintained (Chapters 9 and 13).

The seemingly straightforward relationship of the underlying influences of environment and genotype and their variance components can be complicated by several natural phenomena. The amount of variation due to environment could be different for different genotypes within the same environment. It has been shown that inbred lines show more variation due to environmental causes than do outbred lines of the same species. It is hypothesized that the inbreds are less well buffered against the environment because of their much greater frequency of homozygosity at many loci.

The genotype and the environment may also interact, which may influence the partitioning of the variation. The honey bee represents a very clear example of this correlation. A bee spends a good portion of its life within a colony. This colony environment influences the development of the bee and is in part determined by the genotypes of the related bees who are maintaining the colony. These kinds of correlations may be somewhat controlled in experiments through techniques such as cross-fostering of individuals between colonies or sibling groups. However, generally the influence of correlation between genotype and environment is accepted as being part of the genotypic variation.

Generally, an assumption is made that the environment has the same effect or magnitude of effect on different genotypes, but this is not always the case. Such variation also is generally accepted as part of the variation due to environment and not considered further. With certain experimental designs, it is possible to estimate the influence of an environment on different genotypes by use of a two-way analysis of variance of various genotypes across various environments or treatments. With the statistical analysis, a value can be computed for the magnitude of the interaction between differing environment and differing genotype.

The component of variation having most interest in quantitative genetics is the additive variance or the breeding value (V_A). This factor is the chief cause for resemblance between related individuals and therefore is closely connected to observed genetic properties of a population and responses of that population to selection. It is also the most readily estimated component of a population's variation. Equation (3) might more realistically be expressed as

$$V_P = V_A + V_N + V_E \qquad (4)$$

where the variation of the phenotype (V_P) is equal to the variation due to additive genetic causes (V_A) plus variation due to nonadditive genetic causes (V_N) plus variation due to environmental causes (V_E). The value of additive genetic variation is not usually presented in this form but is expressed as the proportion of variation due to additive genetic causes as a part of the total phenotypic variation. This ratio of additive genetic to total phenotypic variation is referred to as *heritability*. This will be further discussed later.

The presence of additive variation in the variation of a quantitative character does not imply anything about the mode of action of the underlying polygenes. These polygenes may be additive in nature, they may have dominance interactions, and they may have epistatic interactions. Only if all the genetic variation is additive is the mode of action implied by the presence of additive variation. Perhaps this can be looked at as the degree of resemblance between individuals with genes having common ancestry as opposed to the level of dissimilarities between relatives. The dominance variation is due to the association of alleles that show some form of dominance relationship. With multiple loci the epistatic interaction generally is included as part of the genetic variance. However, this is usually very small. Probably it could be further subdivided if one knew the number of loci involved in the polygenes. For further discussion of the epistatic deviation, see Falconer (1981). The additive variance component of the genetic variance does take into account multiple alleles at a single locus through the average effect of all the alleles across the population being studied.

The total phenotypic or observed variation in a population can be partitioned into its several components. Using resemblances between relatives, estimates of the additive genetic variation, or breeding value, are possible, and by using inbred lines the variation of a character due to environmental causes can be estimated. Estimations of the breeding value of an individual are relatively easy to make and are important for the estimation of heritability. Heritability is important for the major objective of predicting success from proposed breeding schemes. This leaves the nonadditive genetic variation, which is included in effects due to dominance and to epistasis. With very large numbers of individuals in a population and more elaborate statistical techniques it is also possible to estimate these two components as separate entities. This is probably not practical for honey-bee quantitative genetics.

V. MORE ON RESEMBLANCE

The previous discussion has dealt with partitioning the variation in a theoretical way. Practically, partitioning of the observable phenotypic variation is begun by grouping individuals into families—that is, groups of individuals with genes that are common by descent. Statistical techniques are then used to compare the similarity of individuals within a group with differences between groups. For groups of siblings, either full sibs (in the case of honey bees, supersibs) or half sibs, the intraclass correlation coefficient t is used, and for the special case of offspring with a parent, or average of two parents (mid-parent), the regression coefficient b is used. These coefficients are estimates of the covariance of related individuals. They will estimate the value of the components of the observed phenotypic variance, with different proportions of these components being estimated based on the type of relationship. For more detailed derivations of these relationships, see Lush (1945), Falconer (1981),and Kempthorne (1955).

Offspring can be grouped by commonality of one parent or an average of both parents. By definition, the average value of offspring is equal to half of the breeding value (V_A) of one of the parents. If mid-parent is used, the relationship will still be true because mid-parent represents the mean of two parents and an estimate of the mean breeding value of the population. Therefore the covariation of offspring on parents is equal to $\frac{1}{2}V_A$. When a regression coefficient b is calculated, it represents the ratio of covariance to total phenotypic variation. Thus, the regression coefficient will estimate $\frac{1}{2}V_A/V_P$, which is one-half the heritability.

For half sibs, those individuals with only one parent in common, the probability is that they will have $\frac{1}{4}$ of their genes by common descent from that one parent, and therefore their covariance is $\frac{1}{4}$ of the breeding value of that parent, $\frac{1}{4}V_A$. For full sibs, the situation is more complicated. They have a probability of having $\frac{1}{2}$ of their genes in common from two parents plus a $\frac{1}{4}$ chance of having the same alleles for a single locus. This probability adds to the covariance, in addition to $\frac{1}{2}$ the breeding value, V_A, a quantity $\frac{1}{4}$ the variation due to dominance, V_D. For honey-bee supersibs, all the genes from the haploid male parent are identical, so the probability is $\frac{3}{4}$ for genes in common and a $\frac{1}{2}$ chance of the same alleles. For all instances using sibs, the intraclass correlation coefficient is calculated and represents the ratio of the covariation to the total variation observed. Therefore, t will be equal to $\frac{1}{4}$ of the heritability for half sibs, a minimum value of $\frac{1}{2}$ of the heritability for full sibs, and a minimum of $\frac{3}{4}$ for supersibs. In the case of full sibs and supersibs, if dominance is not a factor in the variation, the correlation coefficient will be equal to the heritability. If dominance is a factor, the correlation coefficient will overestimate the heritability by $\frac{1}{2}$ the factor of dominance variation.

One other consideration is that relatives frequently occur in a common environment. This common environment increases the similarity between relatives and increases the difference between family groups. This would therefore increase the estimates of heritability where common environment was important. Generally it is ignored except for instances of full sibs where maternal environment plays a part. Probably for honey bees the equivalent common environment could be viewed as the colony. In this case it would influence both calculations with super- sibs and half sibs. It would be up to the individual investigator to judge how important this common environment is, based on a biological understanding of the character under study.

There are a number of assumptions that are made about the basic population from which these calculations have been drawn. We assume that the population is a randomly breeding one having no changes of gene frequency from generation to generation. We also assume no selection and no inbreeding are affecting the gene frequencies in the population. Finally, we assume that there is no differential fitness of genotypes; heterozygotes have the same chance of surviving as either of the two homozygous types. If these assumptions are not true, then the estimates will be correspondingly inaccurate.

VI. HERITABILITY

The term heritability, has two major definitions. The first, a more general definition, is the state or quality of being heritable, inherited, or common by descent. In other words, it is the biological phenomenon of the mode of inheritance of a particular trait. Quantitative geneticists, however, use the term in a much more restrictive sense, as a measure of the degree to which a phenotype is genetically influenced and can be modified by selection (King, 1968). Care must be taken in using this word correctly. There are a number of papers in honey-bee genetic literature that use the word heritability in the title, but are not quantitative in subject.

Heritability, h^2, is one of the most important properties of a quantitative trait. It represents the ratio of the additive genetic variance to the total, or phenotypic, variance:

$$h^2 = V_A/V_P \tag{5}$$

This is the proportion of total variance that is attributable to additive effects, the average effects of all genes affecting a character. The size of h^2 indicates the similarity of related organisms. The most important function of h^2 in the study of quantitative traits is that it can predict how reliable the phenotypic value is as a guide to the individual's actual breeding value. This is so because heritability estimates the proportion of the phenotypic variation that is attributable to genetic causes amenable to selection. This is why it is so important to accurately measure the desired trait. If the environmental variance, V_E, is high due to poor measuring techniques, the estimate of the proportion resulting from additive genetic causes, and thus heritability, would be reduced. A poor measure of the phenotype will result in poor success in a selection program.

The value of heritability ranges from 0 (no genetic influence on the trait) to 1 (all variation of the trait is genetically produced). Traits that are closely related to reproductive fitness, such as egg-laying capacity [$h^2 = 0.16$: Avetisyan and Grankin (1976)], generally have low heritabilities. Higher h^2 values are expected in characters less important to reproductive fitness such as body color, or body-part size [$h^2 = 0.53-0.92$: Avetisyan and Grankin (1976)]. A variety of estimates of heritability for the honey bee are presented in Table 1.

Importantly, heritability is a property not only of a specific character but also of the population and of the environmental circumstances. Thus, heritability value estimates made on one population may be different for that same population in another environment, or for a different population. Environmental variance depends on the conditions of culture or management of the organisms being studied — more variable conditions reduce the heritability, and more uniform conditions increase it. The genetic components of heritability are influenced by the gene frequencies in the population, and these may differ between populations of the same organism because of their different biological histories. Small populations maintained for a long time may become more genetically uniform than do large, randomly mating populations, and they may therefore show lower heritabilities.

A. Special Considerations in Honey Bees

The standard approaches for measuring heritability require comparing the merits of related individuals and estimating heritability from the covariance between them or from a regression or correlation. However, the theoretical and applied work that has been done has largely concentrated on diploid domesticated animals, such as sheep and cows, and the laboratory standby, the fruit fly. The honey bee, a haplo–diploid organism living in colonial aggregations and showing caste differences as well as sex differences, requires slightly modified theoretical bases. Rinderer (1977) and Oldroyd and Moran (1983) have addressed the biological differences for honey bees and the necessary modifications to be made in systems for estimating heritability.

First of all, the haplo–diploid condition changes the relatedness of relatives. Rinderer (1977) discusses the changes in the coefficient of relatedness between sister-workers under three different mating conditions. For a queen mated to a single drone, all the sperm are identical, and the daughter-workers are related at a level of 0.75. When the queen has been mated by many drones, all of them from the same drone-mother, the relatedness between two workers can be either 0.75 or 0.50. Under the more natural situation of a queen multiply mated with drones from many sources, the relatedness is either 0.75 or 0.25 for workers in the same colony. This is contrary to the normal situation in diploid organisms where full sibs (having

TABLE 1. Some Heritability (h^2) Estimates for Honey-Bee Traits

Source	Trait		h^2	Method
Soller and Bar-Cohen (1967)	Winter honey weight		0.57	Analysis of variance
	Spring honey weight		0.60	
	Total honey		0.58	
	Winter brood		0.76	
	Spring brood		0.33	
	Final brood area			
el Banby (1967)	Brood rearing:			Regression: offspring or dams
	winter		0.95	
	citrus season		0.51	
	clover season		0.34	
	cotton season		0.28	
	yearly average		0.90	
	Honey production:			
	clover		1.00	
	cotton		0.75	
Pirchner et al. (1962)	Honey production		0.23	Analysis of variance
	Brood		0.35	
Vesely and Siler (1963)	Honey yield		0.16–0.19	Regression: offspring or midparent
	Brood 6 weeks before flow		0.30–0.41	
Gonçalves and Stort (1978)	Number of hamuli		0.76	Regression: offspring or parents
Collins (1979)	Time to react to isopentyl acetate (IPA)		0.68	Regression: offspring or midparent
Rinderer et al. (1983)	Time to react to IPA		0.03 ± 0.006	Analysis of variance
	Longevity		0.32 ± 0.27	
Oldroyd and Moran (1983)	Number of hamuli		0.68 ± 0.183	Intraclass correlation/ average relatedness
Collins et al. (1984)	Hoarding	E[a]	0.92 ± 0.44	Analysis of variance
	(three day average)	A	0.66 ± 0.69	
	Time to react	E	1.28 ± 0.04	
	to IPA	A	0.31 ± 0.01	
	Time to react to	E	0.31 ± 0.20	
	moving target	A	0.69 ± 0.31	
	Number of stings	E	0.57 ± 0.24	
	in target	A	NE[b]	
	Number of bees	E	0.93 ± 0.03	
	responding	A	0.17 ± 0.01	
	Comb cell size	E	NE	
		A	1.15 ± 0.11	
Milne and Friars (1984)	Pupal weight		0.645 ± 0.065	Analysis of variance
Milne (1985a)	Corbicular area		1.014 ± 0.195	Analysis of variance
Milne (1985b)	Hoarding		0.187 ± 0.029	Analysis of variance
Milne (1985c)	Worker longevity		0.196 ± 0.024	Analysis of variance
Moritz (1985)	Postcapping stage		0.68 ± 0.001	Intraclass correlation/ average relatedness
			0.97 ± 0.06	Regression: offspring or midparent

[a] E, European bees; A, Africanized bees   [b] NE, not estimable.

both parents the same) are related at a probability of 0.50 and half sibs (one parent in common) are related at a level of 0.25. Crow and Roberts (1950) and Laidlaw and Page (Chapter 13) discussed the calculation of the coefficient of relationship for a number of mating systems. A somewhat different approach was taken by Polhemus et al. (1950) and Laidlaw and Eckert (1950) for application of quantitative genetic theory to honey bees. They were able to look at relatedness in the same manner as a diploid system by considering that a mating took place between a dam-queen and a sire-queen.

The second major difference seen in honey bees is the production of castes in individuals that are derived from the same combinations of genetic material (a fertilized egg) but reared under different environmental conditions. In some cases, a trait may be expressed differently, or not at all, by queens and workers. This could be considered an example of variation due to epistasis, where genes controlling caste differentiation have an effect on the expression of genes controlling other characters. In many cases, especially those related to an economic character, a phenotypic value for an individual may be measured on a sibling of a different sex or caste. In these instances the phenotypic value of the reproductive (queens or drones) is based on the phenotype of sisters (workers) with coefficients of relatedness of 0.75, 0.50, or 0.25 in the case of queens, or a coefficient of relatedness equal to 0.50 for drones. This inaccuracy in actually measuring the phenotype of the reproductive individual has consequences for the accuracy of the estimate of heritability.

A third important situation present in honey bees is that they live as a colony of more or less related individuals. In diploid organisms one must be alert to increased covariance due to a common environment. This is most clearly seen in animals that have litters or clutches where there is a strong effect of the common maternal environment. The individual honey-bee investigator must assess the effect of the common colony environment on the trait he is investigating.

Another aspect of the colonial sociality of honey bees is that some of the characteristics that are of most interest are the result of behavior of the individuals within the colony and cannot be readily measured on single individuals. A good example of this is honey production, commonly expressed in terms of weight gain by a colony during a honey production season. A technique proposed by Rothenbuhler (1960) attempted to deal with this problem. He proposed that inbred queens be inseminated by single drones. All the semen from the single drone will be identical, and most of the loci of the inbred queen can be considered to be homozygous. From such a mating a colony of workers of almost identical genotype can be produced. A measurement of the colony behavior, therefore, is the average expression of almost all of these similar genotypes, and a good estimate of their phenotype.

B. Estimation

The simplest way to estimate heritability is to measure a population of mixed genotypes and one of identical genotype in several environments. The first population would provide an estimation of total phenotypic variance. The second would measure only environmental variance, because all genotypes would be identical. The difference between these two phenotypic variances would be the additive genetic value. Heritability could then be directly calculated from the ratio of additive genetic variance to total phenotypic variance.

A common straightforward method for estimating heritability is from the regression of offspring on parents. The data, measurements of parents and the mean values of their offspring, are used to calculate a regression coefficient b. If this is the regression of offspring on one parent, b_{op}, it is a valid measure of $\frac{1}{2}h^2$; if the regression is offspring on midparent, b_{mp}, it actually measures heritability. Some examples of honey bee h^2 values calculated using this method are presented in Table 1. Both Rinderer (1977) and Oldroyd and Moran (1983) caution that this approach is inappropriate for measurements made on different castes. However, it has been used in a number of instances where the queen's phenotypic value was based on measurements made on sister workers (Collins, 1979; Milne, 1985a,b,c; Milne and Friars, 1984; and Moritz, 1985). For characters, such as brood patterns or egg-laying abilities, that are measured only on queens, the situation mimics a diploid one. The midparent then is the average of the phenotypic values of the dam-queen and the sire-queen. Another possibility is to use pooled mixed semen from more than 20 drones, where the resulting segregation of gametes from the sire-queen can be interpreted as a diploid system.

For a trait that can be measured on both queens and workers, an appropriate collection of matings can be used to compare queens with the parental queens, and workers with those same parental queens. Differences in these two regression values will show the magnitude of environmental, dominance, and epistatic effects. Heritability values from the regression of queens on queens would be higher than the values for workers on the same queens. The only limitation on this system is that the phenotypic variance must be the same for both castes for the statistical procedures to be appropriate.

Historically, heritability is most often estimated by sib analysis. Each of several males (sires) is mated to several females (dams), and some offspring

TABLE 2. Calculation of Phenotypic Variance
from Analysis of Variance Mean Squares (MS) for
a Sampling of Sib/Half-Sib Families[a]

Source of variance	Variance	Calculation[b]
Between sires	σ^2_{sire}	$(MS_{sire} - MS_{dam})/dk$
Between dams	σ^2_{dam}	$(MS_{dam} - MS_{within})/k$
Within offspring	σ^2_{within}	MS_{within}
Total population	σ^2_{total}	$\sigma^2_{sire} + \sigma^2_{dam} + \sigma^2_{within}$

[a] For more detail see Falconer (1981) and Henderson (1953).

[b] d, Number of dams; k, number of offspring per dam.

from each female are measured. The individuals measured form a population of half-sib and full-sib families. An analysis of variance divides the phenotypic variance into components attributable to differences in sires, dams mated to the same sires, and among offspring of the same female. The variance components from sires, dams, and the total may be calculated from the mean square values (Table 2). The total variance, or phenotypic variance, is calculated because it is not necessarily equal to the observed variance as estimated from the total sum of squares, though the two seldom differ by much. With these values, estimates of heritability can be made from the sire component, the dam component, or a combination of the two (Table 3).

The use of sib analysis has been carefully examined by Rinderer (1977). In order to really be comparing half-sib and supersib families, the character must be measured on the appropriate caste in both cases, that is, queens or workers but not both. This is because the relationship of workers between colonies of sister queens is that of cousins but not sibs. He proposed a scheme using randomly selected dam-queens mated to randomly selected sire-queens. This is a useful sibling system for estimating heritability. If the

TABLE 3. Calculation of Heritability from
Phenotypic Variances Based on Sib and Half-Sib
Families[a]

Estimate	Calculation
h^2_{sire}	$4\sigma^2_{sire}/\sigma^2_{total}$
h^2_{dam}	$4\sigma^2_{dam}/\sigma^2_{total}$
$h^2_{combined}$	$2(\sigma^2_{sire} + \sigma^2_{dam})/\sigma^2_{total}$

[a] For more detail see Falconer (1981).

matings are single drone inseminations, then the sire component from the analyses of variance is the best source for estimates of h^2. If mixed pooled semen for more than 20 drones from one sire-queen is used, then both the sire and dam component are acceptable for estimates. If the behavior under study is a colony behavior, then only the sire component can be used regardless of the number of drones in each mating.

Oldroyd and Moran (1983) proposed a somewhat different system that has the advantage of not requiring artificial insemination. They use the relationship of intraclass (sib group) correlations (t) for a population and the relatedness (r) of workers in a colony (sib groups). This is expressed as

$$h^2 = t/r \tag{6}$$

For naturally mated queens having a large number of effective matings (eight or more), the value of r approaches $\frac{1}{4}$, the situation in diploid half-sibs. The intraclass correlation value of t is calculated from measurements made on worker offspring. If the trait is one expressed by the colony, such as defense and honey production, the correlation of repeated or replicate measures on groups of the worker offspring would be higher than that from measures from individuals. This would give an upward bias to the heritability estimates. This bias could be reduced by utilizing many replicates or taking measurements across a period of time.

VII. EFFECTS OF SELECTION

Earlier, a number of assumptions were made about the population. One was that the population was not undergoing selection. In many cases, however, we are using quantitative genetics to predict the effect of planned selection programs. Once selection has begun the effects from generation to generation can be monitored. By observing the changes in the population means, variance, covariances, etc., the effects of the underlying changes in gene frequencies brought about by selection can be described. There are three ways that gene frequencies change: (1) by artificial selection of parents for each generation, (2) through differential fertility in parents within a generation, and (3) through differential viability in the offspring. Ways (2) and (3) are major aspects of natural section and are generally always present. Usually these are not correlated with the characters in question. However, their function in the selection system should not be forgotten.

There are two parameters that we can measure in a selection program. One is the response to selection, R. This is the mean deviation of the offspring from the original, or parental, population before selection. This is identical to the breeding value of the selected parents. The other is the selection differential S, which is defined as the deviation of the mean parental value from the mean of the whole parental generation before selection. The relationship of R to S is expressed as

$$R = h^2 S \tag{7}$$

and, where b_{op} is the regression of offspring on the midparent, as

$$R = b_{op} S \tag{8}$$

Assumptions are that there are no nongenetic causes of resemblance between the offspring and parents, such as common environment, and that there is no natural selection.

The heritability value was used to predict the response to the selection. This value of heritability is, in theory, only valid for prediction for one generation because it is estimated from the resemblance between relatives in the parental generation and the genetic make-up of this population changes with selection. In practice, however, the original estimate of h^2 is useful for several generations (five to 10). It can be seen from Eqs. (7) and (8) that it is possible to calculate an estimate of h^2 from the relationship between response to selection and selection differential. This information is obtained from later generations in a selection program. This h^2 is referred to as *realized heritability*. Estimates of realized h^2 are valid in the absence of inbreeding, with reduced environmental effects, a weighting of selection differential to account for natural selection, and no maternal effects. Realized heritability is the most useful way of comparing the effectiveness of selection in different experiments. Because of the many external effects that cannot be totally eliminated from any experiment, the best estimates of realized heritability are those that are predicted over a number of generations of selection. The actual calculation involves plotting the generation means against the cumulative differential and calculating the slope of the regression line. This value, essentially the regression of offspring on the midparent, is the realized heritability. An example of such a plot is presented in Fig. 1 based on data from Rothenbuhler *et al.* (1979), with calculations by Collins. Several other estimates of realized heritability are presented in Table 4.

The selection differential S depends on the proportion of the population selected as parents (a number limited by the minimum number of parents necessary to maintain a viable population and by the variability of the character). Intensity of selection i is a standardized form of S expressed as

$$S/\sigma_p = i \tag{9}$$

where i is dependent only on the proportion selected, and can be deter-

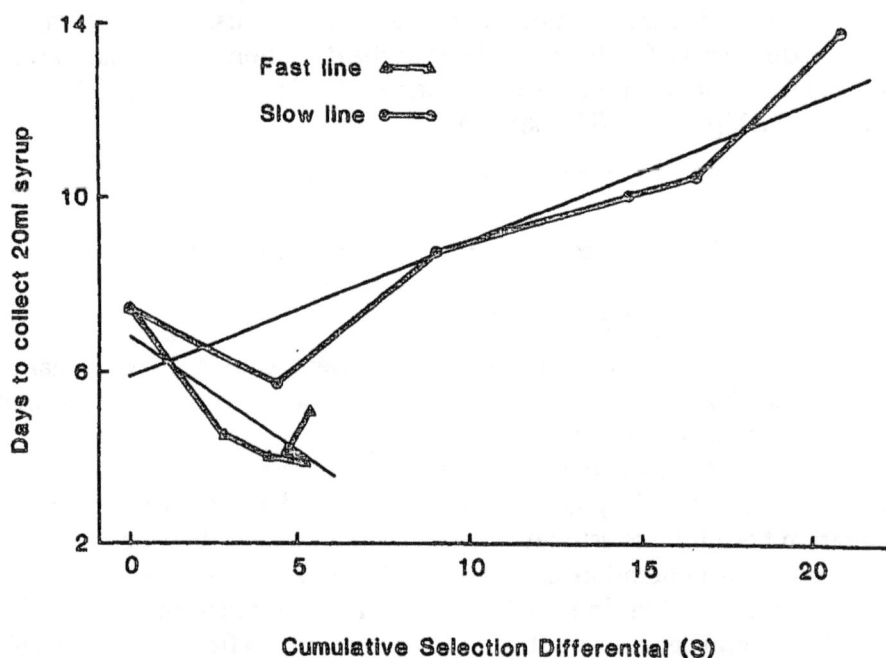

Fig. 1. Two-way selection for hoarding of sugar syrup in honey bees. Response (generation means) plotted against cumulative selection differential (S). The slopes are not significantly different ($p = 0.408$). Data taken from Rothenbuhler et al. (1979). Compare with Rinderer and Collins, Chapter 6, Fig. 4.

mined by the ratio of the value of the trait at truncation to the proportion of individuals selected as parents. It is given in standard deviation units, and if one knows the standard deviation of the trait, S can be predicted. If we then calculate S following selection, weighted by the number of offspring each parent produces, we can calculate an effective value of S. The relative magnitudes of the predicted S and the effective S give us a measure of the importance of artificial and natural selection as it is functioning in our

TABLE 4. Estimates of Realized Heritability (h^2)

Source	Trait	Realized h^2
Soller and Bar-Cohen (1967)	Citrus honey production	0.36
Rothenbuhler et al. (1979)	Hoarding of syrup by caged bees	Fast line, 0.553[a] Slow line, 0.324
Hellmich et al. (1985)	Pollen hoarding	High line, 0.556 ± 0.161 Low line, 0.063 ± 0.190

[a] Calculation by A. M. Collins.

selection scheme. The effective S reflects both, and the predicted S reflects only artificial selection.

There are a number of other phenomena associated with selection that are of interest in a discussion of quantitative genetics of the honey bee. The most important of these, inbreeding depression, has a serious effect on the fitness of selected lines. A more complete discussion of this problem is presented by Cornuet (Chapter 9) and Laidlaw and Page (Chapter 13). Its converse, heterosis, has been used to advantage in a number of honey-bee breeding programs discussed by Kulinčević (Chapter 16). These made use of the greater fitness expressed in certain heterozygotes compared to the respective homozygous conditions. In some cases combinations between certain homozygotes are more fit than other heterozygote combinations, a phenomenon known as combining ability. Both inbreeding depression and heterosis occur because of the dominance interactions within loci, and the amount of heterosis one sees following a cross between two lines or populations is dependent on differences in gene frequencies between the two populations.

VIII. CORRELATIONS

Correlations sometimes occur between characters. These correlations can arise because of commonality, or similarity, in the underlying genotypes. The genetic cause of correlation between characters is largely due to pleiotropy, although linkage of genes on the same chromosome is a transient cause of correlation. Pleiotropy, as discussed earlier, is a property of a gene whereby it effects more than one character. Thus, the degree of correlation between two characters is the extent to which they are controlled by the same genes. Some of these pleiotropic effects may produce negative correlations and others may produce positive correlations.

The correlations that we actually measure are the phenotypic correlations. As with variance, this correlation, or covariance, can be partitioned into its component parts. These are genetic, the correlation of the breeding values for the two traits, and environmental, which includes both environmental and nonadditive genetic correlations. The correlation is calculated by the appropriate covariance divided by the product of the two standard deviations of the characters. The genetic and environmental correlations may be quite different in magnitude and sign. If they are different in sign it means that the genotype and the environment affect the character through different physiological mechanisms.

Given the same kinds of experimental designs for super sib and half-sib families or for regressions of offspring on parents, appropriate calculations

TABLE 5. Significant Correlations between Traits of the Honey Bee

Source	Trait A	Trait B	Correlations	
			Phenotypic	Genetic
Soller and	Spring honey weight	Total honey	0.92	1.02
Bar-Cohen	Winter brood	Final brood area	0.91	1.15
(1967)	Total honey	Winter brood	0.51	1.12
	Winter honey weight	All others	Small and negative	
	Spring honey weight	Winter brood	0.45	1.06
	Spring honey weight	Final brood area	0.37	1.30
	Total honey	Final brood area	0.45	1.32
el Banby (1967)	Honey production on cotton	Brood in cotton season	0.86	1.06
	Honey production for whole year	Whole year brood	0.82	0.77
Kerr *et al.* (1974)	Amount of 2-heptanone	Time to first sting	−0.75	Not calculate
	Amount of 2-heptanone	Time bees were irritated	−0.821	Not calculate
	Amount of 2-heptanone	Number of stings in ball	0.741	Not calculate
	Time bees were irritated	Number of stings in ball	−0.759	Not calculate
Rinderer *et al.* (1983)	Time to react to IPA (cage)	Initial activity level (cage)	−0.61	NE[a]
	Time to react to IPA (cage)	Longevity	−0.06	0.72 ± 0.8
	Response to *Nosema apis*	Longevity	0.76	NE
Collins *et al.* (1984)	Time to react to IPA (cage)	Initial activity level (cage)	E[b] −0.57 A −0.26	1.0 1.0
	Time to react to IPA (cage)	Time to react to moving target	E −0.01 A −0.17	−0.82 ± 0.0 −0.36 ± 0.0
	Time to react to IPA (cage)	Number of stings in moving target	E 0.09 A −0.43	1.0 NE
	Hoarding (13-day average)	Initial activity level (cage)	E 0.05 A 0.54	1.0 1.0
	Hoarding (13-day average)	Time to react to IPA (cage)	E 0.11 A −0.70	1.0 −0.28 ± 0.:
	Hoarding (13-day average)	Time to react to moving target	E 0.15 A −0.32	0.80 ± 0.: −1.0
	Hoarding (13-day average)	Number of stings in target	E −0.05 A 0.61	−1.0 NE
	Time to react to moving target	Initial activity level (cage)	E −0.19 A 0.12	1.0 1.0
	Time to react to moving target	Number of stings in target	E −0.58 A −0.16	−1.0 NE

[a] NE, not estimable.
[b] E, European bees; A, Africanized bees.

to estimate genetic correlations can be made from analyses of covariance. Some examples are presented in Table 5. However, these are very seldom precise. One can estimate correlations from responses to selection in a manner similar to that for realized heritability or, a more useful process, one can predict the change in character Y following selection on character X if one knows the genetic correlation and the heritabilities for the two characters. A discussion of the calculations involved for these estimates is presented by Falconer (1981).

We can also make use of correlations between characters in a process called indirect selection. In indirect selection one selects for a character of secondary importance in order to improve a correlated character of major importance. This can be done if the secondary character has a higher h^2 than the primary character, and if the genetic correlation is high. A possible example involves sex- or caste-limited traits which are correlated to a character expressed in both sexes or castes. On the whole, however, indirect selection is not as effective as simultaneous selection for the two characters, a topic discussed more thoroughly in the next chapter.

ACKNOWLEDGMENT

The chapter was prepared in cooperation with the Louisiana Agricultural Experiment Station.

REFERENCES

Avetisyan, G. A., and Grankin, N. N. (1976). Heritability of some economically useful and some external characteristics of mid-Russian honeybees of the Orel region. *Doklady Timiryazevskoi Sel'skokhozyaĭstvennoĭ Akademii* 221, 177–180.

Banby, M. A. el. (1967). Heritability estimates and genetic correlation for brood-rearing and honey production in the honeybee. *Proc. Inter. Apic. Cong. (Apimondia)* 21, 498.

Collins, A. M. (1979). Genetics of the response of the honeybee to an alarm chemical, isopentyl acetate. *J. Apic. Res.* 18, 285–291.

Collins, A. M., Rinderer, T. E., Harbo, J. R., and Brown, M. A. (1984). Heritabilities and correlations for several characters in the honey bee. *J. Hered.* 75, 135–140.

Crow, J. F., and Roberts, W. C. (1950). Inbreeding and homozygosis in bees. *Genetics* 35, 612–621.

Falconer, D. S. (1981). "Introduction to Quantitative Genetics," 2nd ed. Ronald Press, New York.

Gonçalves, L. S., and Stort, A. C. (1978). Honey bee improvement through behavioral genetics. *Annu. Rev. Entomol.* 31, 197–213.

Hellmich, R. L., II, Kulinčević, J. M., and Rothenbuhler, W. C. (1985). Selection for high and low pollen-hoarding honey bees. *J. Hered.* 76, 155–158.

Henderson, C. R. (1953). Estimation of variance and covariance components. *Biometrics 9*, 226–252.

Hill, W. G., ed. (1984a). "Quantitative Genetics Part I: Explanation and Analysis of Continuous Variation." Van Nostrand Reinholt, New York.

Hill, W. G., ed. (1984b). "Quantitative Genetics Part II: Selection." Van Nostrand Reinholt, New York.

Kempthorne, O. (1955). The theoretical values of correlations between relatives in random mating populations. *Genetics 40*, 153–167.

Kerr, W. E., Blum, M. S., Pisani, J. F., and Stort, A. C. (1974). Correlation between amounts of 2-heptanone and iso-amyl acetate in honeybees and their aggressive behaviour. *J. Apic. Res. 13*, 173–176.

King, R. C. (1968). "A Dictionary of Genetics." Oxford University Press, London.

Laidlaw, H. H., Jr., and Eckert, J. E. (1950). "Queen Rearing." Dadant and Sons, Hamilton, Ill.

Lush, J. L. (1945). "Animal Breeding Plans," 3rd ed. Iowa State University Press, Ames, Iowa.

Mather, K. (1949). "Biometrical Genetics: The Study of Continuous Variation." Methuen and Co., London.

Milne, C. P., Jr. (1985a). An estimate of the heritability of the corbicular area of the honeybee. *J. Apic. Res. 24*, 137–139.

Milne, C. P., Jr. (1985b). A heritability estimate for honey bee hoarding behavior. *Apidologie*. (In press).

Milne, C. P., Jr. (1985c). An estimate of the heritability of worker longevity or length of life in the honeybee. *J. Apic. Res. 24*, 140–143.

Milne, C. P., Jr., and Friars, G. W. (1984). An estimate of the heritability of honeybee pupal weight. *J. Hered. 75*, 509–510.

Moritz, R. F. A. (1985). Heritability of the post capping stage in *Apis mellifera* and its relevance for varroatosis resistance. *J. Hered. 76*, 267–270.

Oldroyd, B., and Moran, C. (1983). Heritability of worker characters in the honeybee (*Apis mellifera*). *Aust. J. Biol. Sci. 36*, 323–332.

Pirchner, F., Ruttner, F., and Ruttner, H. (1962). Erbliche Unterschiede zwischen Ertragseigenschaften von Bienen. *Proc. Inter. Cong. Entomol. 11*, 510–516.

Polhemus, M. S., Lush, J. L., and Rothenbuhler, W. C. (1950). Mating systems in honey bees. *J. Hered. 41*, 151–155.

Provine, W. B. (1971). "The Origins of Theoretical Population Genetics." University of Chicago Press, Chicago.

Rinderer, T. E. (1977). Measuring the heritability of characters of honeybees. *J. Apic. Res. 16*, 95–98.

Rinderer, T. E., Collins, A. M., and Brown, M. A. (1983). Heritabilities and correlations of the honey bee: response to *Nosema apis*, longevity, and alarm response to isopentyl acetate. *Apidologie 14*, 79–85.

Rothenbuhler, W. C. (1960). A technique for studying genetics of colony behavior in honey bees. *Am. Bee J. 100*, 176, 198.

Rothenbuhler, W. C., Kulinčević, J. M., and Thompson, V. C. (1979). Successful selection of honeybees for fast and slow hoarding of sugar syrup in the laboratory. *J. Apic. Res. 18*, 272–278.

Soller, M., and Bar-Cohen, R. (1967). Some observations on the heritability and genetic correlation between honey production and brood area in the honeybee. *J. Apic. Res. 6*, 37–43.

Vesely, V., and Siler, R. (1963). Possibilities of the application of quantitative and population genetics in bee breeding. *Proc. Inter. Apic. Cong. (Apimondia) 19*, 120–121.

CHAPTER 12

Selection

THOMAS E. RINDERER

I. INTRODUCTION

Honey-bee breeding is a small subset of the much larger enterprise of farm-animal breeding and, to some extent, of plant breeding. The comparative enormity and profitability of, for example, the cattle, swine, and poultry industries are the economic underpinnings of population-genetic and stock improvement theory. A fuller review of this theory, which sets the foundations of honey-bee breeding, can be found in Lush (1945), Mather (1949), Li (1955), Lerner (1958), Falconer (1960), and much of this volume. Lerner and Donald (1966) provide a less mathematical approach to the subject that considers the interplay between the technology and the economics of breeding, which should be of interest to both geneticists and practical bee breeders.

Most breeding programs are designed so that the "best" parents are selected and used to produce the next generation. In general terms, the "best" parents should produce the "best" offspring and the average quality of the stock is improved. Such guidelines are deceptively simple, especially with honey bees.

This chapter suggests ways to define and measure honey-bee characteristics so that the "best" parents for honey-bee stock improvement programs can be selected. Home computers make possible the application of fundamental selection theory to the selection programs of small family-owned bee-breeding enterprises. Since these form the majority of bee-breeding enterprises, this chapter speaks primarily to the needs of smaller programs.

II. BEE IMPROVEMENT TOOLS

Humans have been involved in bee husbandry since prehistory (Crane, 1983). However, the breeding of bees and hence their genetic improvement awaited the developments of movable-frame beekeeping equipment (Langstroth, 1853) and methods to produce large numbers of queen bees (Doolittle, 1888). The addition of instrumental insemination techniques for honey bees (Watson, 1927; Nolan, 1932) and their later improvement (Laidlaw, 1944; Mackensen, 1947, 1948; Mackensen and Roberts, 1948; Harbo, Chapter 15) have brought the mating of bees under complete control and "opened a wide door to both bee breeding and genetics" (Cale and Rothenbuhler, 1975).

Throughout the world, apiculturalists have looked through this door and agreed that it is not only desirable but, in some cases, necessary that the discoveries of the past 150 years be used to improve honey-bee stocks or to maintain already improved stocks. However, major differences of opinion exist concerning the definition of improved stock.

III. DEFINING IMPROVEMENT

A. Commercial Interests

World-wide, apiculture is extremely diverse. As an agricultural enterprise it involves corporations that handle many thousands of colonies with modern mechanical technology, small single-family businesses, still smaller side-line businesses used to supplement incomes from other sources, and rural development programs designed to increase income of individuals and groups in developing nations.

These commercial enterprises realize profits mostly from the sale of honey and wax to the general public and of queen bees, colonies, and package bees to other beekeepers. In some parts of the world, colonies of bees are rented for pollination services. Additionally, royal jelly, propolis, and pollen are sold, usually in small quantities. World-wide, some commercial beekeepers make most of their living producing and selling such products.

This wide variety of commercial interests leads to many differing views of what constitutes stock improvement. Improved honey production, handling qualities of bees, colony population growth, wax production, and efficient pollination of specific crops may or may not be desired depending upon the specific origin of a beekeeper's profit. Generally, improved stocks which show increased production in a broad sense with reduced management costs are desired.

B. Recreational Interests

Many beekeepers in much of the world keep bees for recreation. In some areas recreational or hobby beekeepers far outnumber commercial beekeepers. In these areas the commercial beekeepers often earn much of their living by supplying queens, bees, materials, and advice to the hobbyists. Although recreational beekeepers usually value the harvest of at least some honey, characteristics other than productivity often lead their lists of desirable improvements. Some urban hobbyists require especially gentle bees. Others, particularly in Europe, want bees that are typical of the subspecies that evolved in their areas. Still others desire bees having a specific morphological characteristic such as a black or yellow body color.

C. Geographical Range

Beyond this, the range of commercial and hobby beekeeping with *Apis mellifera* extends throughout the world from the tropics to above the arctic circle. This range includes the Old World with its wide variety of ecogeographical subspecies and, through human introduction, the New World areas of the Americas, Australia, and New Zealand, and to some extent the home ranges of *A. cerana, A. dorsata,* and *A. florea.*

This geographical disparity leads to still more variation in bee stock breeding goals. Beekeepers in temperate zones consider overwintering abilities and properly timed spring build-up patterns important. Beekeepers in tropical areas consider colony maintenance through rainy seasons or long droughts and rapid build-up when nectar flows begin important. These examples could head very long lists of stock quality needs, and hence breeding goals arising from the geographical range of beekeeping.

D. Species Diversity

The eastern honey bee, *A. cerana,* is a hive bee and is commercially exploited with much the same bee-husbandry techniques used with *A. mellifera.* Movable frame hives, queen-rearing, and even instrumental insemination (Woyke, 1975) are usable techniques with this species.

Currently, the use of *A. cerana* is restricted to its naturally occurring range (Ruttner, Chapter 12). This is because *A. mellifera* is considered a better honey producer. However, as the pests and parasites of the genus become spread throughout the world there may be specific uses for pollination or hobby beekeeping in other parts of the world where *A. cerana* is not currently found.

Much of the discussion concerning the effects of the commercial and recreational interests and geographic range on *A. mellifera* breeding goals also applies to *A. cerana*. Additional concerns, such as the improvement of pollination effectiveness on crops originating in Europe or Africa, may be breeding goals unique to *A. cerana*.

Additionally, *A. dorsata* and *A. florea* are not "kept" but feral. Colonies are commercially exploited through intensive organized honey hunting. The migratory and absconding nature of these species is a major obstacle to their commercial management. However, a beginning has been made at managed "migrations" of *A. florea* (Dutton and Simpson, 1977). Further management developments may stimulate attempts to breed more commercially desirable stocks of these species.

E. Many Stocks

Thus, what constitutes improvement in bee stock is dependent on a number of considerations. The tremendous variation in the world of beekeeping, including beekeeping interests, existing stocks of bees, local climates and floral resources, complicates the definition of bee stock improvement. Clearly, several answers and stocks are required (Rinderer, 1977), each suited to the needs and desires of specific groups of beekeepers.

IV. BREEDING FOR CONFORMATION

A. Ecogeographical Subspecies

Conformation breeding is the breeding of stock to fit a defined "ideal" phenotype for a breed. Because humans have only recently been able to breed bees, there are no recognized breeds defined by conformation standards as there are in other domestic animals. Indeed, professional bee-breeder associations do not even have mechanisms to set standards and recognize breeds. The ecogeographical subspecies of bees (Ruttner, Chapter 2) replace breeds in the minds of some apiculturalists. Subspecies descriptions, especially those detailing morphological averages, are often misunderstood to be equal to breeder association conformation standards which define the ideal breed "type" and thereby establish the selection criteria for stock breeders.

Breeding bees toward a "type" set by the average of an ecogeographical subspecies is founded on the idea that the "best" bees for an area are those that evolved in that area. A related concept is that since such bees have become "best" for an area through the evolutionary processes of natural selection, selection to improve the stock is unnecessary. The role of artificial selection is to select against "foreign" genes introduced by the importation of bees representing other subspecies.

Certainly, such thoughts are not completely wrong. The evolutionary formation of subspecies is clearly a response to climatic and ecological differences. Ecogeographical subspecies represent the "best" bee for the naturally occurring ecology of an area.

However, neither are such thoughts completely correct. The "best" produced by natural selection is most certainly the best inclusive reproductive fitness suited to an area. It is doubtful that such inclusive fitness includes traits which best coincide with the economic interests of beekeepers, especially in areas extensively changed by modern agriculture.

It is even more doubtful that the economic value of bees can be improved or maintained by breeding programs based on morphology alone. The success of bees, both in nature and as agricultural animals, is too closely tied to behavior. Only when morphology is shown to be highly correlated genetically to desired economically related behavior will it serve as a useful measure in selection. Of course, some apiculturalists value morphological characteristics for their aesthetic qualities. Because of this, even breeding programs strongly aimed at producing stock having superior performance sometimes include a few morphological traits in their selection criteria.

Many apiculturalists wish to preserve local ecogeographical subspecies. In Europe, this desire sometimes is used to support arguments for the exclusive use of "conformation" breed stocks in a subspecies area. The goal of preservation is certainly desirable. However, it must be remembered that the complete description of a subspecies includes the variation of both morphological and nonmorphological characters as well as the average value of characters. The inclusion of bees as protected wild animals in larger nature preserves coupled with the exclusion of beekeeping in such areas would better serve the goals of preservation.

B. Commercial Stocks

In a very few instances, long-term selection by commercial bee breeders has produced surprisingly uniform stock which under the rules of large-animal breeder associations might qualify as a breed. The conformation standards of these stocks have been established by single beekeepers through years of practical beekeeping. Usually, such stocks have combinations of three to five strikingly improved characteristics. In other respects they seem similar to stocks of bees from which they were derived. Breeding goals with such stocks generally involve maintenance of stock quality rather than further significant improvement.

V. BREEDING FOR IMPROVEMENT

The majority of bee-breeding programs involve selection for specific stock-improvement goals. All such programs have resource limitations, and decisions in several areas must be made in order to optimize program success.

A. Genetic Theory

The foundations of selection rest in the mathematical disciplines of quantitative and population genetics. The basic starting point to the understanding of selection is the concept that phenotype (the characteristics of an animal which can be observed) results from the influences of the animal's genetics, the environment in which the animal is found and the interaction between these two factors. This concept can be extended to populations of animals. The variation of phenotypes in a population of animals can be attributed to the variation resulting from the genetics of the population, the variation resulting from the environment of the population and the interactions between these two sources of variation. The variation in phenotypes resulting from genetics can be further subdivided into different types of genetic events. Variance due to additive genetic events is especially important. Additive variance is the chief genetic cause of resemblance between relatives and therefore the chief determinant of how easily a population can be improved by selection (for further discussion see Collins, Chapter 11). One mathematical relationship for populations which has special interest, because it predicts response to selection, is

$$R = h^2 S \tag{1}$$

where R is the predicted improvement or response resulting from selection, h^2 is the heritability of the characteristic under selection in the population, and S is the selection differential. Further, R is the difference between the average phenotypic value (the direct measures) of the parental population and the average value of the offspring of the selected parents, while h^2 is an estimate of the genetic variation in the population which is susceptible to change through genetic selection. As a proportion it ranges from 0 to 1. It serves as a guide to determine the reliability of phenotypic measures as measures of breeding value. The selection differential S is measured in standard deviation units as the difference between the average of the population of potential parents and the average of the selected parents. Since S is expressed in standard deviation units, so is R.

B. Measurement

Regardless of all else, the accurate measurement of characteristics is essential to a selection program's success. Only through accurate measurement can the relative breeding value (Collins, Chapter 11) of potential parents be determined.

Accurate measurement reduces the phenotypic variance. Since h^2 is the ratio of additive genetic variance (V_a) to phenotypic variance (V_p);

$$h^2 = \frac{V_a}{V_p} \tag{2}$$

reductions in V_p have the desirable effect of increasing h^2.

The most important way to increase accuracy is to make measurements on bees which are in a common environment and have had similar management histories. The group of colonies should have had an equal start at some point in their recent histories. Ideally, they then would be measured when apiary or other testing conditions permit a full expression of the colonies' capabilities. For example, honey production measures should be made during nectar-flow conditions which are typical of conditions in which the improved stock is expected to perform.

Second, more precise measures or repeated measures as a method to gain precision may improve accuracy. Whether or not they do can be determined for specific selection programs by comparing the ranking of breeding values obtained from more and less precise methods. Measurement and associated record keeping are expensive, yet they are essential to all breeding programs. Thus, the costs of measurement are important. Small increases in accuracy gained from costly measurement systems are likely valueless. The costs would reduce the number of colonies that can be evaluated. This may force a breeder to decrease the selection differential (Collins, Chapter 11) in an attempt to avoid the problems attending inbreeding (Woyke, Chapter 4; Laidlaw and Page, Chapter 13). Thus, the predicted response to selection may be reduced by using costly and precise measurements which do not greatly improve comparisons of breeding values.

1. Field Measurements

The nature of beekeeping often requires the field measurement of characteristics. Doubtless, field conditions increase environmental variation. Unless attention is given to reducing this variation, measurements may be sufficiently inaccurate that selection yields little or no improvement because of the relationships shown in Eqs. (1) and (2). In effect, the h^2 can be substantially reduced.

Uniform management of colonies is one important approach to reduce environmental variation in field tests. Using queens of equal ages introduced to equal-sized populations of bees, and equal amounts of equipment, diet supplements, drug treatments, etc., are useful. Depending on what is being measured, the time of day and weather conditions might also be important.

Often, commercial breeding programs require several apiary locations because large numbers of colonies will be tested. Different apiaries, even when they are near, will often have average scores that are quite different. There is no way to completely control such between-apiary variance, but its effects on the accuracy of measurement can be reduced. We can assume that the entire difference between the average scores between apiaries is due to location differences. Probably this is not entirely true since it ignores the likely interactions between the bee stock and the variety of local conditions which differ between apiary locations. Nonetheless, if we make this assumption and through using certain statistical tests discussed in most statistics books [see Snedecor and Cochran (1967)] find that the individual colony scores in each apiary are normally distributed, then the process of comparing colonies in different apiaries becomes relatively straightforward. First, each apiary is described in terms of its own mean and standard deviation. Second, the individual responses can then be transformed into standard deviation units and compared, in order to select parents. An individual's position in a population is called its z score and is calculated as

$$z = \frac{X - M}{s} \tag{3}$$

where X is the colony's score, M is the apiary's average score, and s is the standard deviation of the apiary's scores. The effects of Eq. (3) are shown in Fig. 1. A score which equals the apiary average becomes zero, and most scores range from -3 to $+3$. These scores can be compared between apiaries and thus permit the identification of the best parents in a breeding program regardless of the main effects arising from apiary location.

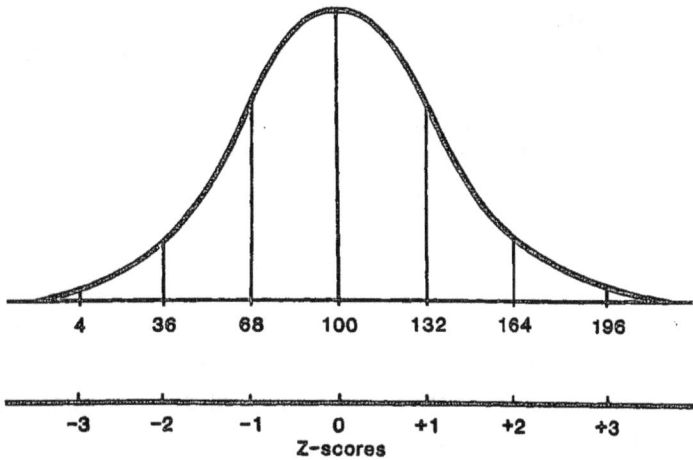

Fig. 1. Illustration of the conversion of a normal distribution of phenotypic scores to z scores, using a normal distribution of honey production values in kilograms.

2. Correlated Measurements

The difficulties of accurately measuring honey-bee characteristics in the field cannot be underestimated. Honey-production measures require nectar-flows; pollination-activity measurements depend strongly on the attractiveness of competing floral sources (Martin and McGregor, 1973); disease-resistance measures depend on the occurrence of epizootics or costly testing. Climate, weather, drifting, unequal colony strengths, and unequal past histories of colonies all make field measurements of colonies less accurate. Also, the direct field observation of characteristics can be costly. If, for example, honey production is measured throughout a year, the cost is measured as only one generation of selection in a year and a lack of accuracy caused by a complete season of environmental variation. A correlated measure might be more accurate or at least be less costly.

The problems attending the measurement of honey production and its cost in generation time have stimulated the development of several measures correlated with honey production. Hoarding behavior (Milne, 1980a), worker-bee longevity (Milne, 1980b), pupal weight (Milne, 1980c), pollen-basket measurements (Milne and Pries, 1984), and a single day's honey production (Szabo, 1982), have all been identified as correlates of overall honey production.

Many other important characteristics of honey bees can be evaluated through correlated measurements. Tests of honey-bee colony defense (Collins, 1979) and response to various diseases (Rinderer *et al.*, 1975; Rinderer and Elliott, 1977; Milne, 1983) have all been developed. Indeed, almost every scientific experiment studying bees uses a measurement system which might be developed into a useful correlated measure of an important bee characteristic.

The development of useful correlated measures or indirect measurement (Falconer, 1960), including laboratory testing (Milne, 1985; Rinderer, 1977), is an important current interest of honey-bee geneticists. There are no widely accepted guidelines for the development of correlated measures for selection programs. Generally, large-animal and poultry breeders have much closer control of environmental sources of variance and can better rely on direct measures of the characters that interest them. Also, they are usually concerned with physiological characters such as weight gain or butterfat production, rather than the results of complex behavioral processes such as honey production or colony defense. Nonetheless, statistics and quantitative genetics contain all the theory necessary to establish guidelines for the development of useful correlated measures.

First, relatively simple studies are required to determine if candidate measurement systems produce results which are correlated with the results of measurements of the actual trait to be improved. Such studies, if properly designed, also will provide estimates of phenotypic means and variation.

The evaluation of such tests involves considering the correlation value, which can range from 0 to 1, the statistical significance of the correlation, the number of colonies used to produce the correlation, extreme data pairs, and the relative costs of the two evaluation systems. At this point, correlative measurement systems which produce low correlations (perhaps 0.2) or prove more costly than direct measurement might be discarded. Statistical significance can only properly be interpreted when the numbers of colonies measured are considered. A correlation of 0.764 from measurements on 10 colonies would be considered highly significant statistically, as would a correlation of 0.081 from measurements on 1000 colonies (Snedecor and Cochran, 1967). However, since the goal is to create ways to accurately measure the breeding value of parents, the first correlation would be considered highly *important* but the second would be considered trivial. Extreme pairs of data from only one or two colonies may create significant and apparently important correlations. The likely sources of single colonies producing scores sufficiently extreme and strong to substantially alter correlation are chance, genetic dominance, or extreme environmental events. Additive genetic variance, the substance of breeding value, is a function of the population generally, and the scores of a single colony should not strongly change an important correlation.

Second, the value of the correlated measure to a selection program must be demonstrated. Experiments can be designed using selected matings which will permit the calculation of h^2 and genetic correlations as well as providing additional information on phenotypic correlations. Such experiments are generally costly with bees, and this expense justifies the use of simple correlation studies.

The results of heritability and correlation studies provide the basis for evaluating the value of correlated measurement systems. Equation (1) is the foundation of this evaluation. Response of desired characteristics (R_d) to selection based on correlated measures is

$$R_d = h_c^2 S r_g \cdot \tfrac{2}{3} \tag{4}$$

where r_g is the genetic correlation between the two measurement systems, h_c^2 is the heritability of the correlated measure, and the $\tfrac{2}{3}$ is the result of drones being one generation behind queens in a population (Moran, 1984).

Since the response of desired characteristics based on direct measures [Eq. (1)] can also be predicted, the ratio of

$$R_d/R \tag{5}$$

compares the relative value of the two measurement systems. However, this assumes that the two measurement systems permit the same sized groups of parent and progeny populations to be measured and that the costs of both measurement techniques are identical. Neither assumption is likely to be valid, and the apparent simplicity of Eq. (5) must be expanded to include the costs of measurement and their real effect on possible changes in the selection differential (S) as it is limited by other selection program constraints. Thus, the ratio of economic responses ER_d/ER is

$$ER_d/ER = [(h_c^2 S_c r_g/M_c)/(h^2 S/M_d)] \tag{6}$$

where M_c and M_d are the total estimated monetary costs of measurement for each program, respectively. In the estimates of M_c or M_d, users of Eq. (6) should account for increased costs incurred by operating more colonies required to improve S or S_c. The numerical value calculated for Eq. (6) is then a comparison of the amounts of population improvement in standard deviation units expected per unit of money spent in stock evaluation.

Third, model selection programs comparing direct and correlated measures or different correlated measures would provide final demonstrations of a correlated measure's efficiency, and additional estimates of h^2, plus genetic and phenotype correlations.

There are several ways that a correlated measure can be better or worse than a direct measure. Environmental variance can be reduced substantially and h^2 will become substantially higher, such that $h_c^2 r_g$ is greater than the h^2 of the direct measurement system. However, $h_c^2 r_g$ might be less than h^2 but the actual mechanics of measurement might be sufficiently simple and relatively inexpensive that S_c can be dramatically improved over S. Alternately, correlated measures might substantially improve R based on a strongly increased h^2 but cost so much that the numbers of progeny and S must be reduced to unacceptable levels.

C. Selection of Stock for Several Characters

Generally, bee breeders would like to improve stock for more than one characteristic simultaneously. This is understandable since the economic value of a stock depends on several characteristics.

1. Tandem Selection

One approach to improving several characteristics in a stock is using sequential or tandem selection. One character after another receives the breeder's attention. The usefulness of this approach depends on the genetics of the characteristic. Quantitative characteristics genetically regulated by several loci and multiple alleles are poor candidates for tandem selection. When selection is relaxed for such characteristics, the average population response tends to return toward original levels. If the improved characteristic was based on selection for additive effects and the average population response does not tend to return toward the original levels, then the additive effects which are useful for the improvement of other characteristics have probably been lost.

There is one valuable use for tandem selection in well-planned honey-bee selection programs. Very frequently a breeder desires specific morphological characteristics in the stock. Often, such characteristics depend upon relatively simple genetic events (Tucker, Chapter 3), and a careful assembly of base stocks prior to embarking on a selection program will produce a parental population very nearly uniform for the desired morphological features and still containing ample additive genetic variance to support the improvement of desired quantitative characteristics.

2. Independent Culling Selection

A second approach to simultaneously improving several characteristics is to evaluate each characteristic separately, determine S for each characteristic, and only accept parents for the next generation which meet the culling standards for all characteristics. This approach to bee stock improvement certainly would work but does present difficulties. The number of characters which can be placed under selection becomes very restricted (Table 1). It quickly becomes apparent that independent culling selection can only be

TABLE 1. Relation Between the Number of Selected Characteristics and the Number of Progeny Which Must be Produced Each Generation[a]

Number of characters	Probability of a single colony being selected	Number of colonies required each generation
1	0.2	500
2	0.04	2,500
3	0.008	12,500
4	0.0016	62,500
5	0.00032	312,500
6	0.000064	1,562,500

[a] These relationships assume that the working breeding population is 100 colonies and that the top 20% of the population meets the culling level for each characteristic.

done for a few characters, and then only by quite large corporations or cooperatives.

3. Selection Index Breeding

Selection index breeding assumes that the breeding value of potential parents can be expressed in a single number (I). Such a number, or selection index score, would be a compound value derived from the colony's individual phenotypic scores, the h^2 of the characteristics, the genetic correlations between these characteristics, and the relative economic value of the characteristics as judged by the bee breeder.

In a selection program, the I for each potential parent of the next generation is calculated and those having the highest ones are used as parents. Hazel and Lush (1942) and Hazel (1943) have compared selection index breeding to other methods of selection for multiple characteristics and demonstrated that it provides the most rapid method of improving the economic value of a stock.

As an illustration of the mechanics of building a simple selection index, consider two characteristics, honey production and colony defensive behavior. Honey production values can be converted to z scores as explained earlier and illustrated in Fig. 1. In the same way, quantitative measure of colony defensive behavior can also be converted to z scores at the apiary level and combined.

Each colony then has two z scores, one for each characteristic. The z-score conversions bring the two measures into the same scale of standard deviation units. This allows the two scores to be weighted for economic value, h^2, and genetic correlations without concern about adjusting for unequal scales of measurement.

A selection index (I) score including economic value may be expressed as

$$I = z_{hp}V + z_d \qquad (7)$$

where z_{hp} is the colony's honey production breeding value, z_d is the colony's defense behavior breeding value, and V is the relative economic value of honey production compared to defensive behavior. Equation (7) assumes that the economic value of increasing the defensive behavior by one standard deviation is the standard of comparison and has a value of 1. V, the relative economic value of increasing honey production by one standard deviation, is then set by the breeder according to his breeding goals. If honey production improvement is considered half as important as defensive behavior improvement then V would be set at 0.5; if twice as important, then V would be set at 2. Equation (7) is the simplest form of a selection index.

Since heritabilities are involved in estimating responses to selection [Eq. (1)], they must be incorporated into the selection index in a way that does not dilute the relative economic value assigned to the two characteristics. The equation

$$I = z_{hp}V(h_{hp}^2/h_d^2) + z_d \qquad (8)$$

adjusts I to accommodate the differential h^2 values of the two characters while maintaining the economic evaluations of the breeder. This formula will favor improvement of the characteristic having the higher h^2. When building a base stock prior to selection, a breeder would probably wish to include a large number of superior colonies for the character with the lower heritability.

The genetic correlation (r_g) between two characteristics is generally the correlation of breeding values. It estimates the proportion of the total additive genetic variance for both characteristics which affect both characters. It can be accommodated into the selection index as

$$I = z_{hp}V(h_{hp}^2/h_d^2) + z_d(1 - r_g) \qquad (9)$$

Where r_g is positive or near zero, selection for both characteristics will predictably improve both. Where r_g is strongly negative, selection has much less chance of simultaneously improving both characteristics. In the example of honey production and defensive behavior, no estimate of r_g exists. However, hoarding behavior, which is a measure related to honey production, is positively correlated to one aspect of defensive behavior and negatively correlated to two other aspects (Collins et al., 1984; Collins, Chapter 11, Table V). Nonetheless, simultaneous selection using a selection index can improve honey production and reduce defensive behavior (unpublished Honey-Bee Breeding, Genetics, and Physiology Laboratory data).

The selection index [Eq. (9)] presented here is limited to two characteristics. More elaborate selection indexes can be developed (Hazel, 1943; Lush, 1948; Falconer, 1960), which use multiple regressions and covariance components to construct the index. Where all the required genetic parameters are known for more than two characteristics, a selection index can be developed from multiple analyses of covariance. When necessary, bee breeders can probably find help in developing formulas specific to their own needs from animal breeders at universities near them.

When genetic parameters are not known, or only some are known, a selection index is still useful. As a minimum, estimates of breeding value based on phenotypic scores can be coupled with the relative economic value of each trait [Eq. (7)]. Such a procedure would be an improvement over independent culling levels, tandem selection, or off-hand field evaluations.

D. Special Constraints

The propagation of bee stock in a selection program provides a challenge to bee breeders. The ease of producing many daughter queens from one or a few colonies tempts a breeder to use too few parents. Certainly, a strong S will predictably improve stock more rapidly. However, one which is too strong will eliminate valuable additive genetic variance from the potential breeders of the next generation. Also, bees suffer more from inbreeding depression than other animals (Woyke, Chapter 4), and special care is required to maintain general genetic heterozygosity while changing the gene frequencies for the trait under selection. Laidlaw and Page (Chapter 13) and Moritz (1984) provide guidance concerning the minimum numbers of parents required to satisfactorily maintain general stock quality.

Certainly no long-term breeding program should use fewer colonies. Probably, most breeding programs employing selection would benefit from using more than the minimum numbers. Economics will dictate the specific numbers desirable for specific programs.

ACKNOWLEDGMENTS

This chapter was prepared in cooperation with Louisiana Agricultural Experiment Station. C. P. Milne, R. F. A. Moritz, and H. A. Sylvester made several useful suggestions for improvements on earlier manuscript drafts.

REFERENCES

Cale, G. H., and Rothenbuhler, W. C. (1975). Genetics and breeding of the honey bee. *In* "The Hive and the Honey Bee" (Dadant and Sons, ed.), pp. 157–184. Dadant and Sons, Hamilton, Ill.

Collins, A. M. (1979). Genetics of the response of the honeybee to an alarm chemical, isopentyl acetate. *J. Apic. Res. 18*, 285–291.

Collins, A. M., Rinderer, T. E., Harbo, J. R., and Brown, M. A. (1984). Heritabilities and correlations for several characters in the honey bee. *J. Hered. 75*, 135–140.

Crane, E. (1983). "The Archaeology of Beekeeping." Cornell University Press, Ithaca, New York.

Doolittle, G. M. (1888). "Scientific Queen Rearing." American Bee Journal, Hamilton, Ill.

Dutton, R., and Simpson, S. (1977). Producing honey with *Apis florea* in Oman. *Bee World 58*, 71–76.

Falconer, D. S. (1960). "Introduction to Quantitative Genetics." Ronald Press, New York.

Hazel, L. N. (1943). The genetic basis for constructing selection indexes. *Genetics 28*, 476–490.

Hazel, L. N., and Lush, J. L. (1942). The efficiency of three methods of selection. *J. Hered. 33*, 393–399.

Laidlaw, H. H., Jr. (1944). Artificial insemination of the queen bee (*Apis mellifera* L.): morphological basis and results. *J. Morphol. 74*, 426–465.

Langstroth, L. L. (1853). "Langstroth on the Hive and the Honey-Bee, a Beekeeper's Manual." Saxton, New York.

Lerner, I. M. (1958). "The Genetic Basis of Selection." Wiley, New York.

Lerner, I. M., and Donald, H. P. (1966). "Modern Developments in Animal Breeding." Academic Press, London.

Li, C. C. (1955). "Population Genetics." University of Chicago Press, Chicago.

Lush, J. L. (1945). "Animal Breeding Plans." Iowa State University Press, Ames, Iowa.

Lush, J. L. (1948). "Animal Breeding Plans." Collegiate Press, Ames, Iowa.

Mackensen, O. (1947). Effect of carbon dioxide on initial oviposition of artificially inseminated and virgin queen bees. *J. Econ. Entomol. 40*, 344–349.

Mackensen, O. (1948). A new syringe for the artificial insemination of queen bees. *Am. Bee J. 88*, 412.

Mackensen, O., and Roberts, W. C. (1948). "A Manual for the Artificial Insemination of Queen Bees." U.S.D.A. Technical Bulletin 250, U.S. Government Printing Office, Washington, D.C.

Martin, E. C., and McGregor, S. E. (1973). Changing trends in insect pollination of commercial crops. *Annu. Rev. Entomol. 18*, 207–226.

Mather, K. (1949). "Biometrical Genetics: The Study of Continuous Variation." Methuen Press, London.

Milne, C. P. (1980a). Laboratory measurement of honey production in the honeybee. 1. A model for hoarding behaviour by caged workers. *J. Apic. Res. 19*, 122–126.

Milne, C. P. (1980b). Laboratory measurement of honey production in the honeybee. 2. Longevity or length of life of caged workers. *J. Apic. Res. 19*, 172–175.

Milne, C. P. (1980c). Laboratory measurement of honey production in the honeybee. 3. Pupal weight of the worker. *J. Apic. Res. 19*, 176–178.

Milne, C. P. (1983). Honey bee (Hymenoptera: Apidae) hygienic behavior and resistance to chalkbrood. *Ann. Entomol. Soc. Am. 76*, 384–387.

Milne, C. P. (1985). The need for using laboratory tests in breeding honeybees for improved honey production. *J. Apic. Res. 24*, 237–242.

Milne, C. P., and Pries, K. J. (1984). Honeybee corbicular size and honey production. *J. Apic. Res. 23*, 11–14.

Moran, C. (1984). Sex linked effective population size in control populations, with particular reference to honeybees (*Apis mellifera* L.). *Theor. Appl. Genet. 67*, 317–322.

Moritz, R. F. A. (1984). Selection in small populations of the honeybee (*Apis mellifera* L.). *Z. Tierz. Züchtungsbiol. 101*, 394–400.

Nolan, W. J. (1932). "Breeding the Honey Bee under Controlled Conditions." U.S.D.A. Technical Bulletin 326. U.S. Government Printing Office, Washington, D.C.

Rinderer, T. E. (1977). A new approach to honey bee breeding at the Baton Rouge USDA Laboratory *Am. Bee J.* 117, 146–147.

Rinderer, T. E., and Elliott, K. D. (1977). Influence of nosematosis on the hoarding behavior of the honeybee. *J. Invertebr. Pathol.* 30, 110–111.

Rinderer, T. E., Rothenbuhler, W. C., and Kulinčević, J. M. (1975). Responses of three genetically different stocks of the honeybee to a virus from bees with hairless-black syndrome. *J. Invertebr. Pathol.* 25, 297–300.

Snedecor, G. W., and Cochran, W. G. (1967). "Statistical Methods." Iowa State University Press, Ames, Iowa.

Szabo, T. I. (1982). Phenotypic correlations between colony traits in the honey bee. *Am. Bee J.* 122, 711–716.

Watson, L. R. (1927). Controlled mating of the honeybee. *Am. Bee J.* 67, 300–302.

Woyke, J. (1975). Natural and instrumental insemination of *Apis cerana indica* in India. *J. Apic. Res.* 14, 153–159.

CHAPTER 13

Mating Designs

HARRY H. LAIDLAW, JR., AND ROBERT E. PAGE, JR.

I. COMPLEXITIES OF HONEY-BEE BREEDING

Mating with its redistribution of genetic material is the central element of genetics and breeding. Therefore, continuing choice of mating designs is basic to breeding.

A. Honey Bees and Breeding

Bee breeding has several inherent complexities that have been responsible, in some measure, for its disappointing progress in the past; these include (1) controlled mating of virgin queens, (2) multiple insemination of queens, (3) sperm storage in the spermatheca and sperm utilization, (4) haplo – diploidy and sex determination, (5) caste determination and differentiation, and (6) sensitivity of bees to environments. These are serious obstacles to honey-bee improvement. Some of them, however, present opportunities for unique approaches to bee breeding that can be exploited.

B. Multiple Mating and Sperm Distribution

It is well established that queens mate with several drones — seven to 20 on one or more mating flights (Oertel, 1940; Roberts, 1944; Triasko, 1951, 1956; Taber and Wendel, 1958; Woyke, 1960, 1964; Kerr *et al.*, 1962; Adams *et al.*, 1977). Varying number of sperm from each mate reach the spermatheca, where they disperse throughout it (Kepeňa and Fl'ak, 1980; Crozier and Brücker, 1981; Laidlaw and Page, 1984; Moritz, R. F. A., 1983; Page *et al.*, 1984). Each queen, therefore, takes into her spermatheca at mating a small sample of the sperm available in the local population. When numerous queens are mated, the sperm stored in the spermathecae of all the mated queens together represents most of the collective genetic pool of the population.

C. Sperm Utilization

As a queen lays fertilized eggs, she withdraws a mixture of sperm in a probable constant volume of spermathecal fluid to fertilize the eggs (Harbo, 1979), and by this means each drone that mated the queen sires a subfamily of diploid progeny, i.e., the worker bees and queens. (Any diploid drones are destroyed as young larvae by the worker bees and can develop as drones only under special circumstances.) All possible subfamilies are probably represented at all times in the colony (Page and Metcalf, 1982; Laidlaw and Page, 1984; Mortiz, 1983; Page *et al.*, 1984.

D. Sex Determination

The method of sex determination in bees is a breeding problem that is almost as intractable as control of mating. Sex is determined by multiple alleles of a single locus (Mackensen, 1943, 1951, 1955; Woyke, 1963a). The normal male, or drone, is haploid and has in his genome any one of the sex alleles. Females are diploid and are heterozygous for any two of the sex alleles. A mature egg, one that has undergone meiosis and has a single genome in the pronucleus, fertilized by a sperm having a sex allele like that of the egg genome will produce a diploid male larva. These larvae are eaten by the worker bees soon after they hatch (Woyke, 1963b). The proportion of brood that is lost because of this varies with the number of different sex alleles in the population and the number of drones mating with a queen. Sex-allele homozygosity can result in actual inviability of up to 50% of the eggs laid. Measured populations have been shown to have from six to 19 sex alleles (Mackensen, 1955; Laidlaw *et al.*, 1956; Kerr, 1967; Woyke, 1976; Adams *et al.*, 1977).

Any inviability of brood can prevent the economic build-up of the worker-bee population (Laidlaw and Eckert, 1950, p. 120; Woyke, 1980, 1981), which increases the difficulty of colony evaluation and breeder queen selection. Breeding programs must be designed to avoid loss of sex alleles in the breeding population, or to eventually restore the alleles should their loss occur.

E. Colony Structure

A normal honey-bee colony consists of the queen as the mother of the colony and the workers as her daughters. Thus there are always two generations. Drones are also present during most of the brood rearing time of the year. Each of the three castes has a different role in reproduction.

1. Queens

Queens are the *primary reproductive* individuals of European races of *Apis mellifera*. The genic composition of all honey-bee gametes, female *and* male, is established in the eggs at meiosis. Spermatogenesis in diploid drones may be an exception. Meiosis assorts the maternal and paternal chromosomes of the queen's genotype and mixes or commingles them in complete haploid sets in the gametes. As a result of meioses, the diploid queen produces many genetically different gametes and a corresponding diversity of drones.

2. Drones

Drones develop as haploids from gametes that have a mixture of maternal and paternal chromosomes of their mother's genotype, and it is apparent that most drones from hybrid queens are themselves hybrids. Drones are usually called queens' sons, though in actuality they represent queens' gametes (Newell, 1915) and are genetically intercalary between mother and mate. It is convenient, nevertheless, to refer to drones as sires or fathers.

Meioses abort in drones as sperm are formed (Snodgrass, 1925), and the sperm of each drone are still genetically the gamete of his mother. Without changing the genome of the gamete that gave rise to him, the drone multiplies it to about 10 million identical ones and reverses the sex of the embodying cells from female to male. His other function is to convey the resulting spermatozoa to the reproductive tract of his mate. Because the drone simply amplifies a genome without adding, subtracting, or changing its genetic composition, except for mutations that may possibly arise, he is merely an agent to propagate one of his mother's gametes. He is a *complementary reproductive* in that he enables the female to function genetically as a male.

3. Workers

Workers retain a reproductive ability that is greatly diminished. They can lay eggs, and these undergo meioses as in the queen. Very few workers of European races lay eggs, though some do that usually develop into drones and rarely into females (Mackensen, 1943, 1951, 1955; Tucker, 1958). Workers of the Cape bee of South Africa are prone to oviposit (Onions, 1909, 1912; Anderson, 1963; Taber, 1983), and the eggs may produce other

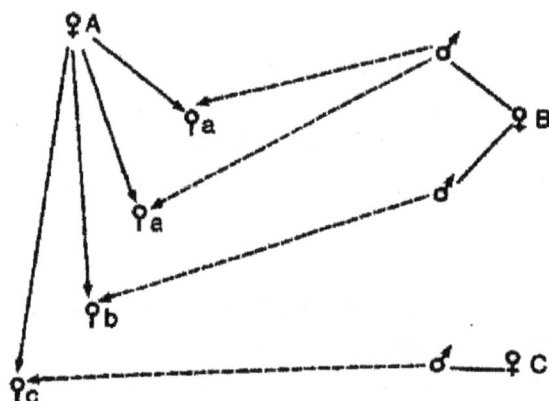

Fig. 1. Family relationships of female bees of a colony: multiple mates, mutiple drone mothers. Each drone mate sires a subfamily of super sisters. a–a, Super sisters. Same queen mother, same drone mother, same physical sire. a–b, Full sisters. Same queen mother, same drone mother, different physical sires. b–c, Half sisters. Same queen mother, different drone mother. ———, Eggs; — — —, sperm. Uncrossed female symbols are workers.

workers. In view of this, worker bees should be considered *facultative para-reproductives.*

4. Subfamilies

With the exception of the queen mother, the female population of a colony consists of subfamilies of highly related super sisters (Rothenbuhler, 1960; Laidlaw, 1974) (Fig. 1). Each subfamily is sired by one drone. The number of subfamilies is determined by the number of mates of the queen mother, but the size of each subfamily depends on the proportion of the total sperm in the spermatheca that came from the subfamily sire, Besides super sisters, there are half sisters between subfamilies that are fathered by drones from different mothers. There may also be full sisters between subfamilies sired by drones from the same mother.

The characteristics of the different subfamilies may differ due to different sires, and the individuals within a subfamily may vary because of segregation during meiosis of the queen's eggs. Colony characteristics are thus a composite of each of these.

F. Laying Workers as Drone Mothers

A zygote heterozygous for the sex alleles can give rise to either a worker or a queen. A laying worker will, therefore, produce the same kinds of drones she would produce if she were a queen.

From a breeding standpoint, laying workers of a colony can be considered multiple, unmated sister queens with impaired reproductive capacity. The array of male genotypes in a colony with laying workers is derived from the genotypes of both the queen mother and her mates and therefore represents greater genotypic variability than the array of drones produced by a single queen.

II. POSSIBLE MATINGS IN HONEY BEES AND THEIR GENETIC CONSEQUENCES

A. General Statements

1. Mating System

Honey bees have a mating system that is similar in some respects to that of hermaphroditic plants with a self-incompatibility system. Honey bees are hermaphroditic in the sense that queens produce eggs which when inseminated with sperm carrying sex alleles *different* from the female gamete produce females, and if not inseminated produce haploid functional males by parthenogenesis. Queens may be genetically mothers and fathers at the same time (Fig. 2a). They are self-incompatible as the result of the genic sex determination mechanism.

2. Reciprocal Matings

Reciprocal matings are not necessarily comparable. A queen will produce the full range of gametes her genotype permits, but each of her mates represents only one of the genomes his mother can produce. Therefore, fertilization of a queen's eggs is limited to the few different genomes of the sperm in her spermatheca. Reciprocal matings can approach those of diploid organisms if semen from many drones of a single drone mother is thoroughly mixed so each egg has a chance to be fertilized by any one of a greater variety of different sperm. In this situation the number of subfamilies is increased but the number of members of each subfamily is reduced and the sisters are more likely to be full sisters than super sisters.

3. Breeding Terminology

Polhemus *et al.* (1950) pointed out the difficulties associated with using physical pairing terminology of diploids for honey bees rather than genetic pairing terminology. Laidlaw and Eckert (1950, pp. 122–124) used the genetic pairing terminology of mating female to female, and showed that full-sib matings, that is, full sister–full sister matings, are possible in bees, but that genetic brother–sister matings are not. Nevertheless, the use of the

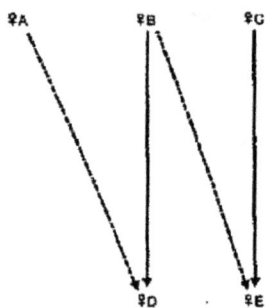

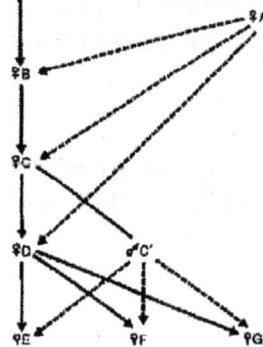

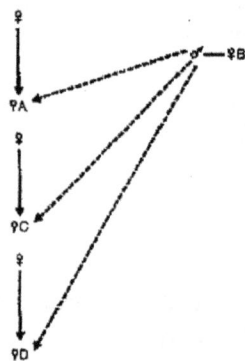

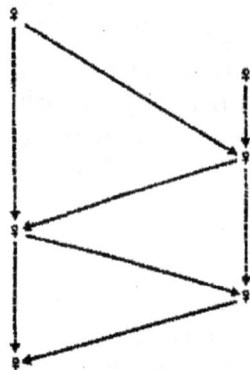

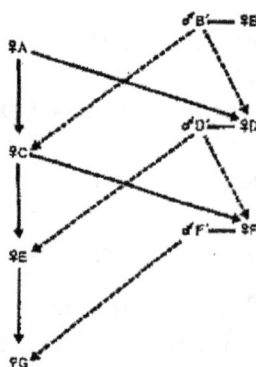

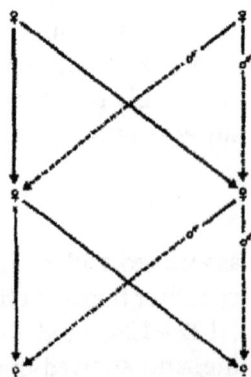

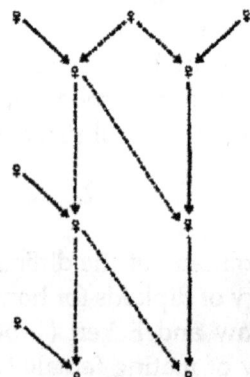

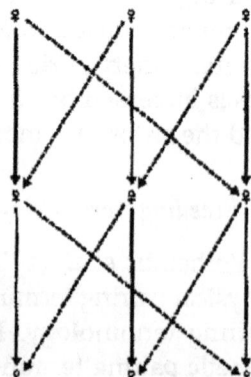

Fig. 2. (a) Hermaphroditic capability of queen bees. A queen is functionally hermaphroditic in that she can genetically and physically be a mother (B × A) and at the same time through her drones be genetically a father (B × C). (b) Irregular pedigree. [From Crow and Roberts, (1950), by permission. Copyright 1950 by Genetics Society of America.] (1) Inbreeding. The equation $F_x = \Sigma(\tfrac{1}{2})^n(1 + F_A)$ is used, where n is the number of female *ancestors* beginning with one parent through a common ancestor (CA) and ending with the other parent. Common ancestors are underlined. Males are passed through without counting.

Path for queen C: $B\underline{A} = (\tfrac{1}{2})^2 = .25$

Paths for queen D: $CB\underline{A} = (\tfrac{1}{2})^3 = .125$

$$C\underline{A} = (\tfrac{1}{2})^2 = .25$$

$$\Sigma = .375$$

Paths for queen E: $D\underline{A}BC = (\tfrac{1}{2})^4 = .0625$

$$D\underline{A}C = (\tfrac{1}{2})^3 = .125$$

$$D\underline{C} = (\tfrac{1}{2})^2(1 + .25) = .3125$$

$$\Sigma = .5$$

(2) Relationship. The relationship between E and G is found with the equation

$$R_{xy} = \Sigma(\tfrac{1}{2})^n \frac{1 + F_A}{\sqrt{(1 + F_x)(1 + F_y)}}$$

where n is the number of *steps* in a path from one of the pair whose relationship is being measured through all ancestors to a common ancestor and back to the other member of the pair. The males, designated by prime, are passed through without counting. The paths are summed. The paths from E to G are

$\underline{C}'G = (\tfrac{1}{2}) = .5$

$\underline{D}G = (\tfrac{1}{2})^2(1.375) = .34375$

$D\underline{C}G = (\tfrac{1}{2})^3(1.25) = .15625$

$D\underline{A}CG = (\tfrac{1}{2})^4 = .0625$

$D\underline{A}BCG = (\tfrac{1}{2})^5 = .03125$

$\underline{C}DG = (\tfrac{1}{2})^3(1.25) = .15625$

$C\underline{A}DG = (\tfrac{1}{2})^4 = .0625$

$CB\underline{A}DG = (\tfrac{1}{2})^5 = .03125$

$\Sigma = 1.34375$

$$R_{EG} = \frac{1.34373}{\sqrt{(1 + .5)(1 + .5)}} = \frac{1.34375}{1.5} = .8958$$

(c) Self fertilization. One or multiple drones. Queen A is mated by her own drone to produce

(continued)

term "brother" when referring to drones or sperm in a physical sense is convenient.

The terminology used here to denote relationship is based on the *origin* of the genomes of gametes, and differs in certain respects from the usual terminology employed with diploid organisms which when applied to haplo–diploids is imprecise. True honey-bee reproductives (queens and workers) are capable of originating genomes of gametes. Complementary reproductives (drones) cannot.

B. Inbreeding

In the United States, bee breeding has been greatly influenced by the methods used in breeding hybrid maize. Inbred lines of bees are developed and crossed to produce genetic hybrids. In order for this to work most effectively, ways are needed to measure the loss of heterozygosity associated with inbreeding when different systems of mating are repeated in successive generations.

1. Coefficient of Inbreeding

Wright's method of path coefficients (1921, 1922, 1933, 1934) was found to be invaluable for determining the amount of homozygosity expected as a consequence of inbreeding. Several of the possible regular systems of inbreeding applicable to haplo–diploidy were studied in detail by Kalmus and Smith (1948), Crow and Roberts (1950), Polhemus *et al.* (1950), Polhemus and Park (1951), and Ruttner (1968). The purposes of all these papers were the development of methods for determination of inbreeding

Fig. 2. (*continued*)

---

female B. Queen B is mated by her own drone to produce female C. (*d*) More than one queen mated to one drone or equivalent (one genome). There is one family of super sisters for each queen, and in this case one drone or equivalent sires three different subfamilies. (*e*) Mother–daughter mating (backcross to queen's mother). Called brother–sister mating. (*f*) Super sister mating. Called aunt–nephew mating. The sisters are sired by one drone. Letters with prime designations signify a male common ancestor that is passed through but not counted. (*g*) Full sister–full sister mating. The sisters are sired by different drones of the same drone mother. (*h*) Paternal half-sister mating. Sisters have same father but different mothers. Drone mothers may be open mated. Half sisters can be identified. (*i*) Half-sister circular mating. Each queen has a hermaphroditic parent and is one in turn. In each generation all progeny of each queen are half sisters to the progeny of each other queen. Multiple or single inseminations may be used, but open mating cannot be used.

coefficients and devising more efficient methods to establish inbred lines that could be crossed to form genetic hybrids. Cruden (1949), Emik and Terrill (1949), and Plum (1954) developed simplified methods for dealing with complex, irregular pedigrees in closed breeding populations of diploids. These methods eliminate the necessity to analyze complete pedigrees for each cross each generation.

The inbreeding coefficient of an individual is the probability that both alleles at any given locus are identical by descent from an ancestor common to both parents. This value, F_X, is identical to the coefficient of consanguinity (or coancestry) of the parents, f_{IJ}, (Malecot, 1948; Falconer, 1960, p. 88; Crow and Kimura, 1970, p. 72; Hartl, 1980, pp. 126–132) and is found by the equation

$$F_X = f_{IJ} = \Sigma(\tfrac{1}{2})^n(1 + F_A) \tag{1}$$

With this equation all possible genetic paths are followed from a designated starting parent to the common ancestor of the path and on to the other parent; n is the number of ancestors in a path including both parents, and F_A is the inbreeding coefficient of the common ancestor of a path. The paths for each separate coefficient are summed.

The equation, developed for diploid organisms, is useful for analyzing pedigrees of haplo–diploids (honey bees) provided the following rules are observed (Fig. 2b):

1. Identify all common ancestors in the pedigree.
2. Count all females but never males in all paths from and including one parent of the individual being measured to and including the common ancestor (CA) of the path and back to and including the other parent.
3. Paths must go through all ancestors in a path, male and female, even though the males are not counted.
4. If a male is the *common ancestor* of a path, pass through him and his mother without counting either.
5. Always go against the arrows in the pedigree from the same parent to the common ancestor and with the arrows returning to the other parent.
6. Ignore the inbreeding coefficient of an individual in a path except when it is the common ancestor of *that specific path*.

In *self fertilization* (Fig. 2c), there is one parent each generation which is also the common ancestor. The equation is

$$F_t = \tfrac{1}{2}(1 + F_{t-1}) \tag{2}$$

2. Coefficient of Relationship

Another regularly used value in breeding is Wright's (1922) coefficient of relationship, R_{xy}. This coefficient is a measure of the average genetic relationship of two individuals, x and y, and is expressed by

$$R_{xy} = \Sigma(\tfrac{1}{2})^n \frac{1 + F_A}{\sqrt{(1 + F_x)(1 + F_y)}} \qquad (3)$$

where n is the number of generation *steps* from one individual through a common ancestor to the other individual. If the common ancestor is male, pass through without counting it. It can also be expressed by

$$r_{IJ} = \frac{2f_{IJ}}{\sqrt{(1 + F_I)(1 + F_J)}} \qquad (4)$$

In the absence of inbreeding this value is exactly twice the value of the coefficient of consanguinity (or coancestry) f_{IJ}, which is equal to the inbreeding coefficient F_x of a *hypothetical* offspring of I and J. The coefficient of relationship is often used as a measure of the proportion of like genes shared by two individuals as a result of common descent. It can be used to assess the genetic relationship of potential mates, and hence the inbreeding coefficient of their offspring, or for determining the average relatedness of individuals within or among colonies.

3. Heterozygosity

The purpose of deliberate inbreeding is to increase homozygosity at specific loci of interest to the breeder. However, the breeder has very little control over which loci become homozygous, and settles instead for an approach to homozygosity of the entire genotype. The proportion H_t of loci remaining heterozygous in any generation t can be determined by

$$H_t = H_0(1 - F_t) \qquad (5)$$

where H_0 is the initial heterozygosity before inbreeding (Crow and Kimura, 1970, p. 66). This is usually assumed to be 0.50 for the *initial* non-inbred generation.

When regular systems of inbreeding are used, it is possible to develop recursion equations to determine the inbreeding coefficient and the remain-

TABLE 1. Recursion Equations for Inbreeding Coefficients of Generation t (F_t) Using Different Regular, Closed Systems of Inbreeding

Mating	Equation
Selfing	$F_t = \frac{1}{2}(1 + F_{t-1})$
Gametic backcross	$F_t = \frac{1}{2}(1 + F_{t-1})$
Maternal mother–daughter	$F_t = \frac{1}{4}(1 + 2F_{t-1} + F_{t-2})$
Paternal mother–daughter	$F_t = \frac{1}{2}(\frac{1}{2} + F_{t-1})$
Super sister–super sister	$F_t = \frac{1}{8}(3 + 4F_{t-1} + F_{t-2})$
Full sister–full sister	$F_t = \frac{1}{4}(1 + 2F_{t-1} + F_{t-2})$
Half sisters (drone mothers super sibs)	$F_t = \frac{1}{32}(7 + 16F_{t-1} + 8F_{t-2} + F_{t-3})$
Half sisters (drone mothers full sibs)	$F_t = \frac{1}{16}(3 + 8F_{t-1} + 4F_{t-2} + F_{t-3})$
Half sister (drone mothers half sibs)	$F_t = \frac{1}{8}(1 + 6F_{t-1} + F_{t-2})$
Aunt–niece (single drone)	$F_t = \frac{1}{16}(3 + 8F_{t-1} + 4F_{t-2} + F_{t-3})$
Aunt–niece (multiple drones)	$F_t = \frac{1}{8}(1 + 4F_{t-1} + 2F_{t-2} + F_{t-3})$
Grandmother–granddaughter	$F_t = \frac{1}{8}(2 + 2F_{t-2} + 3F_{t-3} + F_{t-4})$

ing heterozygosity for any given generation (Table 1). Crow and Roberts (1950) present equations based on Wright's method of path coefficients for expected proportions of heterozygous loci. Equivalent formulas can be derived in terms of inbreeding coefficients using the method of coancestry (Cruden, 1949; Emik and Terrill, 1949; Falconer, 1960, pp. 88–90). This method was used for Table 2, which shows inbreeding coefficients for several generations of regular, closed systems of inbreeding. For these systems of breeding, it is implicitly assumed that multiple drone inseminations result in breeder queens that have different drone fathers, even though this may not always occur. The actual practice of these techniques will depend on chance sampling, with errors in sampling resulting in more rapid inbreeding than indicated. Clearly, the more drones used for inseminations, the less likely it is that errors will occur.

C. Selected Specific Mating Designs

The haplo–diploid condition of honey bees makes many different kinds of matings possible. In some breeding programs, especially those involving intensive inbreeding, reliance is placed on one or very few mating designs. Bee pedigrees can become complicated, and estimates of inbreeding may need to be calculated individually if they are needed at all. However, Table 2 lists some of the different possible mating designs and the inbreeding coeffi-

TABLE 2. Inbreeding Coefficients for Generations of Regular, Closed Systems of Inbreeding[a]

Mating	Generation																	
	1	2	3	4	5	6	7	8	9	10	15	20	25	30	35	40	45	50
Selfing	.500	.750	.875	.938	.969	.984	.992	.996	.998	.999	1.00	1.00	1.00	1.00	1.00	1.00	1.00	1.00
Gametic backcross	.500	.750	.875	.938	.969	.984	.992	.996	.998	.999	1.00	1.00	1.00	1.00	1.00	1.00	1.00	1.00
Maternal mother–daughter	.250	.375	.500	.594	.672	.734	.785	.826	.859	.886	.961	.986	.995	.998	.999	1.00	1.00	1.00
Paternal mother–daughter	.250	.375	.438	.469	.484	.492	.496	.498	.499	.500	.500	.500	.500	.500	.500	.500	.500	.500
Super sister–Super sister	.375	.562	.703	.797	.861	.905	.935	.956	.970	.979	.997	1.00	1.00	1.00	1.00	1.00	1.00	1.00
Full sister–full sister	.250	.375	.500	.594	.672	.734	.785	.826	.859	.886	.961	.986	.995	.998	.999	1.00	1.00	1.00
Half sisters (drone mothers super sibs)	.219	.328	.438	.526	.602	.665	.718	.763	.800	.832	.929	.970	.987	.995	.998	.999	1.00	1.00
Half sisters (drone mothers full sibs)	.188	.281	.375	.457	.527	.589	.642	.689	.729	.764	.883	.942	.971	.985	.993	.996	.998	.999
Half sisters (drone mothers half sisters)	.125	.219	.305	.381	.449	.509	.563	.611	.654	.692	.827	.903	.946	.970	.983	.991	.995	.997
Aunt–niece (single drone)	.188	.281	.375	.457	.527	.589	.642	.689	.729	.764	.883	.942	.971	.985	.993	.996	.998	.999
Aunt–niece (multiple drones)	.125	.188	.250	.312	.367	.418	.465	.508	.547	.584	.726	.820	.882	.922	.948	.966	.978	.985
Grandmother–granddaughter	.125	.250	.281	.359	.430	.476	.527	.575	.614	.651	.789	.873	.923	.953	.972	.983	.990	.994

[a] Values presented were obtained from recursion equations given in Table 1.

cients that can be expected with each, and Table 1 presents the recursion equations for them.

1. Self Fertilization (Physical Pairing of Mother and Son)

By treatment with carbon dioxide, a virgin queen can be stimulated to lay unfertilized eggs. These eggs develop into haploid drones that can then be used to inseminate her (Fig. 2c). The inbreeding of daughters under this system is the same whether a queen is mated wih one or more than one of her drones. When sperm from an infinite number of males is used, the number of subfamilies of a colony is increased proportionately and the member density of subfamilies is reduced. Self fertilization in bees then approaches equivalence to self fertilization in diploid organisms. The brood of self inseminated queens will be only 50% viable because only two different sex alleles are involved.

2. Mating More Than One Queen to a Single Drone or Equivalent

Mating more than one queen to a single gamete is possible with honey bees (Fig. 2d) and can be accomplished in two ways.

1. Semen from a drone that has abundant semen can be taken into a syringe and one-half of it injected into each of two queens, or one-third into each of three queens. Inseminations will be minimal, but enough sperm will usually reach the spermathecae that the required number of female bees can be obtained.

2. Queens in quantity can be fully inseminated with identical gametes, originally derived from a single male, that have been propagated through functional male gynandromorphs (Tucker and Laidlaw, 1966; Kubasek et al., 1980). A virgin queen of a line that produces functional male gynandromorphs is mated by one drone. The male parts of the body of the gynandromorphs that result are derived from identical sperm of the one drone, and the sperm they produce in the testes are (except for mutations) copies of the sperm that developed into male tissue. By this means, one gamete is multiplied greatly. Queens mated to functional male gynandromorphs from a single queen are thus mated with identical gametes.

The same system can be used to backcross to a single genome.

3. Parent–Offspring Mating

a. Mother–Daughter Mating. Physically this is a brother–sister mating. Daughters are mated to their mother's drones. This mating is a common design in breeding programs, resulting in rapid inbreeding and an approach to complete homozygosity of lines (Fig. 2e).

   b. *Father–Daughter Mating.* This is the repeated backcross to one drone mother and is equivalent to repeated backcrosses to one queen mother. It results in a maximum inbreeding of 50%.

4. Super-Sister–Super-Sister Mating (Physically Aunt–Nephew Mating)

Super sisters have the same queen mother and same drone mother and are sired by one drone (Fig. 2f).

5. Full-Sister–Full-Sister Mating (Physically Aunt–Nephew Mating)

Full sisters have the same queen mother and the same drone mother but are sired by different drones, or have hermaphroditic parents that alternate as father and mother (Fig. 2g). Loss of heterozygosity is less rapid than that of the super-sister–super-sister mating and equals that attained by mother–daughter mating.

6. Half-Sister–Half-Sister Mating

Half sisters have one parent in common, and may be *paternal* half sisters if they have fathers from the same drone mother but have different mothers themselves (Fig. 2h), or *maternal* half sisters if the sisters' mother is the same but the fathers are from different drone mothers, (Hartl, 1980, p. 293). To maintain a "closed" half-sister mating scheme, three queens are needed each generation. The maternal half-sister mating scheme is the reverse of the paternal.

   If each half-sister queen has an hermaphroditic parent and is one in turn (Fig. 2i), in each generation all female progeny of each queen are half sisters to the progeny of each other queen. Multiple or single inseminations may be used but open mating may not. The hermaphroditic capability increases possibilities for queen testing because the same queen may be tested zygotically and gametically, that is, by her daughters and by her drones.

7. Cousin Mating

Cousins can be derived in various ways, and their mating gives a slow rate of reduction in heterozygosity each generation. It will eventually result in complete homozygosity, providing the system remains closed.

8. Aunt–Niece Mating

Three queens are needed to originate the line. Single-drone insemination leads to a faster rate of inbreeding than multiple-drone inseminations.

With multiple-drone inseminations, inbreeding is light, being equal to grandmother–granddaughter mating.

9. Grandmother–Granddaughter Mating

Grandmother–granddaughter mating is only slightly more efficient in inbreeding than aunt–niece mating with multiple-drone insemination. It is more likely to be used in a breeding program than aunt–niece mating.

Polhemus and Park (1951) pointed out that the actual rapidity in days of inbreeding depends nearly as much on the time required to rear each generation as on the heterozygosity lost per generation. With all factors considered, mother–daughter mating is the most efficient system in inbreeding. This happens to be the system that has been used most extensively.

III. INBRED-HYBRID BREEDING

Crossing inbred lines to produce hybrids is the method of breeding honey bees that has been employed by all of the organized bee-breeding programs in the United States. Excellent uniform hybrids have been produced that are markedly superior to usual commercial stocks, and beekeeping and breeding methods particularly suitable for bee improvement programs have emerged from these efforts.

A. Desirable Features

Availability of the same hybrid bee each year is one of the major advantages of hybridization. Another feature is that characteristics considered desirable in bees can be combined in a single hybrid colony for particular purposes or beekeeping conditions.

B. Undesirable Features

Hybrid breeding is hindered, however, by the expense of creating suitable inbred lines and the difficulty in maintaining them. Lines quickly become two-sex-allele fraternities as inbreeding progresses, which is accompanied by reduction in brood solidness, and both the females and the drones may exhibit decreased vigor. Complete control of mating is essential.

IV. CLOSED-POPULATION BREEDING

In spite of successes of race crosses and hybrid breeding, the problems associated with them prompted investigations of other methods of bee breeding that would be less difficult or expensive. This led to reexamination of closed-population breeding, which had been rejected in the past in favor of the inbred-hybrid system (Laidlaw and Eckert, 1950, pp. 124–128; Laidlaw, 1979, pp. 156–158, 1981; Kubasek, 1980; Page and Laidlaw, 1982a,b, 1985; Page and Marks, 1982; Page et al., 1982, 1983).

As with inbred lines, closed populations are free from uncontrolled introduction of genetic material, and thus must be maintained by isolation or instrumental insemination but with more flexibility in choice of matings. In contrast to inbred-hybrid breeding, the objective of closed-population breeding is to progressively improve the performance of bees by selective breeding while maintaining high brood viability and genetic variability in the population over the lifetime of the program.

A. Mating Systems

The mating designs employed and the methods of selecting queens and drones as reproductives affect the rate of loss of genetic variability and sex alleles. There are three basic strategies; in each the same breeding queens are both queen mothers and drone mothers.

1. Daughters from all of the breeding queens are each mated to 10 drones selected at random from the entire population of drones or are mated in isolation, and breeder queen replacement daughters are selected at random from among daughters of all of the breeding queens without regard to the daughter's parentage. About 50 breeder colonies must be selected each generation to have a 95% probability of at least 85% viable brood for 20 generations (Fig. 3).

2. With the same method of producing the replacement queens, each queen mother is replaced by one of her own daughters (Page et al., 1983). About 35 breeder colonies are needed to have 95% probability of at least 85% viable brood for 20 generations (Fig. 4). In both cases the next generation breeder colonies are selected on the basis of performance of the queens and their colonies in comparison with sisters and with daughters of all of the breeders.

3. Semen from approximately equal numbers of drones from each of the breeding queens is pooled and thoroughly mixed or homogenized (Laidlaw, 1981; Moritz, 1983). Each daughter of all of the breeding queens is mated to a comparable sample of this semen. By this method, (a) variance in brood viability is reduced among the queens of the closed population, (b) all queens are effectively mated to the entire gene pool, (c) queens can be selected on the basis of the general combining ability of their own genotype, and (d) the loss of sex alleles from a closed population can be reduced significantly by selection for high brood viability (Kubasek, 1980; Page et al.,

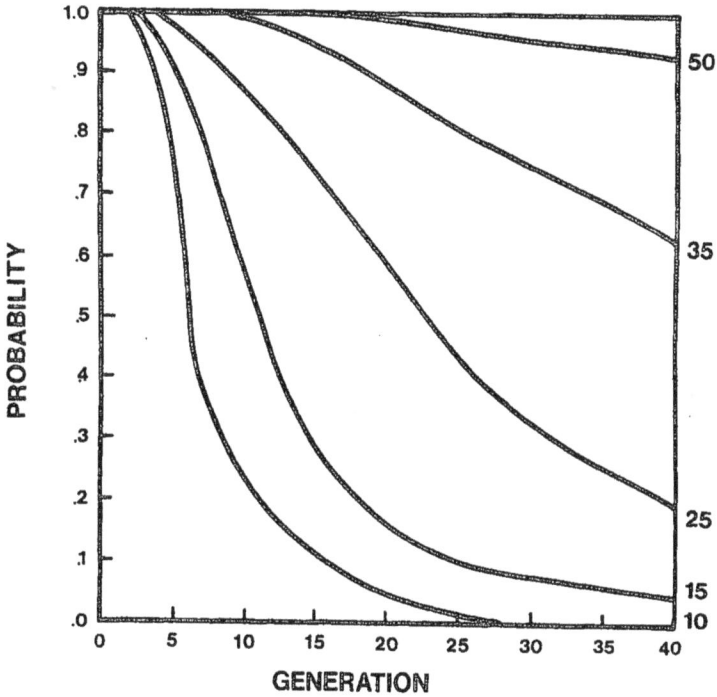

Fig. 3. A computer simulation of the probability that a closed population with 10, 15, 25, 35, and 50 breeder queens will have an average brood viability of at least 85% for 40 generations. It is assumed that each queen is mated to 10 drones. There are 12 sex alleles initially, and breeder queens are selected at random in each generation.

1983), which in substance is selection for rarer sex alleles. Selection of next generation breeder queens is done as in strategy 1 or 2 above.

A fourth mating system may be superimposed upon the above basic ones. Certain queens that are part of the closed population and that head superior colonies may be top crossed through their drones onto the virgin daughters of the closed population breeder queens by two methods (Page *et al.*, 1985).

First, a given proportion (never greater than about 75%) of the virgin daughter queens randomly selected within the population are inseminated only by drones of the selected superior queen and preferably with pooled semen from many of her drones.

Second, a given proportion (never greater than about 75%) of all drones used to inseminate all daughter queens in the closed population come from the selected superior queen. This mixing of semen to inseminate all queens is preferred for maintaining variability at the sex locus. Top crossing by any *given* queen with either of these methods should not be consecutive, nor should top crosses be made more than three times during a given closed breeding program.

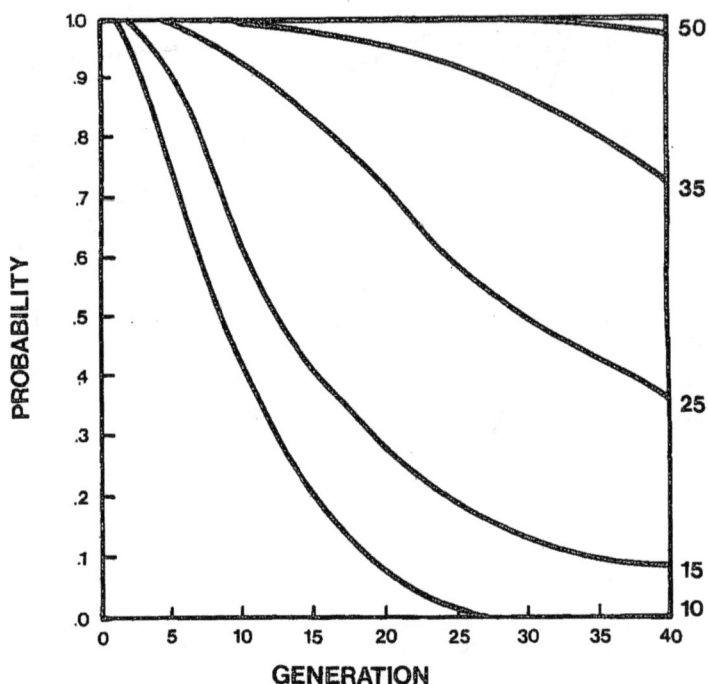

Fig. 4. Same as Fig. 3 except that each breeder queen contributed exactly one daughter queen as a breeder for each subsequent generation.

B. Introduction of New Alleles into Closed Populations

Should it become necessary to restore a minimum number of sex alleles to a closed population, care must be taken to minimize the introduction of undesirable genes. The additional sex alleles should be sought in superior stocks and tests should be made for the presence of different alleles. Testing can be done by inseminating test queens that are outside the closed population with pooled, homogenized semen from drones from all of the breeder queens within the closed population. Queens that have two sex alleles that are different from those within the closed population will have close to 100% brood viability. Only queens with two new sex alleles should be added to the breeder colonies of this closed population. As a rule of thumb, for maximum gene frequency stability, virgin queens and drones from the selected, tested queen should be introduced into the closed population at an approximate frequency of

$$\frac{2(1 - \overline{V})}{3 - 2V} \tag{6}$$

where $\overline{V}$ is the average brood viability of the closed population.

V. GENETIC COMPOSITION AND EXPERIMENTAL COLONIES

Experiments with honey bees often suffer from lack of genetic uniformity of the experimental colonies. By rather simple matings, without inbreeding, highly related queens are produced to head colonies which are unusually highly related to each other (Fig. 5).

1. Queen A is mated with *one* drone of queen B.
2. Daughters, such as C and D, are reared to head the experimental colonies.
3. These daughters are inseminated with thoroughly mixed semen from many drones of unrelated queen E so they will produce normal strong colonies for experiments.

The relationship of the sister queens heading the colonies is 0.75 and between the colony populations is 0.4375.

VI. CONCLUSION

It is obvious that many kinds and combinations of matings can be made with honey bees. Because of this, special experimental stocks can be created for special purposes such as behavioral, feeding, disease, or genetic studies.

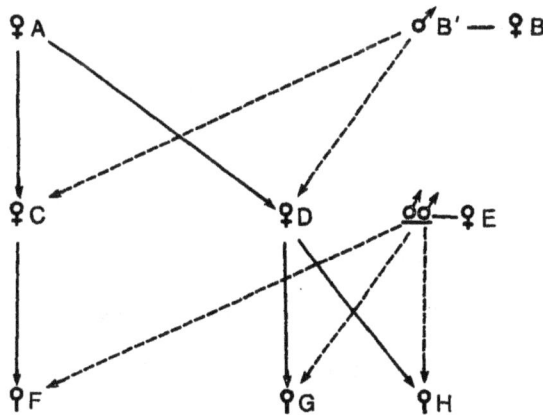

Fig. 5. Closely related but not inbred sisters for use in experiments:

$$r_{CD} = C\underline{A}D(\tfrac{1}{2})^2 + C\underline{B}'D(\tfrac{1}{2}) = 0.75$$

$$r_{GH} = G\underline{D}H(\tfrac{1}{2})^2 + G\underline{E}H(\tfrac{1}{2})^2 = 0.5$$

$$r_{FG} = FC\underline{A}DG(\tfrac{1}{2})^4 + FCB'DG(\tfrac{1}{2})^3 + F\underline{E}G(\tfrac{1}{2})^2 = 0.4375$$

Queen A is mated by one drone of queen B. Queens C and D are mated with a mixture of semen from many drones of queen E.

Drones that have genotypes made up of genes from specified mothers can be produced in enormous numbers when and where needed. The utility of this capability and the potential experimental or economic value of the various mating designs and their combinations are subject to verification and further research.

The potential for introgression of desirable genes into feral bee populations is particularly appropriate for study.

Single-gamete selection, as suggested by Rothenbuhler (1960), may be directly applicable for such traits as longevity by testing longevity in drones, and possibly be indirectly applicable in other cases.

Mating queens with samples of thoroughly mixed semen from many drones is a new breeding technique that appears to have wide application, and it should be tested.

REFERENCES

Adams, J., Rothman, E. D., Kerr, W. E., and Paulino, Z. L. (1977). Estimation of the number of sex alleles and queen matings from diploid male frequencies in a population of *Apis mellifera*. *Genetics 86*, 583–596.

Anderson, R. H. (1963). The laying worker in the Cape honeybee, *Apis mellifera capensis*. *J. Apic. Res. 2*, 89–92.

Crow, J. F., Kimura, M. (1970). "An Introduction to Population Genetics Theory." Burgess Publishing Co., Minneapolis, Minn.

Crow, J. F., and Roberts, W. C. (1950). Inbreeding and homozygosis in bees. *Genetics 35*, 613–621.

Crozier, R. H., and Brückner, D. (1981). Sperm clumping and the population genetics of Hymenoptera. *Am. Nat. 117*, 561–563.

Cruden, D. M. (1949). The computation of inbreeding coefficients in closed populations. *J. Hered. 40*, 248–251.

Emik, L. O., and Terrill, C. E. (1949). Systematic procedures for calculating inbreeding coefficients. *J. Hered. 40*, 51–55.

Falconer, D. S. (1960). "Introduction to Quantitative Genetics." The Ronald Press Company, New York.

Harbo, J. R. (1979). The rate of depletion of spermatozoa in the queen honeybee spermatheca. *J. Apic. Res. 18*, 204–207.

Hartl, D. L. (1980). "Principles of Population Genetics." Sinauer Associates, Inc., Sunderland, Mass.

Kalmus, H., and Smith, C. A. B. (1948). Production of pure lines in bees. *J. Genetics 49*, 153–158.

Kepeňa, L., and Fl'ak, P. (1980). [Relationship between number of drones mating (with the queen) and the morphological characteristics of the progeny in honeybee colonies]. Počet trúdov pri párení a morfologické znaky potomstva včiel (*Apis mellifera*). *Pol'nohospodárstvo 26*, 566–573. Abstracted in *Apic. Abstr.* 889/83.

Kerr, W. E. (1967). Multiple alleles and genetic load in bees. *J. Apic. Res. 6*, 61–64.

Kerr, W. E., Zucchi, R., Nakakaira, J. T., and Butolo, J. E. (1962). Reproduction in the social bees. *J. N.Y. Entomol. Soc. 70*, 265–270.

Kubasek, K. J. (1980). Selection for increased number of sex alleles in closed populations of the honey bee. An investigation via computer simulation. Masters thesis, Louisiana State University, Baton Rouge, La.

Kubasek, K. J., Rinderer, T. E., and Lee, W. R. (1980). Isogenic sperm line maintenance in the honey bee. *J. Hered.* 71, 278–280.

Laidlaw, H. H., Jr. (1974). Relationships of bees within a colony. *Apiacta* 9, 49–52.

Laidlaw, H. H., Jr. (1979). "Contemporary Queen Rearing." Dadant and Sons, Hamilton, Ill.

Laidlaw, H. H. (1981). (Honey bee genetics and its application to pollinator breeding). *Honeybee Science* 2, 1–4 (in Japanese, English summary).

Laidlaw, H. H., Jr., and Eckert, J. E. (1950). "Queen Rearing." Dadant and Sons, Hamilton, Ill.

Laidlaw, H. H., Jr., and Page, R. E., Jr. (1984). Polyandry in honey bees (*Apis mellifera* L.): sperm utilization and intracolony genetic relationships. *Genetics* 108, 985–997.

Laidlaw, H. H., Jr., Gomes, F. P., and Kerr, W. E. (1956). Estimations of the number of lethal alleles in a panmictic population of *Apis mellifera*. *Genetics* 41, 179–188.

Mackensen, O. (1943). The occurrence of parthenogenetic females in some strains of honey bees. *J. Econ. Entomol.* 36, 465–467.

Mackensen, O. (1951). Viability and sex determination in the honey bee (*Apis mellifera* L.). *Genetics* 36, 500–509.

Mackensen, O. (1955). Further studies on a lethal series in the honey bee. *J. Hered.* 46, 72–74.

Malecot, G. (1948). "Les Mathematiques de l'Heredité." Massonet Cie, Paris.

Moritz, R. F. A. (1983). Homogeneous mixing of honeybee semen by centrifugation. *J. Apic. Res.* 22, 249–255.

Newell, W. (1915). Inheritance in the honeybee. *Science N.S.* 41, 218–219.

Oertel, E. (1940). Mating flights of queen bees. *Glean. Bee Cult.* 68, 292–293, 333.

Onions, G. W. (1909). Workers and queens from eggs laid by workers. *Agric. J. Cape of Good Hope* 35, 527.

Onions, G. W. (1912). South African fertile-worker bees. *Agric. J. Union S. A.* 3, 720.

Page, R. E., Jr., and Laidlaw, H. H., Jr. (1982a). Closed population honeybee breeding. I Population genetics of sex determination. *J. Apic. Res.* 21, 30–37.

Page, R. E., Jr., and Laidlaw, H. H., Jr. (1982b). Closed population honey bee breeding. II Comparative methods of stock maintenance and selective breeding. *J. Apic. Res.* 21, 38–44

Page, R. E., Jr., and Laidlaw, H. H., Jr. (1985). Closed population honeybee breeding. *Bee World* 66, 63–72.

Page, R. E., Jr., and Marks, R. W. (1982). The population genetics of sex determination in honey bees: random mating in closed populations. *Heredity* 48, 263–270.

Page, R. E., Jr., and Metcalf, R. A. (1982). Multiple mating, sperm utilization, and social evolution. *Am. Nat.* 119, 263–281.

Page, R. E., Jr., Erickson, E. H., Jr., and Laidlaw, H. H., Jr. (1982). A closed population breeding program for honey bees. *Am. Bee J.* 122, 350–351, 354–355.

Page, R. E., Jr., Laidlaw, H. H., Jr., and Erickson, E. H., Jr. (1983). Closed population honey-bee breeding. 3. The distribution of sex alleles with gyne supersedure. *J. Apic. Res.* 22, 184–190

Page, R. E., Jr., Kimsey, R. B., and Laidlaw, H. H., Jr. (1984). Migration and dispersal o spermatozoa in spermathecae of queen honey bees. (*Apis mellifera* L.) *Experientia* 40, 182–184.

Page, R. E., Jr., Laidlaw, H. H., Jr., and Erickson, E. H., Jr. (1985). Closed population breeding 4. The distribution of sex alleles with topcrossing. *J. Apic. Res.* 24, 38–42.

Plum, M. (1954). Computation of inbreeding and relationship coefficients. *J. Hered.* 25, 92–94

Polhemus, M. S, and Park, O. W. (1951). Time factors in mating systems for honey bees. *J. Econ Entomol.* 44, 639–642.

Polhemus, M. S., Lush, J. L., and Rothenbuhler, W. C. (1950). Mating systems in honey bees. *J Hered.* 41, 151–155.

Roberts, W. C. (1944). Multiple mating of queen bees proved by progeny and flight tests. *Glean. Bee Cult. 72*, 255–259, 303.

Rothenbuhler, W. C. (1960). A technique for studying genetics of colony behavior in honey bees. *Am. Bee J. 100*, 176, 198.

Ruttner, F. (1968). Intra-racial selection or race-hybrid breeding of honey bees. *Am. Bee J. 108*, 394–396.

Snodgrass, R. E. (1925). "Anatomy and Physiology of the Honeybee." McGraw-Hill, New York.

Taber, S., III. (1983). Bee behavior. Queen rearing with the Cape bee—it's much different. *Am. Bee J. 123*, 435–437.

Taber, S., III., and Wendel, J., Jr. (1958). Concerning the number of times queen bees mate. *J. Econ. Entomol. 51*, 786–789.

Triasko, V. V. (1951). (Signs indicating the mating of queens.) *Pchelovodstvo 28*, 25–31, (in Russian). Abstracted in *Apic. Abstr. 5/53.*

Triasko, V. V. (1956). Polyandry in honey bees. *XVI Int. Beekeeping Cong. Prelim. Sci. Meet.* Abstracted in *Apic. Abst. 233/56.*

Tucker, K. W. (1958). Automictic parthenogenesis in the honey bee. *Genetics 43*, 299–316.

Tucker, K. W., and Laidlaw, H. H., Jr. (1966). The potential for multiplying a clone of honey bee sperm by androgenesis. *J. Hered. 57*, 213–214.

Woyke, J. (1960). Naturalne i sztuczne unasienianie matek pszczelich. [Natural and artificial insemination of queen honeybees]. *Pszczel. Zesz. Nauk. 4*, 183–275. Summarized in *Bee World* (1962), 43, 21–25.

Woyke, J. (1963a). Drone larvae from fertilized eggs of the honeybee. *J. Apic. Res. 2*, 19–24.

Woyke, J. (1963b). What happens to diploid drone larvae in a honeybee colony. *J. Apic. Res. 2*, 73–75.

Woyke, J. (1964). Causes of repeated mating flights by queen honeybees. *J. Apic. Res. 3*, 17–23.

Woyke, J. (1976). Population genetics studies on sex alleles in the honeybee using the example of the Kangaroo Island Bee Sanctuary. *J. Apic. Res. 15*, 105–123.

Woyke, J. (1980). Effect of sex allele homo-heterozygosity on honeybee colony populations and on their honey production. 1. Favourable development conditions and restricted queens. *J. Apic. Res. 19*, 51–63.

Woyke, J. (1981). Effect of sex allele homo-heterozygosity on honeybee populations and on their honey production. 2. Unfavourable development conditions and restricted queens. *J. Apic. Res. 20*, 148–155.

Wright, S. (1921). Systems of mating. *Genetics 6*, 111–178.

Wright, S. (1922). Coefficients of inbreeding and relationship. *Am. Nat. 56*, 330–338.

Wright, S. (1933). Inbreeding and homozygosis. *Proc. Natl. Acad. Sci. U.S.A 19*, 411–420.

Wright, S. (1934). The method of path coefficients. *Ann. Math. Statist. 5*, 161–215.

CHAPTER 14

Storage of Germplasm

ANTONIO CARLOS STORT AND LIONEL SEGUI GONÇALVES

I. INTRODUCTION

Apiculture has been gaining importance in many countries since it represents a source of income not requiring large investments. Because of this worldwide interest in apiculture, many people are working on the improvement of bee subspecies, strains, and stocks that will better fulfill scientific and commercial needs. This work includes studying differences in behavior between existing subspecies and strains and also developing new bee stocks for both beekeepers and research laboratories. A fundamental aspect of such work is the storage and maintenance of germplasm.

Germplasm is the genetic material used in the production of new stocks. In *Apis mellifera* bees, the germplasm mainly includes spermatozoa and eggs. With the development of tissue cultures, it may eventually include the tissues that originate these gametes.

Thus far, germplasm maintenance has been done through the queens, which store their own eggs and the spermatozoa of the drones they mate in their spermathecae, or by preserving spermatozoa in liquid nitrogen at a temperature of $-196°C$. Germplasm storage by propagating queens and mating them has been greatly advanced by instrumental insemination. This permits full mating control, which does not occur as efficiently in controlled matings carried out on islands or in mountain valleys.

In this chapter, we emphasize the methods for the production and storage of drone spermatozoa, the effect of extreme temperature on sperm formation, and stock propagation and storage.

II. SEMEN COLLECTION

The drone reproductive structures are located inside the abdomen. During copulation, the penis undergoes an eversion process that ultimately causes ejaculation. In nature, this phenomenon is triggered when the drone is stimulated by the queen, which results in a strong contraction of the abdominal muscles. The copulating organ is ejected and the seminal vesicles and mucus glands suffer strong pressure that causes them to eject their contents: semen is ejected first, followed by mucus (Woyke and Ruttner, 1958).

When semen is collected in the laboratory for immediate use or for storage, the eversion of the drone's penis must be induced. According to Laidlaw (1977), exposure of the drone to chloroform fumes, contact with weak electric current, or exposure to other chemicals, heat, or cold will usually induce partial eversion and ejaculation. However, the most commonly used method is to manually apply pressure on the drone's thorax and abdomen. The semen and mucus are expelled through the orifice of the ejaculating duct and remain on the outer surface of the penis bulb. The semen, which is light brown in color, does not mix with the mucus although it stays in contact with it. Semen can then be collected with a special needle and syringe under the microscope and stored, for example, in a capillary tube.

III. SEMEN COMPOSITION

The semen is a cream-colored fluid consisting of two main elements: sperm and seminal plasma. The pH of ejaculated semen is 6.8–7.0. The sperm : seminal-plasma ratio is 1 : 1 to 1 : 2 depending on climatic conditions. The former ratio prevails at the beginning of the year (Taber, 1977). The spermatozoa removed from the seminal vesicles show a slow collective motion, but become intensely mobile when diluted with saline solution. This effect can be obtained with many diluents and may reflect an increase in oxygen availability to the sperm.

The chemical composition of semen is quite complex. The entire male reproductive system contains three types of sugars: fructose, glucose, and trehalose. These sugars appear in seminal plasma and probably are a source of energy for the spermatozoa. Fructose is rapidly metabolized by the spermatozoa: 40 min after ejaculation, fructose levels are reduced to minimal amounts. The semen also contains many ions such as magnesium, calcium, sodium, copper, iron, and manganese. Only copper, iron, and magnesium are detected in sperm cells (Blum *et al.*, 1962).

Analysis by paper chromatography has shown free and bound amino acids in both seminal plasma and spermatozoa. Many amino acids were detected: tyrosine, methionine, leucine, cystine, isoleucine, tryptophan, lysine, phenylalanine, arginine, glutamic acid, glycine, alanine, aspartic acid, serine, and threonine (Novak *et al.*, 1960).

Many enzymes of the dehydrogenase group have been found in spermatozoa, such as α-glycerophosphate, lactate, glucose 6-phosphate, isocitrate, succinate, malate, glutamate, β-hydroxybutyrate, $NADH_2$, and $NADPH_2$ (Blum and Taber, 1965). Free fatty acids, phospholipids, tryglycerides, and esterols have also bee extracted from the semen (Blum *et al.*, 1967).

IV. SPERMATOZOA

A. Structure

Mature *Apis mellifera* spermatozoa have been described from light microscope studies by Toedtmann (1914) and Kurennoi (1954) and more recently from electron microscope studies by Bretschneider (1950), Rothschild (1955), Hoage and Kessel (1968), Cruz-Hofling *et al.*, (1970), and Lensky *et al.* (1979). Spermatozoa are thin elongated cells 250 μm long and 0.7 μm wide. This diameter is maintained throughout the entire cell, except for the two ends. Anatomically, only the head, which is sickle-shaped, and the tail can be recognized. When stained with acetic carmine, the nucleus can be clearly differentiated from the acrosome, which is thinner and shorter than the nucleus. After treatment with orange acridine, examination by fluorescence microscopy shows a green fluorescent nucleus and a slightly red tail. Using the electron microscope, the tail appears to be composed of two long mitochondrial derivatives parallel to the flagellum. If the spermatozoon region containing the mitochondria is defined as an intermediate part, these spermatozoa have a long intermediate part and a short tail. The mitochondrial derivatives extend along approximately 80% of the spermatozoon length and are of different sizes. Externally they exhibit transversal striae (cristae), and internally a partially crystallized matrix. The axial filament has the common structure observed in flagella and cilia.

The nucleus has no special structures, except in the regions that are in contact with the acrosome and tail. The acrosome has an internal acrosomal

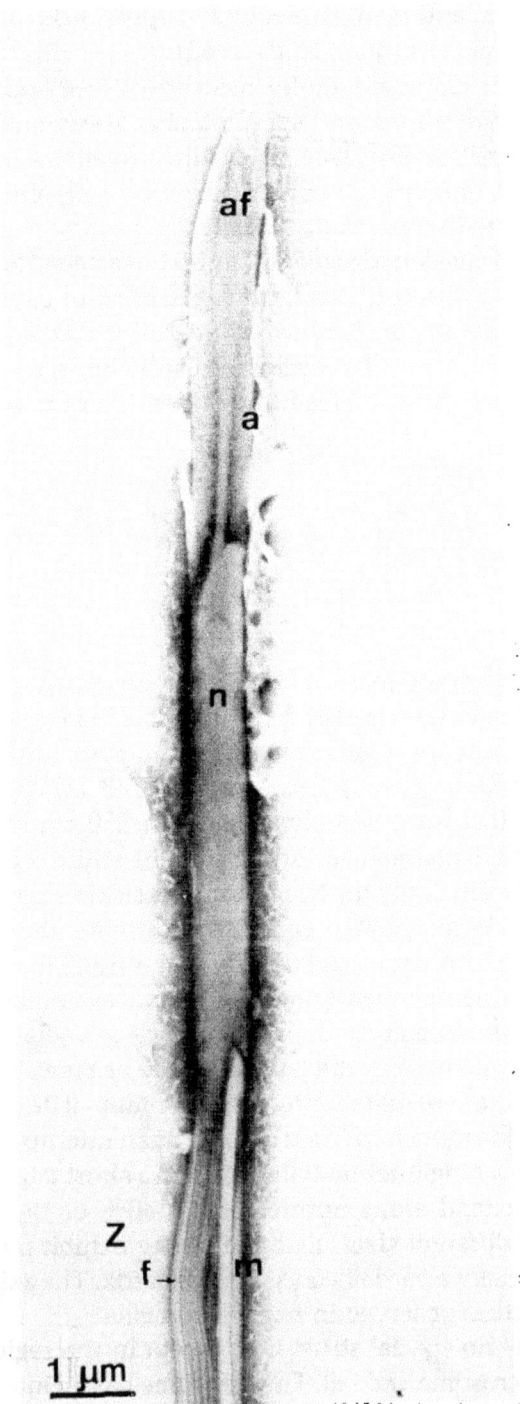

filament that extends from the internal part of the nucleus almost all the way to its end (Fig. 1).

B. Temperature and Spermatogenesis

Excellent studies are available on the structures of spermatozoa, as well as of the cells that originate them, when submitted to variations in temperature during their formation. Tarelho (1981a,b) carried out an extensive study on the influence of low and high temperature. Drones at the pupa and prepupa stage were submitted to nine different temperatures (from 5 to 45°C), each one 5°C higher than the preceding one. No changes were observed in the external phenotype of the adult drones or in the testes of prepupae and pupae submitted to low temperatures. At higher temperatures (40 and 45°C) than the control (35°C), however, the individuals became flaccid, as did their testes, which also became larger and transparent. Tarelho found many chromosomal aberrations during meiosis in males submitted to low temperatures (5–25°C) (Figs. 2 and 3). The list includes metaphases with reduced haploid chromosome number, polyploid metaphases, duplicated chromosomes, lost chromosomes, nonmigrating chromosomes which became isolated, laggard chromosomes in anaphase (one or more chromosomes are delayed in their migration to the poles of the daughter cells), spermatids with lost chromosomes, one or more chromosomes not included in the daughter cells, binucleate spermatids (cells in which no cytokinesis occurred and thus having two nuclei), binucleate spermatozoa (having two nuclei but only one flagellum and probably originating from binucleate spermatids), and normal spermatozoa with one nucleate bud. Among the aberrations observed during meiosis in males submitted to high temperatures (45°C), Tarelho found giant cells and vacuoles in the cytoplasm of normal-size spermatocytes and the adherence of chromatin to the nuclear membrane (Fig. 4).

The frequency of aberrations caused by low temperatures was high among pupae and higher among prepupae. There were indications that most aberrations were due to the effect of temperature variation at the spindle level, by interference with the depolymerization mechanism of the microtubules that compose it as well as with the other reactions involved in chromosome motion and distribution during division. Thus, the production

---

Fig. 1. Longitudinal view of sperm head. Negatively stained whole mount showing the acrosome (a) containing the acrosomal filament (af) and nucleus (n). In the lower part, one of the mitochondrial derivatives (m) and the axial complex (f) can be seen. [From Cruz-Höfling *et al.* (1970), with permission. Copyright 1970 by Academia Brasileira de Ciências.]

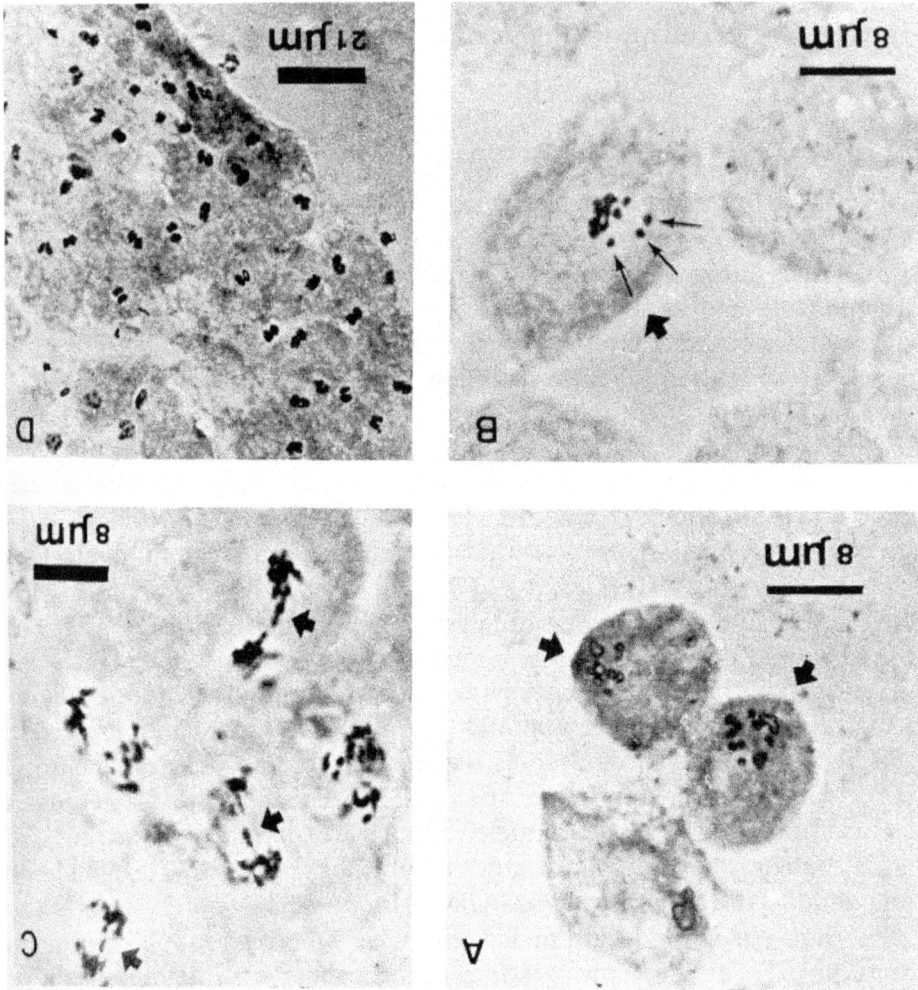

Fig. 2. (A,B) Spermatocytes of *Apis mellifera* L. pupae, aged 15 days, submitted to 10°C for 320 min: (A) reduced metaphase, $n/2 = 8$ (arrows); (B) metaphase with lost chromosomes (thick arrow) and isolated chromosomes (thin arrows). (C,D) Spermatogenesis in 15-day-old *Apis mellifera* L. pupae exposed to 5°C temperature for 20 min: (C) aberrant anaphases, late chromosomes attached to the remaining ones (arrows); (D) cyst showing several binucleate spermatids. [From Tarelho (1981a,b), with permission.]

of binucleate spermatids and spermatozoa suggests the interference of low temperature with cytokinesis, i.e., with the reactions that culminate in the formation of the dividing groove and in the separation of the two daughter cells.

Thus, a change in normal temperature (35°C) during spermatogenesis in *Apis mellifera* causes several types of aberrations.

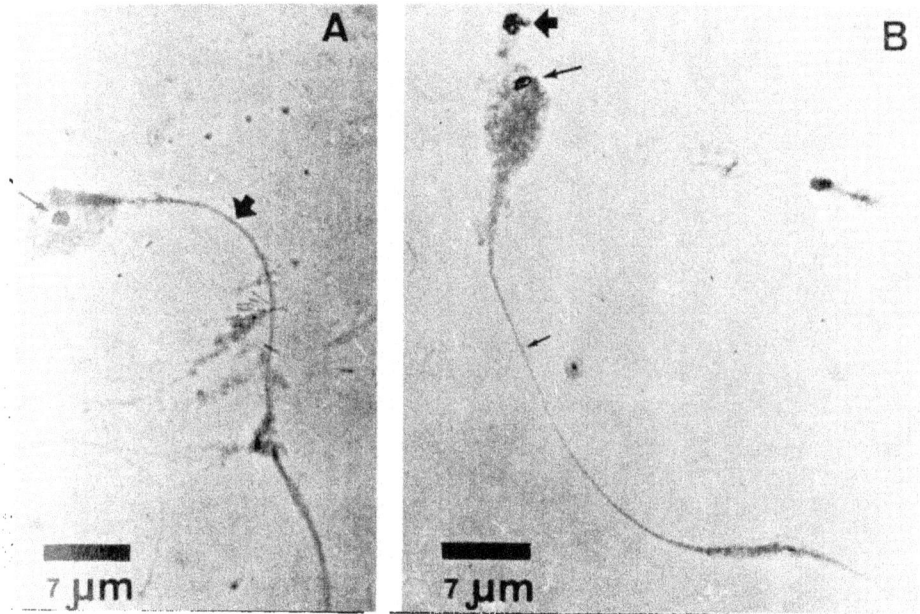

Fig. 3. Spermatozoa of 15 day old *Apis mellifera* L. pupae exposed to 10°C for 320 min. (*a*) Anomalous young spermatozoan with two nuclei (thin arrow). The thick arrow points to the flagellum. (*b*) Normal young spermatozoan with the nucleate bud (thick arrow) attached to the larger cell (long thin arrow) by a cytoplasm bridge. The short thin arrow shows the flagellum. [From Tarelho (1981b), with permission.]

C. Semen Storage

1. Nonfrozen Storage

Honey-bee males are not always available, especially in countries with cold winters. Even in tropical-climate countries such as Brazil it is difficult to obtain drones in the winter months, especially in the south. If semen could be stored for several months, instrumental matings could be made throughout the year. Transfer of genetic material between countries in the form of semen might avoid problems of transmission of diseases and parasites of honey bees.

Semen is easily collected, and the potential advantages of storing semen for long periods of time have stimulated researchers to carry out several studies related to semen storage. Taber and Blum (1960) studied the effect of several factors such as different diluents, replacement of atmospheric air inside the capillary tubes with other gases, temperature, etc. on the viability of sperm stored in capillary tubes with heat-sealed ends. The results showed that semen diluted with hemolymph or jelly coagulated. When the semen

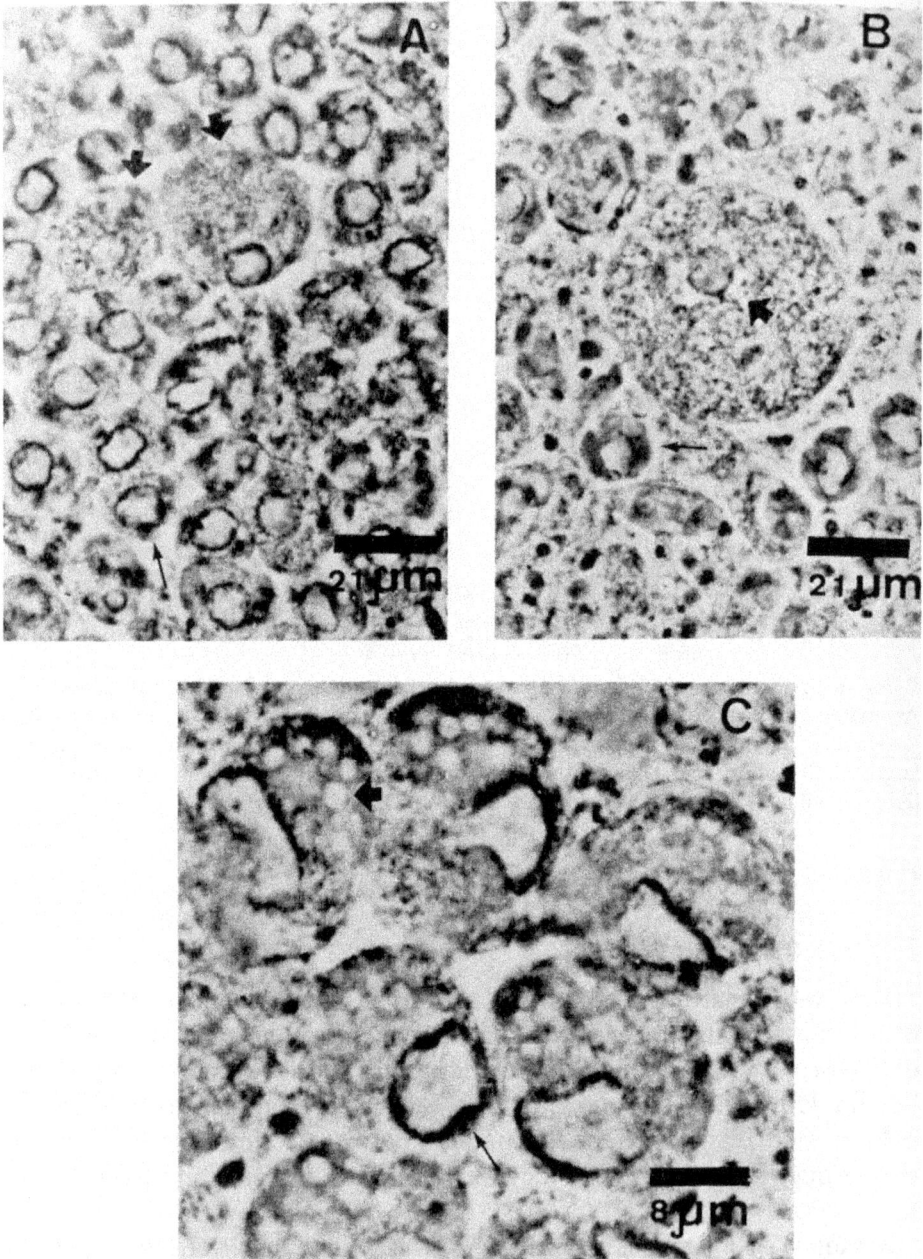

Fig. 4. Spermatocytes of 15-day-old pupae of *Apis mellifera* L. submitted to 45°C temperature for 80 min. (*a*) Spermatocytes of normal size (thin arrows) and in two growth phases (thick
arrows). (*b*) Giant spermatocytes (thick arrow pointing at the nucleus) and spermatocytes of
normal size (thin arrow). (*c*) Normal-size spermatocytes showing vacuoles in the cytoplasm
(thick arrow) and the excentric nucleus with chromatin close to the nuclear membrane (thin
arrow). [From Tarelho (1981a), with permission.]

was diluted with Ringer solution or with Ringer solution plus fructose, partial coagulation occurred. All queens that received semen from tubes where the atmospheric air had been replaced with CO_2 died. A few tubes that were stored for 2 weeks or more showed contamination with microorganisms, and the semen they contained became discolored or more viscous than normal.

For the study of the effect of temperature, sealed tubes containing semen were maintained at room temperature, 35°F, and 90°F from 27 to 68 days with chlortetracycline added as a preservative. When the semen in these tubes was used for instrumental insemination, no fertile eggs were obtained from queens fertilized with semen stored at 35 or 90°F. Spermatozoa stored for 4 weeks or more (up to about 68 days) at room temperature remained viable. Fertilized eggs were obtained from queens inseminated with this semen.

Poole and Taber (1969) approached the problem with better success. The use of better sterilization techniques and treatment with different antibiotics reduced contamination with microorganisms. They stored semen in capillary tubes at 57°F with several antibiotics. The addition of streptomycin and tetracycline prolonged the period of time that stored semen retained its viability. As a result, these investigators were able to successfully store semen for 3–4 months.

Jaycox (1960) tested several saline solutions in an attempt to activate bee spermatozoa that had lost their motility after storage in capillary tubes, and obtained moderate motility after 10 days when semen was diluted with saline solution containing glycine.

According to Poole and Edwards (1970), simple oxygenation of the semen–saline mixture does not induce sperm motility and does not change the folded configuration of the spermatozoa. In contrast, when 3 drops of any saline solution–sugar diluting solution are added to the semen–saline mixture, oxygenated or not, vigorous spermatozoon motion is immediately observed although many spermatozoa remain folded. These results suggest that sugars at low (0.016%) concentration can induce sperm motility in mixtures of semen plus 0.85% saline.

In a more complete study, Poole and Edwards (1972) demonstrated the role played by different sugars in inducing sperm motility when added to the saline solution. As expected, glucose, fructose, and sucrose induced motility. Trehalose, which is a normal constitutent of seminal plasma, was not as effective. Galactose and melibiose, which are toxic to bees, were not toxic to spermatozoa and actually induced motility. Mannose, raffinose, and rhamnose, which are also toxic to bees, did not induce motility. Camargo (1975) investigated several diluents with the objective of finding one that would maintain living spermatozoa for as long as possible and thereby increase the percentage of spermatozoa stored in the queen's spermatheca after instrumental insemination. Ringer, Locke, Tyrode, and Jaycox solutions were not satisfactory as diluents because the spermatozoa survived for less than 1 hr. However, good results were obtained when semen was diluted with coconut water plus dehydrostreptomycin (pH 7.0) at 10°C. The use of an antibiotic is indispensable, since diluents are easily contaminated with bacteria.

2. Frozen Storage

Intense research has been carried out on storage of bee semen in liquid nitrogen, and a considerable body of information has been obtained. Spermatozoa cannot survive at −196°C when semen alone, packed in capillary tubes, is stored in liquid nitrogen. A compound must be added to act as a cryoprotectant. Thus, using hemolymph as a cryoprotectant, Melnichenko and Vavilov (1976) were able to successfully store semen in liquid nitrogen. In 1971, Brown et al. were able to store some insect cells in liquid nitrogen without problems by using dimethyl sulfoxide (DMSO) as a protector. On the basis of their data, Harbo (1977, 1979a) tested DMSO on honey-bee spermatozoa and demonstrated its effectiveness. According to Harbo, three variables must be taken into consideration when spermatozoa are frozen: freezing rate, chemical compounds mixed with the semen, and the semen–diluent ratio. He used a hydraulic syringe connected to an apparatus that accurately measured the collection, dilution, and delivery of semen, and a thermocouple inserted into the semen that recorded the freezing and thawing rates. From collection to insemination, the storage process only caused a 6.6% loss of the semen mixture.

Sperm survived when a constant cooling rate was used, such as 25°C per minute, whereas they did not survive a freezing rate of 300°C per minute. Thawing, however, must be rapid, since a slow thawing process may cause intercellular ice crystals to grow or agglomerate and damage the cells (Mazur, 1966). Rapid thawing was used in Harbo's study (1979a): about 1100°C per minute, with 12 sec needed for semen to thaw.

Out of all the ratios of diluent (saline solution), semen, and cryoprotectants, the one that gave best results was a mixture containing 60% semen, 30% saline solution, and 10% DMSO. Queens inseminated with this mixture that had been stored in liquid nitrogen retained 41% of the spermatozoa received in their spermathecae when compared to controls inseminated with fresh semen. Glycerol provided some protection, but less than DMSO. Glycerol in association with DMSO was of no benefit.

More than 50 inseminations were carried out that produced workers.

However, many queens produced more drone than worker brood, showing that this system of storage in liquid nitrogen is not yet perfect. Harbo (1979b) observed that queens inseminated with spermatozoa stored in liquid nitrogen produced eggs with a lower hatching rate than the controls ($82 \pm 11\%$ and $95 \pm 2\%$, respectively).

Later research moved one step further in the study of the consequences of using semen stored in liquid nitrogen. *Apis mellifera* queens were inseminated with previously frozen semen and produced daughters (F_1 queens) that in turn were allowed to lay eggs (F_2), which were studied for viability. The mortality rate of the F_2 eggs was significantly higher ($5.6 \pm 5\%$) than it was in the controls ($2.6 \pm 1.5\%$). This is evidence that genetic damage may have occurred in the F_2 progeny of spermatozoa that had been stored in liquid nitrogen, although the data showed no persistence of deletrious effects in the F_2 generation (Harbo, 1981).

Other observations have also shown an apparent inability by spermatozoa from stocks preserved in liquid nitrogen to join the pronucleus of the egg, as well as the formation of drones with mosaicism (Harbo, 1980).

The main objective of low-temperature ($-196°C$) sperm storage is to use this sperm for queen insemination such that the performance of these queens will be as good as that of queens inseminated with fresh sperm. If this goal is reached, frozen spermatozoa could be widely utilized in commercial queen production to obtain multiple hybrid combinations at any time of year (Harbo, 1979a). This goal, however, has not yet been reached, and a series of difficulties remains to be overcome.

No cytological or histological analysis have been made of spermatozoa from semen stored at very low temperatures.

V. STOCK MAINTENANCE

A. Propagation

Stocks can be maintained by the continued production of queens or drones and their maintenance in the field by colonies they produce. Few methods have been reported in the literature for stock production. The best known ones are line breeding and hybrid breeding. The first involves the mating and selection of queens and drones from the best hives in a small population and requires the use of instrumental insemination or controlled open mating in islands, valleys, or clearings. In the line-breeding method, inbreeding is practically unavoidable. Inbreeding results in low viability rates (Woyke, Chapter 4). The hybrid breeding method involves mating queens and drones from several selected stocks. In general, the resulting hybrid progenies are superior to their parents because of hybrid vigor or heterosis. In general, hybrid breeding involves more than three inbred lines. However, regardless of the type of mating involved, it is important to know the degree of inbreeding and relationships among the individuals in the stock (see Laidlaw and Page, Chapter 13). Roberts (1974), studied inbreeding and stock production and developed several mating systems which avoid unbreeding.

B. Queen Storage

Several studies have been conducted in attempts to maintain large numbers of queens. A study by Harp (1969) deserves special mention. He used a special frame containing 27 independent compartments with one queen in each and brood combs inside to permit egg laying. The queen compartments were in contact with the colonies' workers through an opening covered with excluding netting. Thus, 40 queens were maintained for 5 months in a single colony and functioned normally when they were introduced individually into new colonies.

Levinsohn and Lensky (1981) also obtained satisfactory results with queen storage. The queens were maintained individually in wooden, 4.5 × 4.5 × 2.3 cm high boxes covered on both sides with excluding copper wire netting and attached to a frame. The frame was placed in hives called reservoir colonies that might or might not have a queen. The confined queens had contact with the workers. In queenless reservoir colonies, 80% of stored fertilized queens survived for 5 months. In reservoir colonies containing a queen, 78% of stored fertilized queens survived for 2 years.

Storage of queens totally isolated from workers has also been studied. However, solitary confinement of queens, workers, or drones for long periods of time is difficult (Park, 1949). Nelson and Roberts (1967) built a box containing eight round compartments, each 34 mm in diameter, covered with excluding netting, and containing one queen. The queens were fed diluted honey mixed with a little royal jelly. Fertilized queens survived about 14 days.

Fertilized queens were also confined individually in mailing cages supplied with water and food containing honey and royal jelly. The queens were maintained in incubators at 27°C with 40–50% relative humidity. Approximately 84% of the queens survived 80 days in confinement, and 17% survived 180 days (Shehata, 1982).

C. Packing Semen for Transport

Legal obstacles exist in several countries against bee importation, because of fear of disease introduction. In view of these laws, new genetic material can only be obtained by receiving mailed semen.

Several researchers have sent bee semen to other countries through the mails using capillary tubes (75 mm in length and 1.45–1.65 mm in diameter) with sealed ends. This semen has been successfully used to inseminate queens.

According to Taber (1977), it is important that the capillary tubes contain more semen than air. Tubes containing more air than semen must frequently be discarded because of bacterial or fungal growth.

Semen is collected with a normal insemination needle and syringe and is transferred to capillary tubes that are centrifuged, sealed, and sent to other research centers or stored. For use, the capillary tube is broken and the semen is withdrawn with an inseminating needle. Harbo (1974) proposed a new method in order to simplify this process by eliminating the steps involving the transfer of semen from the needle to the capillary and centrifugation: a modified capillary tube was used both for storage and as an inseminating needle. After semen collection, the capillary tube is disconnected from the special syringe, both ends are sealed with petrolatum, and the tube is ready for storage or shipment. When semen is used, the tube is reapplied to the syringe, the petrolatum plug is removed from the sharp end, and insemination is performed.

More recently Kaftanoglu and Peng (1980) described a new process in which the Mackensen-type tip of the syringe was modified so that a storing capillary could be fitted to its base. A special syringe was set up for precise control of semen motion. The capillary tubes containing semen can be removed from the syringe and readapted to it without interfering with the semen column because the system is in equilibrium.

D. Packing Queens for Transport

For queen shipment, the box normally utilized is a Benton standard box. It is essentially a wooden block containing three round interconnected cavities, one used for storage of food in the form of a soft "candy" made of powdered sugar and inverted syrup. The other two compartments are for the queen and young worker-bee attendants. The top of the Benton box is covered with netting. Battery cages may also be used, which consist of a container in which queens can be maintained without attendants or candy. Individual cages are maintained on racks inside a traveling container that also contains several thousand workers. A can containing liquid food is attached to the apparatus. This arrangement has the advantage of providing better care for the queen because of the excessive amount of workers. There are also disadvantages with some handling restrictions when battery cages are shipped by plane (York, 1979).

ACKNOWLEDGMENTS

The authors are very grateful to Dr. Zuleice V. S. Tarelho (UNESP-Araçatuba) and Dr. Carminda da Cruz-Landim (UNESP-Rio Claro) for the originals of the figures. Thanks to Mrs. Elettra Greene for translating the text.

REFERENCES

Blum, M. S., and Taber, S., III. (1965). Chemistry of the drone honey bee reproductive system. III. Dehydrogenases in washed spermatozoa. *J. Insect Physiol.* 11, 1389–1501.

Blum, M. S., Glowska, Z., and Taber, S., III. (1962). Chemistry of the drone honey bee reproductive system. II. Carbohydrates in the reproductive organs and semen. *Ann. Entomol. Soc. Am.* 55, 135–139.

Blum, M. S., Bungarner, J. E., and Taber, S., III. (1967). Composition and possible significance of fatty acids in lipid classes in honey bee semen. *J. Insect Physiol.* 13, 1301–1308.

Bretschneider, J. H. (1950). Elektronen-mikroskopische Strukturuntersüchungen and Spermien. *Proc. K. Ned. Akad. Wet.* 53, 531–544.

Brown, B. L., Nagle, S. C., Lehman, J. D., and Rapp, C. D. (1971). Storage of *Aedes aegypti* and *Aedes albopictus* cells under liquid nitrogen. *Cryobiology* 7, 249–251.

Camargo, C. A. (1975). Biology of the spermatozoon of *Apis mellifera*. I. Influence of diluents and pH. *J. Apic. Res.* 14, 113–118.

Cruz-Hofling, M. A. da, Cruz-Landim, C., and Kitajima, E. W. (1970). The fine structure of spermatozoa from the honey bee. *Anais Acad. Bras. Ciênc.* 42, 69–78.

Harbo, J. R. (1974). A technique for handling stored semen of honey bees. *Ann. Entomol. Soc. Am.* 67, 191–194.

Harbo, J. R. (1977). Survival of honey bee spermatozoa in liquid nitrogen. *Ann. Entomol. Soc. Am.* 70, 257–258.

Harbo, J. R. (1979a). Storage of honey bee spermatozoa at −196°C. *J. Apic. Res.* 18, 57–63.

Harbo, J. R. (1979b). Egg hatch of honey bees fertilized with frozen spermatozoa. *Ann. Entomol. Soc. Am.* 72, 516–518.

Harbo, J. R. (1980). Mosaic male honey bees produced by queens inseminated with frozen spermatozoa. *J. Hered.* 71, 435–436.

Harbo, J. R. (1981). Viability of honey bee eggs from progeny of frozen spermatozoa. *Ann. Entomol. Soc. Am.* 74, 482–486.

Harp, E. R. (1969). A method of holding large numbers of honey bee queens in laying condition. *Am. Bee J.* 109, 340–341.

Hoage, T. R., and Kessel, R. G. (1968). An electron microscope study of the process of differentiation during spermatogenesis in the drone honey bee (*Apis mellifera* L.) with special reference to centriole replication and elimination. *J. Ultrastruct. Res.* 24, 6–32.

Jaycox, E. R. (1960). The effect of drying and various diluents on spermatozoa of the honey bee (*Apis mellifera* L.). *J. Econ. Entomol.* 53, 266–269.

Kaftanoglu, O., and Peng, Y. S. (1980). A new syringe for semen storage and instrumental insemination of queen honeybees. *J. Apic. Res.* 19, 73–76.

Kurennoi, N. M. (1954). Structure and viability of drone spermatozoa. *Pchelovodstvo 30*, 19–25, (in Russian).

Laidlaw, H. H., Jr. (1977). "Instrumental Insemination of Honey Bee Queens." Dadant and Sons, Hamilton, Ill.

Lensky, Y., Ben-David, E., and Schindler, H. (1979). Ultrastructure of the spermatozoon of the mature drone honeybee. *J. Apic. Res. 18,* 264–271.

Levinsohn, M., and Lensky, Y. (1981). Long term storage of queen honeybees in reservoir colonies. *J. Apic. Res. 20,* 226–233.

Mazur, P. (1966). Physical and chemical basis of injury in single celled microorganisms subjected to freezing and thawing. *Cryobiology 2,* 213–315.

Melnichenko, A. N., and Vavilov, Y. L. (1976). Many years keeping of drone semen when freezing in liquid nitrogen. *Dokl. Vses. Akad. Skh. Nauk. 1,* 25–26.

Nelson, E. V., and Roberts, W. C. (1967). Storage of queen honey bees, *Apis mellifera* (Hymenoptera: Apidae), in solitary confinement. *Ann. Entomol. Soc. Am. 60,* 1114–1115.

Novak, A. F., Blum, M. S., Taber, S., III., and Liuzzo, J. A. (1960). Separation and determination of seminal plasma and sperm amino acids of the honey bee, *Apis mellifera. Ann. Entomol. Soc. Am. 53,* 841–843.

Park, O. W. (1949). The honey bee colony-life history. *In* "The Hive and the Honey Bee" (R. Grout, ed.), pp. 21–78. Dadant and Sons, Hamilton, Ill.

Poole, H. K., and Edwards, J. F. (1970). Induction of motility in honey bee (*Apis mellifera* L.) spermatozoa by sugars. *Experientia 26,* 859–860.

Poole, H. K., and Edwards, J. F. (1972). Effect of toxic and non-toxic sugars on motility of honey bee (*Apis mellifera* L.) spermatozoa. *Experientia 28,* 235.

Poole, H. K., and Taber, S., III. (1969). A method of in vitro storage of honey bee semen. *Am. Bee J. 109,* 420–421.

Roberts, W. C. (1974). A standard stock of honeybees. *J. Apic. Res. 13,* 113–120.

Rothschild, L. (1955). The spermatozoa of the honey bee. *Trans. R. Entomol. Soc. Lond. 107,* 289–294.

Shehata, S. M. (1982). Long term storage of queen honeybees in isolation. *J. Apic. Res. 21,* 11–18.

Taber, S., III. (1977). Semen of *Apis mellifera:* fertility, chemical and physical characteristics. *In* "Advances in Invertebrate Reproduction," Vol. I (K. G. Adiyodi and R. G. Adiyodi, eds.), pp. 219–251. Peralam-Kenoth, Karivellur, Kerala, India.

Taber, S., III, and Blum, M. S. (1960). Preservation of honey bee sperm. *Science 131,* 1734–1735.

Tarelho, Z. V. S (1981a). Effects of low and high temperatures on the spermatogenesis of *Apis mellifera* Linne, 1754. I. Effect of temperature on metaphase, survival and development. *Rev. Bras. Genét. 4,* 193–212.

Tarelho, Z. V. S. (1981b). Effects of low and high temperatures on the spermatogenesis of *Apis mellifera* Linne, 1754. II. Effects of temperature on anaphase, telophase and spermiogenesis. *Rev. Bras. Genét. 4,* 383–397.

Toedtmann, W. (1914). Die Spermatozoen von *Apis mellifica. Bios 2,* 67–74.

Woyke, J., and Ruttner, F. (1958). An anatomical study of the mating process in the honeybee. *Bee World 39,* 3–18.

York, H. F., Jr. (1979). Production of queens and package bees. *In* "The Hive and the Honey Bee" (Dadant and Sons, eds.), pp. 559–578. Dadant and Sons, Hamilton, Ill.

CHAPTER 15

Propagation and Instrumental Insemination

JOHN R. HARBO

I. INTRODUCTION

A. Genetic Control of a Colony

The ability to control mating is basic to any breeding program. However, before the mating of the honey bee could be controlled it was necessary to learn a little about bee reproduction and colony manipulation. This knowledge first became available in the mid nineteenth century, and shortly thereafter (about 1870) there was a sudden interest in controlled mating.

1. Basic Reproductive Biology

Work by Dzierzon and Langstroth led to early attempts to control mating. First came Dzierzon's theories. Dzierzon (1845) stated that female bees develop from fertilized eggs and males develop from unfertilized eggs. There is more to sex determination (Woyke, Chapter 4), but for practical bee breeding Dzierzon's model is adequate. Dzierzon also learned that a queen lays all the eggs in a colony and that once egg laying begins a queen will not mate again. The discovery of bee space and the development of the movable frame hive by L. L. Langstroth in 1851 enabled beekeepers to examine their bees without seriously disrupting the colony. Thus, by the 1850s beekeepers knew enough about bee reproduction to want to control the queen, and with the movable frame hive they could replace queens with relative ease.

2. Queen Rearing

Queen rearing became a prime area of interest as Dzierzon's and Lang-stroth's discoveries became known to beekeepers. Doolittle (1889) developed a method of queen rearing that involved the transfer of young larvae from worker cells into beeswax cups that were made about the size of natural queen cups. Doolittle's methods, with little or no modification, are still used in the beekeeping industry.

3. Controlled Mating

Natural mating (NM) of the queen (Koeniger, Chapter 10) can be controlled. However, one needs to exclude all drones from an isolated area except those of the type desired to mate with the queens. The isolated area need not be an island, but islands have been successful in obtaining pure matings. Numerous attempts to mate queens in cages or other confined areas have failed or have not been repeatable (Harbo, 1971).

Instrumental insemination (II) is an alternative to island mating and requires much less space. II permits controlled matings in a place that is not geographically isolated, and it allows the bee breeder to use many different drone types at one place on the same day. This feat would otherwise require a different isolated area for each drone type used.

II also enables breeders to make matings that are impossible with natural mating, for example, (1) mating a queen to a single drone or to a few specific drones, (2) mating mutant queens and drones, and (3) mating a queen to her own male offspring (selfing).

B. History of Instrumental Insemination

The basic principles of II were developed between 1926 and 1947. Lloyd Watson first demonstrated a successful technique in 1926 (Cale, 1926). The success of Watson's technique was confirmed by Nolan (1929), who also developed holding hooks and an insemination stand that are similar to those presently in use (Nolan, 1932). Laidlaw (1944) vastly improved the success rate of II by learning to insert the insemination tip past a flaplike structure (the valvefold) that covers the entrance of the median oviduct. He depressed the valvefold and injected the semen directly into the median oviduct. Mackensen (1947) used carbon dioxide to immobilize queens during insemination. This made it easier to insert the tip into the median oviduct. Moreover, he found that CO_2 narcosis caused queens to begin laying eggs sooner after insemination. The first comprehensive manual for II of queens was written by Mackensen and Roberts (1948).

The major use of II has been in research. It has been used to develop inbred lines, maintain mutant markers, and make specific matings for genetic research such as backcrosses and single drone inseminations. II has been used very little in commercial breeding programs, and attempts to market II queens for use in field colonies have ended in failure.

C. Chapter Objective

For instrumental insemination, this chapter describes only the equipment and procedures that I now use. The techniques are similar to those of most other workers with two exceptions. First, I collect large quantities of semen before beginning inseminations; others collect semen between inseminations. Second, I use glass rather than plastic insemination tips. The techniques of other workers are described in Mackensen and Tucker (1970), Ruttner (1976), and Laidlaw (1977). The procedure begins with queen and drone production and ends with laying queens.

II. QUEEN PRODUCTION

A. Natural Queen Production

Queen and worker honey bees are genetically identical; both are females. They differ only in the way workers feed and care for the larvae. Queens can be produced from worker larvae that are $3\frac{1}{2}$ days old or less ($6\frac{1}{2}$ days from egg laying) (Becker, 1925), so any colony with young worker brood has the potential to produce queens. However, most often they choose not to do so.

The reason a colony does not constantly produce queens can be traced to the resident queen. A mated queen inhibits the workers from producing queen cells, and this inhibition is effective only if the queen is able to travel freely over the brood area (i.e., not caged or restricted to a portion of the broodnest) (Butler, 1957). Lensky and Slabezki (1981) found that a pheromone is produced on the tarsi (feet) of a queen and deposited on the comb surface by her foot pads. When applied to the lower edges of a brood comb in an overcrowded colony, this chemical, in combination with secretions from the mandibular gland, inhibits the production of queen cups. When used alone, neither of these secretions inhibits construction of cups. However, if queen cells have already been started, a laying queen apparently does not suppress the rearing of queens (Lensky, 1971).

The natural periods for a colony to rear queens are when they are about to swarm, when they are replacing a failing or poor queen, and when they are replacing a queen that has been removed or accidentally killed. All these conditions could be attributed to inadequate queen movement over the brood area, for a crippled, old, or poorly laying queen might be unable to get to all parts of the broodnest as the colony demands. Those queens therefore fail to inhibit cell production, and new queens are produced. Queen rearing under crowded, swarm conditions may also be caused by the immobility of the queen. According to Lensky and Slabezki (1981), the crowding of worker bees at more than 2000 bees per liter of actual space (hive volume minus comb volume) restricts the movement of a queen so that queen cells are produced.

Under conditions of swarming and supersedure, a queen usually will lay eggs in queen cups. Therefore, their restricted movement does not completely restrict them from where the queen cups are produced.

In the case of sudden queen loss or emergency supersedure, eggs are not laid into queen cups but queen cells are produced from larvae in worker cells. Örösi Pál (1957) found that emergency queen cells are started on cells with larvae, not on cells with eggs.

B. Management for Queen Rearing

To produce queens, a beekeeper must manage a colony of bees so that the workers will rear certain larvae, chosen by the beekeeper, into queens. There are many ways to do this (Laidlaw, 1979; Morse, 1979), but most methods rely on putting a colony into a natural queen-rearing state and then adding the young larvae that are to be reared into queens. These larvae are taken from worker cells and put into human-made beeswax cups that simulate natural queen cups (Fig. 1, *a* and *b*).

1. Cell-Building Colony

A colony that is managed to rear queen cells is called a cell builder. Sometimes these colonies have a laying queen (usually confined to a section of the colony away from the cells), sometimes they are queenless, and sometimes a queen producer uses two cell builders, one to start the cells and one to finish them. I have used both queenright and queenless cell builders, and recommend queenless cell builders for all but those who produce thousands of queens per year. Thus I will discuss only the management of a queenless cell builder.

There are some general qualities that a cell-building colony should have. It should have a larger rather than a small population of worker bees (> 20,000). The bees should be crowded into a colony with 10, or at the most 20, frames (Fig. 1*b*). The colony should be fed sugar syrup, and also pollen if no natural pollen is available (feeding pollen often causes disease, so beware of the pollen source). Finally, young brood in the colony should be arranged

Fig. 1. Queen rearing and drone storage. (*a*) A queen cell cup with a drop of royal jelly and a larva floating on the jelly. The end of a wire grafting tool extends into the cup. (*b*) Cups are arranged on two bars that are made to slip in and out of a frame that goes into a cell builder. (*c*) Capped queen cells put into vials that are arranged on a wooden rack that goes into an incubator (35 °C). Each vial has artificial sponge material and a small amount of queen candy (made from honey and powdered sugar) at the bottom of the vial. A newly emerged queen can live about 2 days on the candy, and the sponge gives her a dry, nonslippery walking surface. (*d*) Drones stored in cages. The cages have excluder material on one side and screen (8 mesh per 25 mm) on the other. Cage size and shape can vary, but cages should not be high and narrow because the drones tend to gather at the bottom. If the drones become more than 3 or 4 cm deep at the bottom, some may die. Therefore, these cages measuring (16 × 19 × 2 cm) are given no more than 125 drones.

next to where the queen cups are to be placed, because young brood attracts nurse bees.

The cell-building colony must be managed on a regular schedule. If properly managed, the same colony can be used to rear cells for an entire season. A weekly schedule is the easiest to keep. Begin the weekly management by harvesting the queen cells (now capped) and put them into an incubator (Fig. 1c) or into colonies. Since the colony is queenless, it needs to be given brood or bees to maintain its population. About two good frames of brood (one with uncapped cells) usually fill this need and can replace two broodless combs in the cell builder. Finally, destroy any queen cells that may be reared on the brood comb, because any virgin queen that emerges in the cell builder will destroy the other cells.

2. Breeder Queens

A breeder queen is the mother of the queens to be reared in the cell builder. A breeder queen need not be kept in a large colony, yet she needs to be laying in a colony that will adequately feed the newly hatched larvae.

3. Basic Grafting

The process of taking larvae from the worker cells to queen cups is called grafting (Fig. 1a). Of course, the larvae come from the breeder queen and are put into the cell-building colony after being transferred to the cups. Grafting should be done after rather than before the weekly management of the cell builder.

A grafting tool is used to transfer a larva from the bottom of a worker cell to a queen cup. This tool can be purchased but is often homemade by bending and filing a wire or carving a green twig. A moistened (chewed) toothpick is often used because it is convenient, not because it is best.

Grafting is best done inside a building under a bright light. Four items are needed: the bars of cell cups (detached from the frames), royal jelly (diluted 1:1 with water), a grafting tool, and a frame of young brood from the breeder queen. Before grafting, place a drop of royal jelly into each cup on a bar of cups. Royal jelly is not necessary, but it is easier to get the larvae off the grafting tool if jelly is present. (Royal jelly is collected from queen cells before they are capped. Remove the queen from a colony and return in 3 days to collect jelly, or graft as usual and then harvest royal jelly 3 days later. Royal jelly stores well in the freezer.)

Choose larvae for grafting that are as young as possible, for the youngest larvae produce the best queens (Woyke, 1971). Weiss (1974a) found that the major decline in queen quality comes when larvae are over 48 hr old at the time of grafting and that there is very little difference in queens when the grafted larvae are 0–36 hr old.

4. Double Grafting

This is a grafting technique that some queen producers use in an effort to produce larger and presumably better queens. Graft as described above, then after 24 hr remove the cells from the cell builder and prepare to graft again. Discard the larvae from the cells (retain the jelly), graft new larvae into those cells, and then return the cells to the cell builder.

There are conflicting opinions as to the value of double grafting. Weiss (1974b) found that when grafting larvae 24 hr old, there was no difference in queen quality when the larvae were double or single grafted. However, when grafting larvae that were 36–48 hr old, the double-grafted larvae produced superior queens.

5. Grafting Eggs

This is a technique that was developed by Örösi Pál (1960). It is a form of double grafting, but instead of putting another larva into a cell, a 2- to 3-day-old egg together with a 3-mm disk of beeswax at its base is set into the queen cup. The jelly surrounds the beeswax base but does not touch the egg.

To graft eggs, one needs eggs of known age and a way to transfer a wax disk and egg to the cell cup. Örösi Pál (1958) recommends a simple punch made by wrapping a thin piece of metal (10 $\times$ 60 mm) around a 3-mm nail. The circle of the punch does not close completely, so that after the disk and egg are in the punch, a pin can reach through the side of the punch onto the disk to remove the disk and egg from the punch. An opening 2 $\times$ 5 mm is filed about 3 mm from one end and opposite the slit so that one can see inside.

6. Seasonal Effects

The best time to produce queens is when nutritious pollen is available to the bees. This usually corresponds to the natural growth period of a colony and will vary among different parts of the world. In general, if drone brood is being reared, queens can be produced. Queens can be produced during suboptimal periods if one is willing to feed large quantities of pollen (Taber and Poole, 1974).

III. DRONE PRODUCTION

A. Rearing Drones

Drones are usually reared in their home colony. Thus, there are no cell builders for drones as there are for queens.

However, as with cell-building colonies, a good drone-producing colony has certain requirements. Probably the most basic need is a good pollen supply. Also, a large population is more apt to produce drones than a smaller population, and of course more drones will be produced if a colony has some drone comb available.

If one needs drones from a particular queen that is not producing any or is producing too few, the queen can be moved to a colony that does produce drones. Find a colony that is actively producing drones, remove the queen from the latter colony, kill the capped drone cells by scraping them with a hive tool, and introduce the queen from which drones are desired. Ten days later, kill the capped drone cells again (these are still drones from the previous queen). The workers will usually rear drones for the new queen as well as they did for their mother.

One can force a young queen to produce drones by not allowing her to mate. Such a queen is induced to lay eggs by giving her 3 min of CO_2 narcosis on each of 3 consecutive days or simply by letting her age (caged) for 5 or 6 weeks. The unfertilized eggs produced by these queens develop into small drones if the eggs are laid in worker-sized cells. These small drones produce viable semen that can be collected and used in instrumental insemination. However, if drone-sized cells are available in the colony, the unmated queens seem to prefer to lay their fertilized eggs in them, and the result will be normal-sized drones.

B. Storing Drones

Drones will not yield semen until they are 6–12 days old. Therefore, they must be aged by storing them somewhere. Moreover, drones are often stored for longer than 6–12 days simply for the convenience of the operator or to coordinate semen collection with queen maturity. The problem is to age the drones while keeping the group free of unwanted drones that may mingle with the desired drones.

One way to do this is to mark the drones. Individual drones can have a dot of paint put on their thorax, or they can carry a visible genetic marker. Of course the genetic marker should not be present in adjacent colonies, and if the marker is an eye marker, the drones must be kept from flying so that they will not be lost. A group of drones can be marked with spray paint if the nozzle has its orifice enlarged to deliver droplets that are slightly larger than used in painting. These painted drones can then be released into a colony and collected as needed.

A second technique is to confine the drones to a single brood chamber. W. C. Roberts used this technique when continuously rearing and storing many drones from a single queen. The drone-rearing colony serves as the drone storage colony. An empty super and two excluders are brought to the colony that has drone brood of the desired stock. All the bees are brushed from the frames of drone brood as well as the other frames that are put into the new super (uncapped worker brood, pollen, etc.). When full of combs, this super containing no adult bees is placed above the brood chamber with an excluder above and below. Of course, adult workers quickly move into this super (drone chamber), so it is free of adults for a very short time. These colonies can be managed on a weekly schedule and should be opened only in mornings so that the drones do not fly out and unwanted drones do not fly in. During management, the empty drone and worker combs are moved down to the brood chamber, and combs of drone brood, worker brood, and pollen are brushed free of bees and moved into the drone chamber.

A third method of storing drones is to put them into cages (Fig. 1*d*). The cages should have excluder material on one side so that workers can pass into the cage to feed the drones. Drones survive best in populous colonies between frames of brood.

IV. INSEMINATION PROCEDURE

Simply stated, II is the instrumental transfer of semen from one or more drones to the oviducts of a queen (see Fig. 6). In both NM and II, the semen is deposited in the lateral oviducts. Once in the oviducts, 4–25% of the spermatozoa migrate to the spermatheca. The percentage of migration depends on the amount of semen given (discussed later). Sperm migration is usually complete within 24 hr (Bishop, 1920; Woyke 1983b), and the insemination process is complete at that point.

A. Equipment

The complete apparatus, except for the CO_2 tank and regulator, is pictured in Fig. 2. The insemination stand and hooks are described in detail by Mackensen and Tucker (1970). They suggest that the queen holder block be adjusted so that it leans 30° from the vertical and makes a 10° angle with the syringe (the syringe being 10° more to the horizontal) (Fig. 2). The adjustment of the stand varies with the operator, but the 10° angle between the queen and syringe should be maintained.

The syringe and tip are shown in Fig. 3. The design in Fig. 3 is an improved version of a syringe designed for use in semen storage (Harbo, 1979). The present version has a removable storage tube, but the storage tube is also used during routine inseminations when no semen storage is planned. Boiled saline within the tubing and micrometer produce an air-free hydraulic system throughout the syringe and make the system responsive to slight

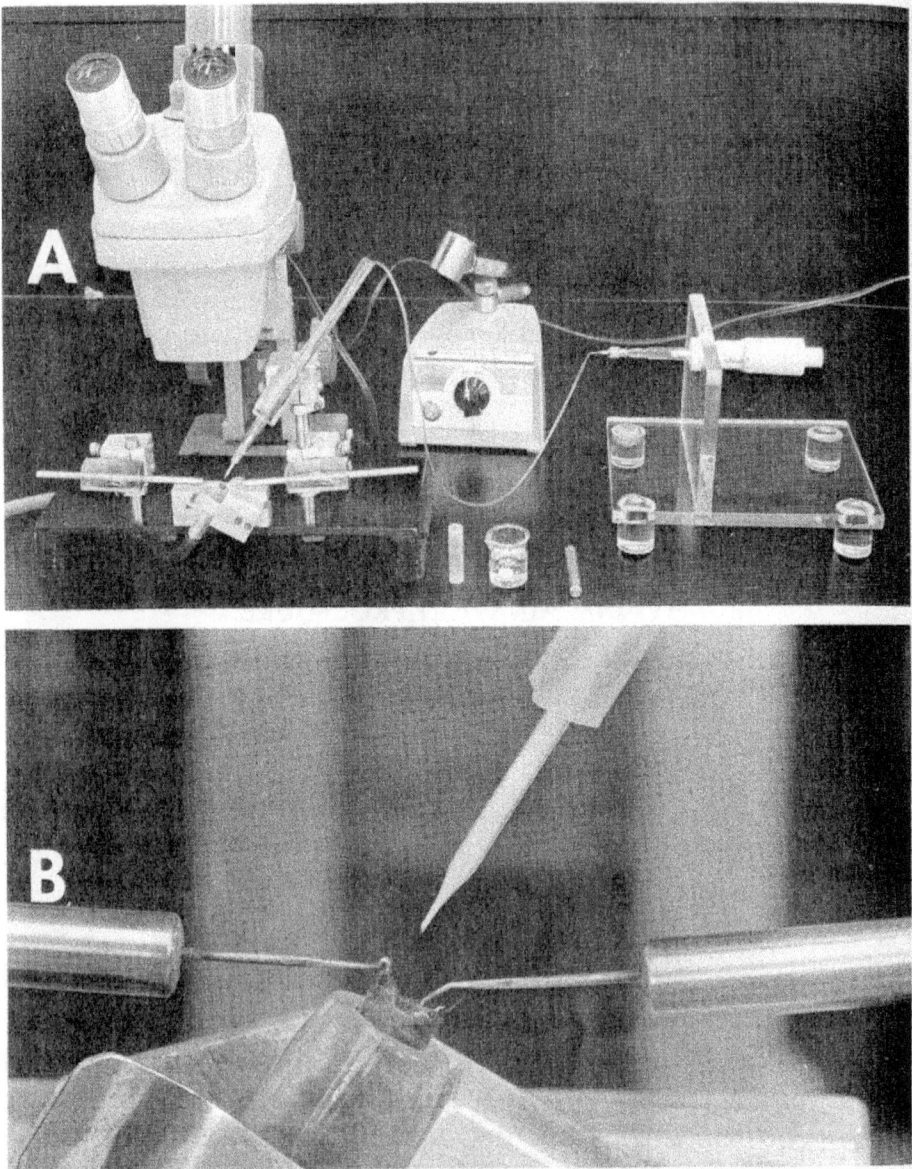

Fig. 2. Insemination equipment in use. (*a*) A Mackensen insemination stand (Mackensen and Tucker, 1970) holds the syringe shown in Fig. 3. The micrometer for the syringe (200 μl capacity) is mounted in a stand made of acrylic plastic. (*b*) Side view of a queen ready for insemination. The top of the queen holder is aligned with the top of the block. About $2\frac{1}{2}$ abdominal segments protrude beyond the end of the holder (hind legs should not protrude. [From Harbo (1985). Copyright in public domain.]

10X enlargement of tip

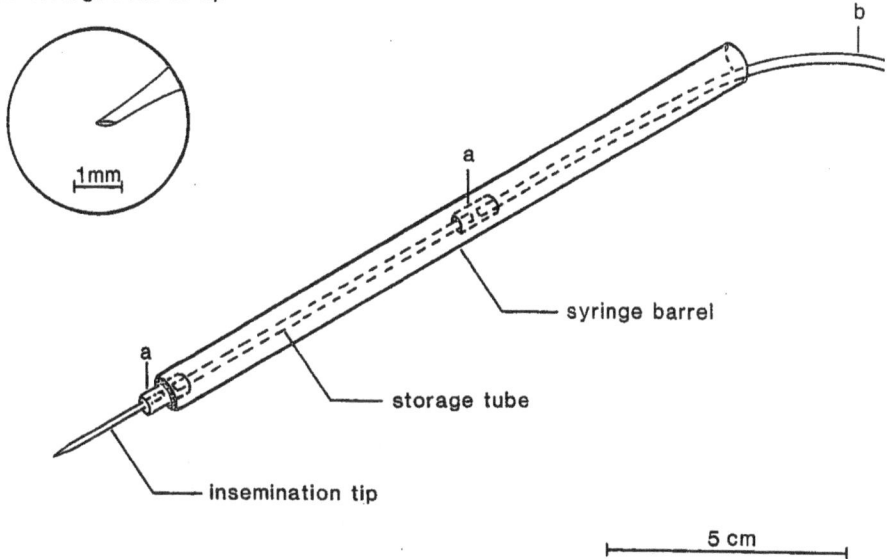

Fig. 3. The barrel and tip portions of the insemination syringe. The tip is a glass capillary tube that was drawn to a point with heat and polished to an angular point (see enlargement). The ID at the orifice of the point is 0.18 ± 0.03 mm; the OD at the orifice is 0.28 ± 0.03 mm. The storage tube is a glass or plastic tube (usually about 1 mm ID) that receives semen when large amounts are collected and that can be removed and stored, if desired. The syringe barrel is glass tubing with an ID 5–6 mm and an OD 7–8 mm. The opening at the tip end of the barrel is reduced to a diameter of ca. 4.2 mm. This reduced orifice forms a tight fit around the latex connector that holds the tip. Thus the tip is held firmly for inseminations and yet is flexible enough to avoid breaking if bumped. The connectors (a) are sections of latex tubing (ID 1.2, OD 4.5 mm). Tube b (polyvinyl tubing; ID 0.5, OD 1.5 mm) forms the hydraulic connection to the syring micrometer shown in Fig. 2a. To assemble the parts, the storage tube, followed by tube b, is pushed into the barrel until the storage tube protrudes out the end where the tip attaches. The tip and the latex connector are attached to the storage tube, and then these parts are pushed back into the barrel until the connector fits as shown. The syringe and plastic tubes are filled with boiled saline. It is boiled to remove dissolved air, thus keeping the hydraulic system free of air and responsive. However, the saline in the storage tube and tip and that used during insemination is not boiled. [From Harbo (1985). Copyright in public domain.]

movements of the micrometer dial (semen is metered with an accuracy of ±0.1 µl).

B. Semen

1. Physical Properties

The color of honey-bee semen is light tan in young drones, changing gradually to dark tan with age. This pigmentation is probably in the plasma fraction of the semen, because the spermatozoa appear white when the plasma is removed. Each drone produces about 10 million spermatozoa (Mackensen, 1955), and there are about 7.5 million spermatozoa per micro-liter of semen (Woyke, 1960; Mackensen, 1964). Spermatozoa are filamen-tous, without a distinct head, and 221–270 micrometers long (mean length = 242 μm) (Woyke, 1983a). Each cell has a volume of about 21 cubic micrometers (measured with a Coulter Counter®*). The specific gravity of semen is 1.077 g/ml (based on five samples totaling 260 μl). Verma (1973) reported osmolarity of honey bee semen to be 467 milliosmolar. See Taber (1977) for an extensive review of the chemical composition of semen and for metabolism of spermatozoa.

2. Obtaining Semen from Drones

Drones are not sexually mature when they emerge as adults. Kurennoi (1953) found that spermatozoa begin to move from the testes to the seminal vesicles when a drone is about 3 days old. Transfer is complete in 3–6 days. The age at which drones become mature (when all the spermatozoa are in the seminal vesicles) varies from 6 to 12 days. Spermatozoa remain in the seminal vesicles until the time of mating.

As a source for semen, it is best to choose drones that have aged 10–21 days after emergence. Drones younger than 10 days are often not yet sex-ually mature, and those older than 21 days are more likely to cause disease in the queens (Mackensen and Tucker, 1970) or leave a residue of semen in the oviducts (Woyke and Jasinski, 1978). Both conditions will kill a queen before she begins to lay eggs. The disease problem comes only from caged drones (Mackensen and Tucker, 1970), and the semen residue affects only caged queens that receive an insemination dose > 4 μl (Veselý, 1970). Al-though there seems to be an ideal age for drones, a breeder can still use older drones if a few precautions are taken: antibiotic in the semen, reasonable sanitation, and smaller insemination doses should eliminate problems caused by older drones.

For semen collection, mature drones should be placed in a small cage that is next to the insemination device. The cage should be large enough for drone flight and hinged at the top so that a cage of drones can be easily placed inside. My cage is 32 × 28 × 22 cm. It should be screened on the sides and top, and fitted with a cloth front so that one can insert a hand to collect the drones. It is easier to see inside the cage if the screen is painted

---

* Mention of a trademark, proprietary product, or vendor does not constitute a guarantee or warranty of the product by the U.S. Department of Agriculture and does not imply its approval to the exclusion of other products or vendors that may also be suitable.

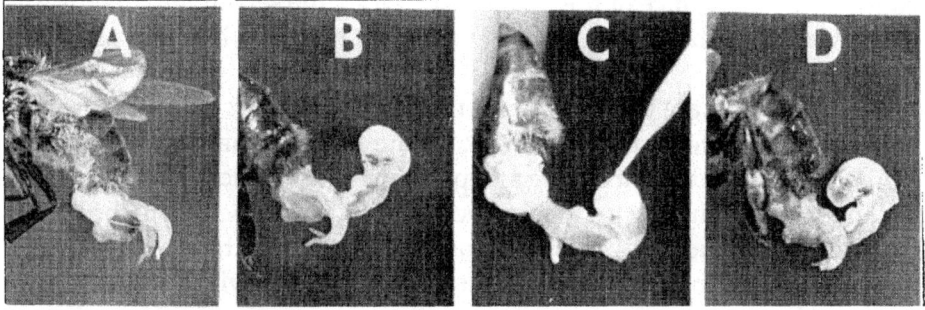

Fig. 4. Obtaining semen from mature drones. (*a*) The first stage of drone eversion caused by a dorso-ventral squeeze of the head and thorax. If the abdomen is turgid at this stage, the drone will likely yield semen. (*b*) The second stage of eversion. This is produced by a lateral squeezing of the turgid abdomen. Semen and mucus are discharged at this stage, making it ideal for semen collection. (*c*) Semen collection from a second stage eversion. (*d*) Very active drones may evert to this third stage when handled or after squeezing the thorax or abdomen. Semen can also be collected from the third stage, but the semen is sometimes discharged onto the abdomen or wings of the drone. [From Harbo (1985). Copyright in public domain.]

black. Release ca. 30 drones at a time so that they do not exhaust themselves before they are ready to be used. The procedure for squeezing drones to yield semen is shown in Fig. 4.

3. Collecting Semen into the Syringe

Assembly of the syringe is described in Fig. 3. The syringe tip and storage tube should be clean and should contain fresh saline before semen collection begins. Saline in the micrometer and plastic tube need not be replaced as often, since it does not come in contact with semen. I use 0.85% NaCl and 0.25% dihydrostreptomycin sulfate (Mackensen and Tucker, 1970). This saline solution is simple and adequate. However, Ruttner and Tryasko (1976) describe other salines ("Kiev" and "Hyer") that give excellent results. Between uses, the tips are stored in a 5% solution of sodium hypochlorite (common laundry bleach). The bleach is rinsed from the tip by forcing water through it with a plastic squeeze bottle.

Before semen enters the syringe, an air-free column of saline should exist from the tip through the tubing and micrometer. Move this column back about 3 μl from the tip so that the incoming semen will be separated from the saline by an air bubble. Focus the microscope at the end of the tip and have the tip orifice facing downward (as in the tip enlargement in Fig. 3). With the drone held in such a way that the ejaculated semen is at the top of the everted genitalia, move the semen to the tip from below. Semen is the light tan material on the surface of a white globe of mucus. Touch the tip with the semen and then pull the semen away slightly, but do not lose contact with the orifice of the tip (Fig. 4c). Draw the semen into the tip. The cohesiveness of the semen will help separate it from white mucus, which can plug the tip.

No air bubbles are put between semen from different drones. However, after collecting semen from a drone, keep the semen column about 1–2 mm from the point to prevent drying and plugging at the orifice. As the next load of semen is brought toward the tip, move the column of semen to the tip and make direct contact between the semen in the tip and that to be collected. Dip the tip in saline and wipe with tissue if crusty material accumulates on the outside.

C. Queen Preparation

The storage of queens from emergence through insemination can be done in three ways. The simplest way is to have each queen free in a small nucleus colony that has a queen excluder over the entrance to prevent the queen from escaping. For insemination, the queens are found, caged, inseminated, returned to their colonies, and released. Be sure that the wings on one side are clipped (about half) so the queen does not fly away when the colony is opened. A second technique keeps up to 70 queens together in a larger colony in cages such as shown in Fig. 5. One trip to the colony brings all the queens in for insemination and a second trip returns them. This technique requires less time transporting queens. A third technique, described by Woyke and Jasinski (1979), eliminates the colony. The queens are kept in an incubator in separate cages with each queen attended by 150 workers. I have never used this last method, so further discussion will involve only the first two.

Queens should be inseminated when they are young. They should be at least 24 hr but not more than 5 or 6 weeks old. However, older queens can be inseminated (Section IV,F). When inseminating with 8 μl of semen, Woyke and Jasinski (1976) had the most spermatozoa enter the spermatheca when queens were 4–8 days old. They recommended that queens be inseminated at age 5–14 days.

A virgin queen can be inseminated after she has laid eggs (Section IV,F). However, if a queen has been previously mated (II or NM), allowed to lay eggs, and then reinseminated by II, my experience has been that the queen dies.

D. Insemination

Semen does not need to be used immediately after collection. It can be stored in the syringe at room temperature (20–25°C) for as long as 2 days with little or no loss of viability.

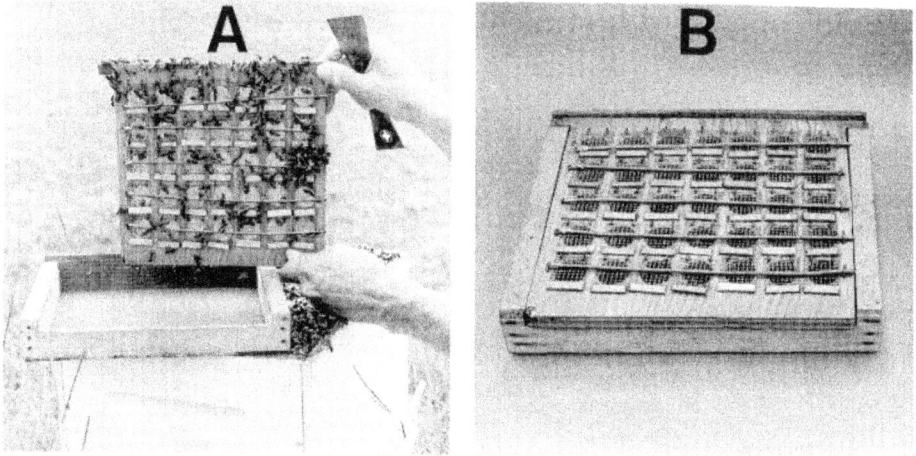

Fig. 5. A system for storing queens together in a colony. (a) The board used for holding queens in a colony is made of plywood (13 mm thick) that was cut to be the same size as a frame in the colony. Queens are stored in holes measuring 24 mm in diameter and covered with a permanent screen (10 mesh per 25 mm) on the back side and with hinged screens (8 mesh per 25 mm hardware cloth) on the front. The two hinges on each screen consist of 10 mm staples driven about 8 mm into the wood. Rigid horizontal wires (also hinged with staples) keep the screens from opening. Metal tags glued to each screen serve as labels and as weights to keep the screens closed when the horizontal wires are unlatched. Before a cage of queens is brought inside to be inseminated, the box (shown on the colony) is taken to the colony, worker bees are brushed or shaken into the box, and the board of queens is placed on the box to form a bee-tight seal. (b) Caged queens during insemination. Attendant workers trapped below enable queens to remain out of the colony for over 8 hr if necessary.

The flow rate of carbon dioxide should be adjusted before queens are put into the queen holder. The rate (ca. 35 ml/min works well if one has a flow meter) does not need to be too precise. The flow should be barely detectable with moistened lips. Increase the rate slightly if the queens are not completely motionless 15 sec after entering the queen holder. The purposes of CO_2 are (1) to immobilize the queens to make inseminations easier and (2) to cause the queens to begin egg laying earlier. Something in the natural mating process stimulates egg laying, but II queens without CO_2 treatment will not begin laying eggs any sooner than virgin queens.

I collect $0.5-1.0$ μl of saline into the insemination tip between inseminations, leaving no air space between semen and saline. The saline precedes the column of semen into the oviducts of the queen. The functions of the saline are (1) to prevent the semen from drying out and plugging the tip, (2) to lubricate the tip for easier insertion, and (3) to add antibiotic for disease control.

The next step is to get the queen into the queen holder and place the queen holder in the insemination stand. Force the queen to crawl into a dead-

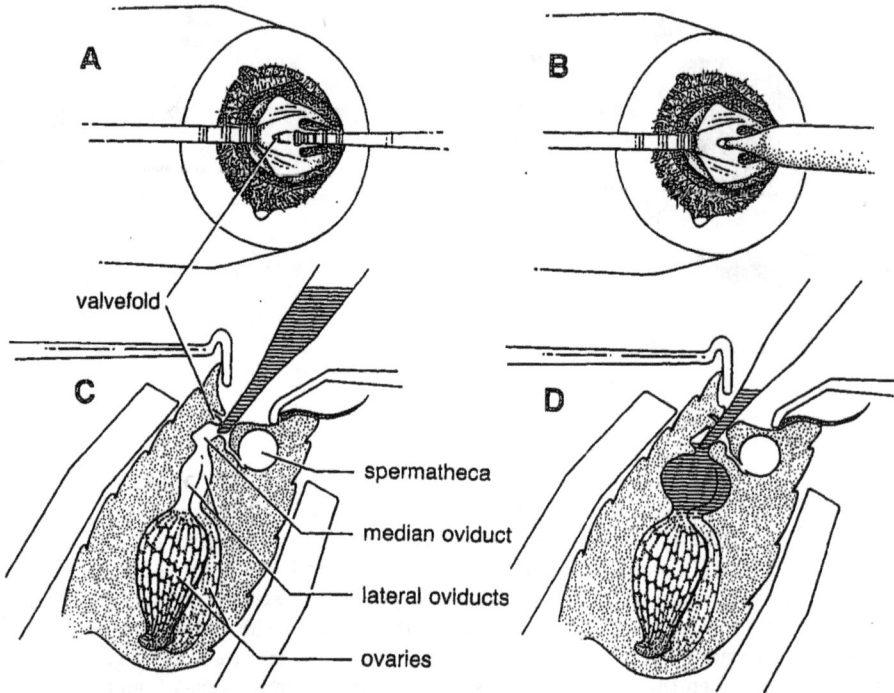

Fig. 6. The insertion of the insemination tip. (*a*) The operator's view of a queen ready to be inseminated. (*b*) The same view as (*a*) but showing the first stage of tip insertion. The tip has been inserted about 0.5 mm (about twice as far as the tip is wide). (*c*) A side view of (*b*). (*d*) Final placement of the tip in the median oviduct. After the first stage of tip insertion (*b* and *c*) the tip is moved ventrally (left) about 0.5 mm to bypass the valvefold, and then inserted another 0.75 mm. The total insertion depth is about 1.25 mm. Semen that was discharged from the tip has caused the lateral oviducts to swell. [From Harbo (1985). Copyright in public domain.]

ended tube (shown on end next to the saline in Fig. 2*a*). Then place an open tube (the queen holder shown in Fig. 2, *a* and *b*) end to end with the open end of the dead-ended tube. When the queen reaches the end of the dead-ended tube, she will back into the queen holder until her abdomen protrudes from the constricted end (Fig. 2*b*). The plug at the end of the CO_2 hose is inserted into the tube to hold the queen in position. See Fig. 2 for queen adjustment.

The sting hook and ventral hook separate the sting apparatus from the ventral plate to expose the vagina (Fig. 6). These hooks should be moved at the handles with the little-finger edge of the operator's hands resting on the table. The first step is to separate the dorsal and ventral plates. Then hold the sting down with a sting depressor (a pin-sized wire with a handle) and reposition the sting hook under the base of the sting. The hook is shaped to fit under the base of the sting. While the sting is pulled dorsally, the other hand is on the queen holder, turning it slightly to keep the sting shaft aligned with the sting hook.

Insertion of the tip into the oviduct is shown in Fig. 6. Just before insertion, lubricate the tip by dipping it in saline.

With the tip in the median oviduct, discharge the semen. The volume given to a queen will be discussed in Section VI,A. Equalize pressure before removing the tip by moving the micrometer about 0.5 μl beyond the intended mark then return to the mark. This prevents semen from running out of the tip as the syringe is withdrawn from the queen.

Tips sometimes get plugged during semen discharge or between inseminations. If it plugs during semen collection or between inseminations, the operator has usually forgotten to retract the semen from the point or to collect saline into the tip after the last insemination. If this happens, soak the tip in saline, wipe it with tissue, and apply normal suction with the syringe. If the tip plugs during semen discharge, a mucus plug is usually the cause. This is caused by careless semen collection, for mucus is more apt to cause a plug when it is discharged than when it is collected. One can often see the mucus plug in the neck of the tip. This may require tip removal and cleaning.

When inseminations are complete, return the queens to their colonies. Queens free in a nucleus colony can remain out of their colony for at least 20 min, so one can collect two or three queens per trip to save time; they can be returned narcose. Caged queens can stay out of the colony for many hours as long as worker bees are in the box below (Fig. 5).

E. Oviposition

To stimulate egg laying, queens are given three treatments of CO_2. Carbon dioxide during insemination counts as a treatment, so if a queen was inseminated twice, she needs only one more treatment. The following is one of many ways to treat queens with CO_2. Put caged queens into a plastic bag and eliminate as much air as possible. Inflate the bag with CO_2 by running a CO_2 hose into the bag while the bag is held shut. When the queens are motionless, turn off the CO_2 and leave the queens narcose in the sealed bag for about 3 min.

Queens free in the colony should be treated with CO_2 once per day on the 2 days after insemination. Expect egg laying to begin about 4 days after the last CO_2 treatment if the queens were 1 week old at the time of insemination.

Queens stored together in a colony do not need CO_2 treatments on the days immediately following insemination. For convenience, I often give queens their final treatment on the day that they are transferred from the queen bank to a small colony to lay eggs. For best success, keep the queens in the storage colony until they are at least 2–3 weeks old. Queens that are over 2 months old when they are put into colonies to lay eggs do not need CO_2 treatments.

Acceptance of an II queen by worker bees is sometimes a problem. To reduce the frequency of queen death, have the II queens lay in a small colony for at least 4 days before putting them into large field colonies. Place each caged queen into a small colony consisting of 2000–5000 worker bees. These colonies may be newly made without brood. Release the queens after 3 days if the workers are not biting the cage. About 10 min after release, check some of the queens to see if any are being chased or bitten by one or more workers. Those that are should be recaged for 2 or 3 more days. If some of the queens in a group are being chased or bitten, all should be rechecked and perhaps recaged. Queens 3–4 weeks old usually begin to lay 1–2 days after release.

F. Special Techniques

1. Single-Drone Insemination

This is a common technique in research. Since all the gametes from one drone are identical, queens mated with semen from a single drone will produce daughters that are all more closely related than full sisters. These daughters, known to have a gamete in common, are called super sisters.

The procedure is similar to that of a regular insemination except one must rinse the tip between inseminations to ensure that the tip contains only semen from one drone. Before collecting semen, collect 0.5 μl of saline just after the air space. This saline is discharged into the queen with the semen and reduces the amount of semen that is left coating the sides of the insemination tip. For best results, choose a drone that has a large semen load.

2. Inseminating Many Queens with Identical Spermatozoa

This technique enables one to mate identical gametes to different queens. As with single-drone inseminations, each queen produces a family of super sisters. Moreover, the families produced by the different queens have a unique relatedness to each other.

There are two ways to do this. The first way is simpler, but more limited since one merely chooses a drone that yields a large quantity of semen. Collect semen as for a single-drone insemination and then inseminate each queen with 0.2 μl of semen. Before each insemination, collect 0.2 μl of saline which will be injected into the oviducts with the semen. Young queens inseminated in this manner have, in my experience, produced worker brood. I have inseminated as many as 14 queens from a single drone (ca. 0.5 million spermatozoa per queen), and they averaged 84% worker brood.

The second technique was demonstrated by Kubasek *et al.* (1980) and requires the use of queens that produce gynandromorph progeny. The advantage of this technique is that isogenic spermatozoa are produced in large quantities and can be collected throughout the season. The inseminations can be like multiple inseminations because the male tissues of many individuals produce identical gametes.

3. Inseminating Queens with a Uniform Mixture of Spermatozoa

Among other possibilities, this technique can be used to maintain genetic heterogeneity in a breeding program or to produce uniform, but genetically heterogeneous, inseminations in a group of queens. These inseminations include genetically diverse spermatozoa, but they are uniform because spermatozoa for each insemination are taken from the same large, thoroughly mixed pool of spermatozoa. Kaftanoglu and Peng (1980), have shown that spermatozoa can be highly diluted, mixed, and then recovered by centrifugation. Queens have been successfully inseminated with this "washed spermatozoa" (Kaftanoglu and Peng, 1980; Williams and Harbo, 1982; Mortiz, 1983, 1984). Moritz (1983) showed that this technique produces uniform mixing of cells.

4. Collecting Spermatozoa from the Seminal Vesicles of a Drone

This technique is used when one desperately needs semen from a particular drone and the drone has failed to ejaculate semen after being squeezed in the usual manner. Because of the time of migration of spermatozoa from the testes to the seminal vesicles, drones must be at least 4 days old before there is much hope for success. Drones aged more than 1 week should have sufficient numbers in their seminal vesicles to perform a minimal insemination.

The seminal vesicles are the smaller of the paired, sausage-shaped organs in the abdomen of a drone. Mackensen and Ruttner (1976) describe a technique where they remove a seminal vesicle, pinch it with a forceps at the testes end, and thereby start peristaltic contractions which force the spermatozoa out the other end where they are collected with a syringe. A less elegant technique of dropping the seminal vesicles into a small glass cone, adding a small amount of saline, and then partially macerating them to free the spermatozoa has also proved satisfactory.

5. Using Spermatozoa from the Spermatheca of One Queen to Inseminate Another

This technique is used to make gamete backcrosses (Cale and Gowen, 1964) or simply to recover spermatozoa that have been stored in other queens. Thus, vigorous queens can be used as storage banks for spermato-zoa.

Take the spermatheca from the donor queen(s), and remove the tracheal covering. Place the spermatheca on a smooth wax surface that has a slight depression and puncture it with a sharp needle. Then insert the insemina-tion tip into the spermatheca, collect the contents into the syringe, and inseminate as usual.

6. Inseminating Very Old Queens

Queens more than 8 weeks old are considered old for insemination, but insemination is still possible for at least 5 months. There may be no age limit. An old queen should be placed in a mailing cage with five to eight workers for about 3 days to reduce the size of her abdomen. The insemination volume should be 2 μl or less to enhance survival. After insemination, place the queen in a colony and release her in 2 or 3 days. No CO_2 treatments are necessary to induce egg laying. The percentage of spermatozoa entering the spermatheca is lower for old than for young queens, and many queens inseminated when aged > 4 months lay a high proportion of unfertilized eggs.

7. Inseminating a Queen with Semen from Her Own Drones

This is a breeding scheme called self-fertilization that was first described by Mackensen (1951). A virgin queen is treated with CO_2 to get her to lay unfertilized eggs that will develop into drones. The queen is not insemin-ated at this time. If some drone-sized cells are available, these "drone-lay-ing" queens will usually lay eggs in them, and the resulting drones will be larger than if the queens are forced to lay in worker-sized cells. When one or more of these drones are about 2 weeks old, their semen can be used to inseminate the queen. The procedure is identical to that of inseminating a very old queen (described above).

V. EVALUATION

A. Spermatozoa in the Spermatheca

A necessary step in the mating of honey bees is to get spermatozoa into the spermatheca of a queen. A queen retaining more spermatozoa in her spermatheca is considered to be better mated than one with fewer.

Estimating the number of spermatozoa in the spermatheca enables quan-tification of success beyond simply mated or not mated. One method is to count spermatozoa in a hemacytometer (a cell counting chamber). A spermatheca can contain as many as 7 million spermatozoa, so the contents should be dispersed in at least 10 ml of solution. More concentrated spermatozoa are difficult to count.

A much faster technique uses light absorbance to estimate numbers of spermatozoa. Harbo (1975) found a strong linear correlation ($r = 0.96$) between absorbance at 230 nm and density of spermatozoa in a 0.5 M NaCl solution. During the past 4 years, I have compiled a new correlation using a 10-ml rather than a 5-ml dilution (same diluent and wavelength) and including samples from a wider range (0 – 8 million cells/10 ml). Based on 194 samples, the regression formula was $Y = 0.034 + 0.061X$ (Y = absorbance units, X = millions of spermatozoa); $r = 0.95$. Rearrangement changed the formula to $X = 16.47Y - 0.56$, which converts absorbance (Y) to millions of spermatozoa (X). This formula was adjusted slightly because it consistently estimated high at the high end of the range (usually about 0.2 million higher than the counted mean) and slightly low at the low end. The line was adjusted to pass more accurately through the observed zero intercept but still through the mean coordinates. The adjusted line is within the 0.95 confidence interval for the line and the slope; it is $X = 15.72Y - 0.41$.

Figure 7 shows the correlation between the amount of semen given and the number of spermatozoa entering the spermatheca of young queens (queens <3 weeks old). Results are variable, but this represents an average expectation.

Even under seemingly uniform conditions, the number of spermatozoa entering the spermatheca is variable. Among sister queens inseminated in sequence with the same amount of semen, the coefficient of variation (CV) is 26% for 8-μl inseminations (Woyke et al., 1974), 35% for 2.5-μl inseminations (Mackensen and Roberts, 1948), and 49% for inseminations with 2 million cells (Bolten and Harbo, 1982). With two inseminations of 2 million cells, the CV dropped to 31%, and for three inseminations of 2.5 μl the CV was 16%. In general, uniformity increased with larger inseminations and with multiple inseminations.

During natural mating, queens retain about 5.7 million spermatozoa in their spermatheca (CV = 18%). An average that approaches this (5.5 million) can be achieved with II, but it requires four inseminations with 2.5 μl (CV = 10%) (Mackensen and Roberts, 1948).

Both genetic and environmental factors affect the number of spermatozoa reaching the spermatheca. Woyke et al. (1974) showed that different geographic races differed in the number of spermatozoa entering the spermatheca after inseminations with 8 μl of semen. Woyke and Jasinski (1976) found that younger queens (<3 weeks old) retained more spermatozoa in their spermathecae after an insemination of 8 μl than did older queens.

John R. Harbo

Insemination (microliters)

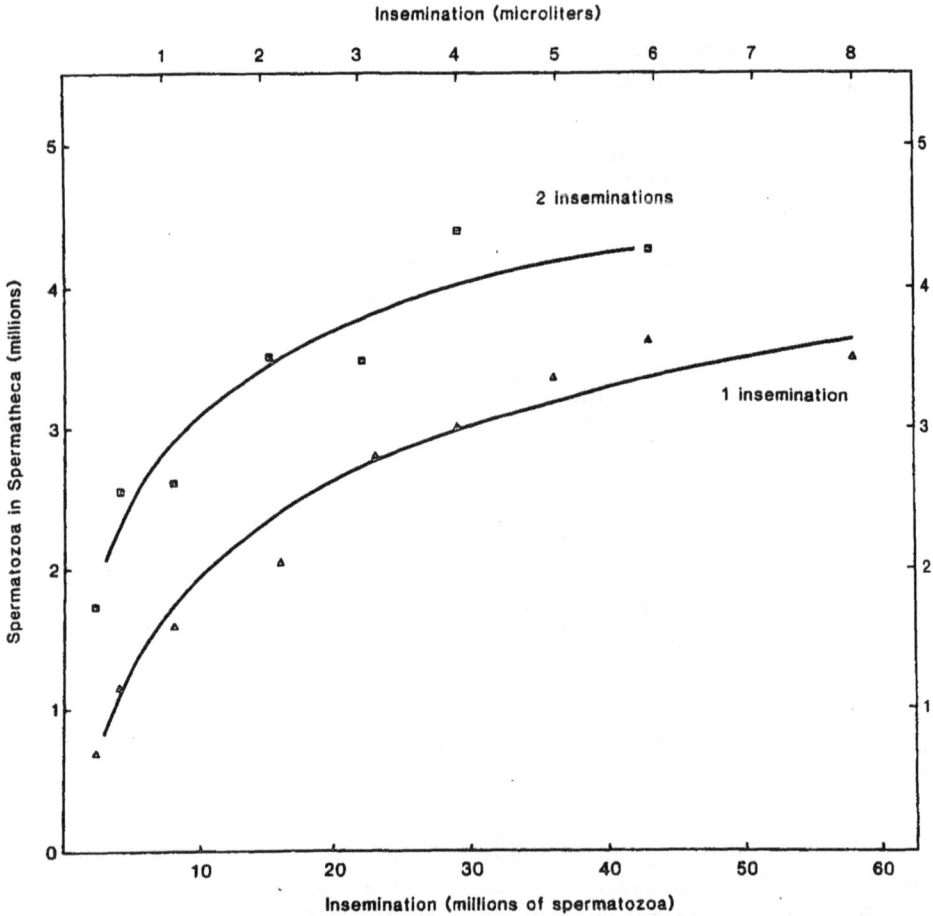

Fig. 7. The effect of insemination dosage on the number of spermatozoá entering the spermatheca. Data were compiled by Woyke (1960, 1971), Mackensen (1964), Veselý (1970), Woyke and Jasinski (1973), Bolten and Harbo (1982), and J. R. Harbo (unpublished). Regression formulas were derived from the means (plotted points) for each dose. Total n is 351 queens for one insemination and 94 queens for two. Regression formulas, $Y_1 = -0.24 + 0.95(\ln X_1)$ and $Y_2 = 1.1 + 0.86(\ln X_2)$, predict the number of spermatozoa entering the spermatheca (Y) from the number of spermatozoa given (X) with one and two inseminations, respectively ($r_1 = 0.98$, $r_2 = 0.96$). X and Y in the formulas are always in millions (e.g., 4,400,000 is entered and recalled as 4.4). [From Harbo (1985). Copyright in public domain.]

Queens running free in a mating hive retained more spermatozoa in their spermathecae than did those caged in a mating hive when each was inseminated with 6 μl of semen (Veselý, 1970). However, he found no significant difference between free and caged queens when insemination volumes were 2 or 4 μl.

To produce 100% worker brood, a young queen needs to have about

100,000 spermatozoa in her spermatheca. The 100,000 estimate is based on two groups of unrelated queens. The first group (10 queens) was inseminated with 0.2 ± 0.1 µl of semen each. Five queens had 111,000–575,000 spermatozoa in their spermathecae, and those all produced >99% worker brood. The remaining five had 1000–39,000 in their spermathecae, and they produced 2–92% worker brood. The second group of queens received about 0.05 µl of semen (about 250,000 cells). Five had 45,000–88,000 spermatozoa in their spermatheca, and those produced 91–99% worker brood. The remaining five had 6000–34,000, and those produced 12–54% worker brood. These results show that young queens produce some worker brood from even the smallest inseminations. Thus, if a young queen produces only drone brood, it is usually not because she was given an inadequate amount of semen during insemination; more likely the semen was not injected into the oviducts.

The presence or absence of spermatozoa in the spermatheca affects the egg-laying behavior of a queen. Queens with no spermatozoa in the spermatheca do not lay as many eggs per day and do not position eggs as uniformly as do queens with spermatozoa in their spermathecae (Harbo, 1976). However, among queens that are producing worker brood, I failed to find a correlation between egg-laying rate and the number of spermatozoa in the spermatheca.

B. Queen Survival

Queen loss after insemination can be a serious problem. Probably the two most common causes are (1) a semen residue remaining in the oviducts for more than 1 or 2 days after insemination and (2) worker aggression when II queens are first introduced to colonies.

Insemination with large volumes of undiluted semen and keeping queens caged after insemination tend to cause semen residue in the oviducts. Normally, semen is discharged from the oviducts and some of the spermatozoa enter the spermatheca. If insemination doses are high (≥ 6 µl), a residue of semen is often left in the oviducts of caged queens (Veselý, 1970). This residue may serve as a medium for the growth of microorganisms, which may be the cause of death. Woyke and Jasinski (1978) found that after inseminating caged queens with 8 µl of semen, a residue of semen was found more often when semen came from older drones (those >4 weeks old) than when semen came from drones aged 2–3 weeks. Veselý (1970) concluded that queens free in a colony do not seem to have a problem with semen residue in the oviducts, nor do caged queens inseminated with <4 µl of semen.

Queen loss when first introducing an II queen to a colony is common because a young II queen that has never laid eggs is not accepted by workers as readily as a laying queen. A procedure for introducing II queens is described in Section IV,E. In general, older II queens are easier to introduce than younger ones, and smaller colony populations seem to accept them more often than larger populations.

C. Production Costs

The equipment needed to produce II queens is minimal. Excluding the beekeeping equipment, which is not included in this analysis, the one-time cost of the insemination equipment, CO_2 regulator, binocular microscope, and light is about $1000 (1985). The only parts that need replacement are the CO_2 supply, the insemination tip, and the sting hook. Their costs are less than $40 per year unless the operator is careless.

Table 1 lists the labor costs of producing II queens when the queens are separately caged in a large colony. There are controllable variables in the procedure: the amount of semen given and the number of times a queen is inseminated. An uncontrollable variable exists in semen collection because of variability in drones. Sometimes a group of drones averages much less than 0.5 μl of semen per drone, and at other times they may average as much as 0.9 μl.

The labor cost of producing II queens is higher when each queen is running free in a small colony. About 2 min more per queen is required because three trips are necessary to find and bring the queens in for insemination or CO_2. Moreover, a producer needs to operate more of these small colonies when the young queens are allowed to run free because free-running queens spend about twice as much time in a colony as do queens that are only put into colonies when they are ready to lay eggs. A free queen needs a colony for 2 weeks; a 3-week-old banked queen can be put into a small colony, released on day 3, and be laying on day 4 or 5.

VI. RECOMMENDATIONS

A. Semen Dosage

Caged queens need multiple inseminations if they are to average more than 3 million spermatozoa in the spermatheca, because caged queens often cannot tolerate inseminations of undiluted semen that are >4 μl. I suggest two inseminations of 3 μl. If one wants to approach the sperm count of a naturally mated queen, three inseminations of 2 or 3 μl are needed. Multiple

TABLE 1. Time Required to Produce Instrumentally Inseminated Queens in a Group of 40 or More[a]

Step	Time required
Producing, caging, and storing drones of known parentage[b]	10 sec/drone
Collecting semen[c]	50 sec/μl
Total cost of semen (at 0.5 μl/drone)[d]	1.2 min/μl
Rearing queen cells[b]	4 min/cell
Marking and storing queens[b]	2 min/queen
Cost of a mature, virgin queen	6 min/queen
Insemination (includes transporting caged queens from and to the storage colony)[e]	3 min/queen
Additional CO_2 treatments	15 sec/queen
Total cost of queens inseminated once with:	
3 μl of semen[f]	13.1 min/queen
8 μl of semen[f,g]	19.1 min/queen
Total cost of queens inseminated twice with 3 μl of semen[f]	19.7 min/queen

[a] Queens produced in one cell builder and kept in one queen storage colony.

[b] K. Tucker, J. Harbo, and J. Bishop; unpublished data presented orally at the annual meeting of the American Bee Breeders Assn., 1974.

[c] Mackensen (1964) reported that collecting 1 μl of semen required 45 sec. J. Harbo (unpublished) measured 47 sec/μl. If drones are sluggish, too young, or for some other reason not prone to ejaculate, the time could easily be doubled.

[d] Drones that yield semen usually produce more than 0.5 μl. The estimate was based on a typical group of 203 drones; 93 yielded no semen, and 102 μl was collected from the others.

[e] Estimate came from 126 inseminations that were timed in groups of 4–28. The mean $\pm$ SD = 2.46 $\pm$ 0.4 min per queen. This included record keeping, but not semen collecting or transporting queens. Time was increased to 3 min to allow for transporting the group of queens from and to their storage colony.

[f] Cost does not include queen introduction and assumes no queen mortality after insemination.

[g] Doses of 8 μl are not recommended for caged queens (see semen dosage). Add ca. 2 min per queen to find and bring each queen in for insemination and CO_2 treatments.

inseminations of caged queens should each be 3 μl or less and should be spaced 2 or 3 days apart.

Woyke (1960) has long recommended one insemination of 8 μl or two inseminations of 4 μl for queens that are free in a small colony. Mackensen (1964) concluded that two inseminations of 3 μl may be an adequate minimum if queens are to be used in large field colonies. Multiple inseminations of free-running queens are given on consecutive days. Use Table I and Fig. 7 to decide on dosage, but consider (1) that a single insemination as large as 8 μl does not seem to be detrimental if the queen is running free in a small colony and (2) that multiple inseminations give more uniform results (lower

CV). There are practical limits where the cost of more inseminations or of using more semen yields diminishing returns in queen performance.

Sometimes one simply needs mated queens, and long-term egg laying is not important. In these cases, I simply inseminate the queen once with about 2 μl of semen. Such queens seem to perform well for many months, and some even survive more than a year.

B. Learning Insemination

When first learning to inseminate, use about 15 caged virgin queens. Two microliters of semen per queen is enough to practice the insertion and injection. If 2 μl is discharged into a queen without backing out around the tip during injection, either the tip is in the correct position or the operator has punctured the body cavity. Check for success on the following day by removing the spermatheca from each queen. No microscope or dissecting tools are needed. Use fingernails or forceps and pull away the last one or two abdominal segments. A heavy tracheal network makes the spermatheca look like a ball of string about 1 mm in diameter. The presence of spermatozoa does not seem to affect the size or the turgidity of the spermatheca. Remove the tracheae by rolling the spermatheca between thumb and forefinger. If the spermatheca is crystal clear, the insemination failed; if it is white and opaque, it succeeded.

After having some success with practice queens, start serious inseminations with queens that are free in small colonies. Inseminate each queen with at least 2 but no more than 8 μl of semen. Follow with CO_2 treatments on the next 2 days and watch for worker brood.

I suggest starting with queens free in colonies for two reasons. First, very good results can be obtained with only one large insemination, and fewer insertions reduce the probability of queen injury. Second, queen death from various causes can be expected, especially for beginners, and it is best for novice inseminators to take these losses (queen death in storage colonies and nonacceptance by workers when released in colonies) before the insemination step.

ACKNOWLEDGMENTS

The preparation of this review and the newly reported research were done in cooperation with Louisiana Agricultural Experiment Station. Robert Spencer prepared Figs. 1 and 5, and the graphics in Fig. 7 were prepared by Sandra Kleinpeter.

REFERENCES

Becker, R. (1925). Die Ausbildung des Geschlechtes bei der Honigbiene. *Erlanger Jahrb. Bienent.* 3, 163–223.

Bishop, G. H. (1920).Fertilization in the honey-bee II. Disposal of the sexual fluids in the organs of the female. *J. Exp. Zool.* 31, 267–286.

Bolten, A. B., and Harbo, J. R. (1982). Numbers of spermatozoa in the spermatheca of the queen honeybee after multiple inseminations with small volumes of semen. *J. Apic. Res.* 21, 7–10.

Butler, C. G. (1957). The process of queen supersedure in colonies of honeybees (*Apis mellifera* L.). *Insectes Soc.* 4, 211–223.

Cale, G. H. (1926). The first successful attempt to control the mating of queen bees. *Am. Bee J.* 66, 533–534.

Cale, G. H., Jr., and Gowen, J. W. (1964). Gamete backcross matings in the honey bee. *Genetics* 50, 1443–1446.

Doolittle, G. M. (1889). "Scientific queen-rearing." Thomas G. Newman, Chicago, Ill.

Dzierzon, J. (1845). Gutachten über die von Hr. Direktor Stöhr im ersten und zweiten Kapital des General-Gutachtens aufgestellten Fragen. *Eichstadt. Bienenzeitung.* 1, (11) 109–113, (12) 119–121.

Harbo, J. R. (1971). "Annotated Bibliography on Attempts at Mating Honey Bees in Confinement." Bibliography No. 12, Inter. Bee Research Assn., Gerrards Cross, Bucks., England.

Harbo, J. R. (1975). Measuring the concentration of spermatozoa from honey bees with spectrophotometry. *Ann. Entomol. Soc. Am.* 68, 1050–1052.

Harbo, J. R. (1976). The effect of insemination on the egg-laying behavior of honey bees. *Ann. Entomol. Soc. Am.* 69, 1036–1038.

Harbo, J. R. (1979). Storage of honey bee spermatozoa at $-196°C$. *J. Apic. Res.* 18, 57–63.

Harbo, J. R. (1985). Instrumental insemination of queen bees–1985. *Am. Bee J.* 125, 197–202, 282–287.

Kaftanoglu, O., and Peng, Y. S. (1980). A washing technique for collection of honeybee semen. *J. Apic. Res.* 19, 205–211.

Kubasek, K. J., Rinderer, T. E., and Lee, W. R. (1980). Isogenic sperm line maintenance in the honey bee. *J. Hered.* 71, 278–280.

Kurennoi, N. M. (1953). When are drones sexually mature? *Pchelovodstvo* 30, 28–30, (in Russian).

Laidlaw, H. H., Jr. (1944). Artificial insemination of the queen bee (*Apis mellifera* L.): morphological basis and results. *J. Morphol.* 74, 429–465.

Laidlaw, H. H., Jr. (1977). "Instrumental Insemination of Honey Bee Queens, Pictorial Instructional Manual." Dadant and Sons, Inc., Hamilton, Ill.

Laidlaw, H. H., Jr. (1979). "Contemporary Queen Rearing." Dadant and Sons, Inc., Hamilton, Ill.

Lensky, Y. (1971). Rearing queen honeybee larvae in queenright colonies. *J. Apic. Res,* 10, 99–101.

Lensky, Y., and Slabezki, Y. (1981). The inhibiting effect of the queen bee (*Apis mellifera* L.) foot-print pheromone on the construction of swarming queen cups. *J. Insect Physiol.* 27, 313–323.

Mackensen, O. (1947). Effect of carbon dioxide on initial oviposition of artificially inseminated and virgin queen bees. *J. Econ. Entomol.* 40, 344–349.

Mackensen, O. (1951). Self fertilization in the honey bee. *Glean. Bee Cult.* 79, 273–275.

Mackensen, O. (1955). Experiments in the technique of artificial insemination of queen bees. *J. Econ. Entomol.* 48, 418–421.

Mackensen, O. (1964). Relation of semen volume to success in artificial insemination of queen honey bees. *J. Econ. Entomol.* 57, 581–583.

Mackensen, O., and Roberts, W. C. (1948). "A Manual for the Artificial Insemination of Queen Bees." U.S.D.A. Bureau of Entomol. and Plant Quar. ET-250.

Mackensen, O., and Ruttner, F. (1976). The insemination procedure. *In* "The Instrumental Insemination of the Queen Bee" (F. Ruttner, ed.), pp. 69–86. Apimondia, Bucharest, Romania.

Mackensen, O., and Tucker, K. W. (1970). "Instrumental Insemination of Queen Bees." U.S.D.A. Agric. Handbook No. 390. U.S. Government Printing Office, Washington, D.C.

Moritz, R. F. A. (1983). Homogeneous mixing of honeybee semen by centrifugation. *J. Apic. Res.* 22, 249–255.

Moritz, R. F. A. (1984). The effect of different diluents on insemination success in the honeybee using mixed semen. *J. Apic. Res.* 23, 164–167.

Morse, R. A. (1979). "Rearing queen honey bees." Wicwas Press, Ithaca, N.Y.

Nolan, W. J. (1929). Success in the artificial insemination of queen bees at the Bee Culture Laboratory. *J. Econ. Entomol.* 22, 544–551.

Nolan, W. J. (1932). "Breeding the Honey Bee under Controlled Conditions." U.S.D.A. Tech. Bull. No. 326.

Örösi Pál, Z. (1957). Succession in starting queen cells. *Méhészet* 5, 223–225.

Örösi Pál, Z. (1958). Results of queen rearing with egg transfer. *Méhészet* 6, 133–134, (in Hungarian).

Örösi Pál, Z. (1960). Experiments on queen rearing. Part II. *Kisérletügyi Közl.* 1, 31–79, (in Hungarian).

Ruttner, F. (1976). "The Instrumental Insemination of the Queen Bee." Apimondia, Bucharest, Romania.

Ruttner, F., and Tryasko, V. V. (1976). Anatomy and physiology of reproduction. *In* "The Instrumental Insemination of the Queen Bee" (F. Ruttner, ed.), pp. 11–24. Apimondia, Bucharest, Romania.

Taber, S., III (1977). Semen of *Apis mellifera*: fertility, chemical and physical characteristics. *In* "Advances in Invertebrate Reproduction," Vol. I (K. G. Adiyodi and R. G. Adiyodi, eds.), pp. 219–251. Peralam-Kenoth, Kerala, India.

Taber, S. III, and Poole, H. K. (1974). Rearing and mating of queens and drone honey bees in winter. *Am. Bee J.* 114, 18–19.

Verma, L. R. (1973). Osmotic analysis of honey bee (*Apis mellifera* L.) semen and haemolymph. *Am. Bee J.* 113, 412.

Veselý, V. (1970). Retention of semen in the lateral oviducts of artificially inseminated honeybee queens (*Apis mellifera* L.). *Acta Entomol. Bohemoslov.* 67, 83–92.

Weiss, K. (1974a). Zur Frage des Königinnengewichtes in Abhängigkeit von Umlarvalter und Larvenversorgung. *Apidologie* 5, 127–147.

Weiss, K. (1974b). Neue Untersuchungen zum "doppelten Umlarven." *Apidologie* 5, 225–246.

Williams, J. L., and Harbo, J. R. (1982). Bioassay for diluents of honey bee semen. *Ann. Entomol. Soc. Am.* 75, 457–459.

Woyke, J. (1960). Natural and artificial insemination of queen honey bees. *Pszczel. Zesz. Nauk.* 4, 183–275, (in Polish, English summary).

Woyke, J. (1971). Correlations between the age at which honeybee brood was grafted, characteristics of the resultant queens, and results of insemination. *J. Apic. Res.* 10, 45–55.

Woyke, J. (1983a). Lengths of haploid and diploid spermatozoa of the honeybee and the question of the production of triploid workers. *J. Apic. Res.* 22, 146–149.

Woyke, J. (1983b). Dynamics of entry of spermatozoa into the spermatheca of instrumentally inseminated queen honeybees. *J. Apic. Res.* 22, 150–154.

Woyke, J., and Jasinski, Z. (1973). Influence of external conditions on the number of spermatozoa entering the spermatheca of instrumentally inseminated honeybee queens. *J. Apic. Res.* 12, 145–151.

Woyke, J., and Jasinski, Z. (1976). The influence of age on the results of instrumental insemination of honeybee queens. *Apidologie 7*, 301–306.

Woyke, J., and Jasinski, Z. (1978). Influence of age of drones on the results of instrumental insemination of honeybee queens. *Apidologie 9*, 203–212.

Woyke, J., and Jasinski, Z. (1979). Number of worker bees necessary to attend instrumentally inseminated queens kept in an incubator. *Apidologie 10*, 149–155.

Woyke, J., Jasinski, Z., and Smagowska, B. (1974). Comparison of reproductive organs and effects of artificial and natural insemination of honey bees of different races and their hybrids. *Pszczel. Zesz. Nauk. 18*, 53–75, (in Polish, English summary).

CHAPTER 16

Breeding Accomplishments with Honey Bees

JOVAN M. KULINČEVIĆ

I. INTRODUCTION

The honey bee (*Apis mellifera* L.) and the silkworm (*Bombyx mora* L.) are the only two insects contributing beneficially to agriculture to the extent that they support worldwide industries. Both have been domesticated and bred by humans because of the value of their products. However, the worldwide value of the honey bee as an agricultural animal far exceeds that of the silkworm. Not only does the honey bee directly produce useful agricultural products, but also large portions of other agricultural enterprises are dependent on the pollination activities of honey bees.

The honey bee's economic value has attracted considerable scientific and technical interst. For geneticists and breeders this interest is increased by the special scientific and technical challenges inspired by the unusual genetical and behavioral attributes of honey bees. This scientific interest and work has contributed to substantive breeding accomplishments with honey bees which have enhanced their role in modern agriculture.

II. GOALS IN HONEY-BEE BREEDING

A. General Considerations

Successful honey-bee breeders have first examined the specific needs of the particular region in which the stock they develop is used. Only in this way have breeding programs started with useful goals. This examination includes identifying the expected sources of receipts from beekeeping, the adverse beekeeping conditions in the area, and the apparent shortcomings of existing stocks of bees.

Well-founded breeding programs have emphasized goals appropriate to the economic needs of specific regions. Generally, income in beekeeping arises from the sale of honey, beeswax, queens, breeding stock, package bees, pollen, and in some parts of the world royal jelly, bee venom, and propolis. These are direct sources of economic return. An indirect source of income for an increasing number of commercial beekeepers comes from the rental of bees for pollination services (McGregor, 1976). There are some regions in the world, for instance in California, where some beekeepers earn nearly half of their income from pollination fees.

In addition to the potential sources of income, the biological adversities of a beekeeping area often have been important to breeding programs. Absence of bee diseases and parasites are valued factors in a successful honey-bee industry. In practice, honey-bee diseases and parasites are usually controlled by the application of chemical agents and antibiotics. However, such control is not always successful, is costly, and often requires regulatory controls to prevent product contamination with drugs and chemicals. Other adverse conditions, such as severe winters, limited periods of nectar availability, and for some areas even a complete absence of pollen are additional difficulties. The production of stock which is resistant to diseases and parasites and adapted to adverse regional conditions has often been a goal for breeding programs.

An equally desirable goal has been the elimination of certain bee traits not conducive to modern beekeeping. Such traits would include excessive stinging, swarming, running on the comb, and absconding.

B. Types of Breeding Programs

Based on these general considerations, breeding programs have been designed to improve bee stocks. Specific projects generally have fallen into two categories. First, small pioneering projects, usually undertaken at a university by one or more investigators, have produced information concerning how single traits of interest are inherited. These programs provide detailed information such as the heritability of traits, methods of measuring them, and an understanding of their genetic basis. Other small projects have explored ways to breed bees to avoid inbreeding, to reduce between colony variation, or to maintain a desirable stock. Such programs have not been designed to produce bee stocks. Rather, they have been designed to provide genetic and breeding information which is useful to commercial breeding programs. These pioneering programs are characterized by having small base populations, and usually involve control stocks which permit precise evaluations of success.

The second category of breeding project, usually but not always undertaken by a commercial beekeeping company, has produced the stocks of bees actually used by beekeepers. The more successful commercial breeding programs often have taken advantage of information gained by the small pioneering projects. Commercial breeding programs are characterized by having large numbers of colonies in base populations and selected generations and lack costly control stocks. As such, the stock improvement attained by such programs, while often obvious, is more difficult to document.

III. BREEDING FOR RESISTANCE TO HONEY-BEE DISEASES AND PESTS

A. Natural Variation

Variation in resistance to several pathogens and pests exists in the honey bee (Rothenbuhler et al., 1968). Practical beekeepers have often noticed that some colonies of honey bees are more resistant to various diseases and pests than others. One commercial beekeeper who very early recognized the significance of such observations and took advantage of them in commercial breeding is Charles Mraz (Park, 1936).

It is known that the North European dark bee (*Apis mellifera mellifera*) which was originally brought to the United States is much more susceptible to the wax moth (*Galleria mellonella* L.) than is the yellow Italian subspecies (*A. m. ligustica*). These same subspecies have similar differential responses to European foulbrood (EFB). The susceptibility of *A. m. mellifera* to EFB resulted in massive requeening programs with Italian bees in North America during 1910–1920 (Sturtevant, 1920). Additionally, American foulbrood (AFB) (see Gochnauer et al., 1975) has been investigated quite closely, and variation in resistance to this infectious disease has been often demonstrated (Eckert, 1951; Park, 1936, 1937; Rothenbuhler, 1958).

B. Resistance to American Foulbrood

Breeding for resistance to AFB has been reviewed by Rothenbuhler (1958), Rothenbuhler et al. (1968), and Cale and Rothenbuhler (1975).

Park et al. (1937) started a selective breeding program to produce stock resistant to AFB with 25 colonies (Italian, Caucasian, Carniolan, and some crossbred stock) from all over the United States. These colonies were thought to have some degree of natural resistance to AFB. A group of six

TABLE 1. Improvement in Resistance to American Foulbrood through 15 Selected Generations[a]

Genera-tion	1	2	3	4	5	6	7	8	9	10	11	12	13	14
Number of colonies	25	27	114	111	148	89	59	90	89	55	66	101	37	68
Percentage diseased	72	67	29	19	8	22	14	27	24	2	2	8	19	29

[a] From data presented by Rothenbuhler (1958) in review of published and unpublished work by O. W. Park and associates beginning in 1935.

susceptible colonies was included in this breeding program as controls. Testing of the colonies was done by inserting a piece of comb containing dead diseased brood with at least 75 scales into one of the brood nest combs of each colony. Combs used in making inoculations were taken from a single source. In the first year, all 31 colonies were inoculated. AFB became well established in all six control colonies and they never recovered. Of the 25 supposedly resistant colonies, only seven were later found to be apparently free from disease. This experiment confirmed the existence of variation in resistance to AFB in the honey bee.

A new generation of queens and drones was reared from the most resistant colonies. An isolated apiary was established for naturally mating these queens and drones. Following inoculation with scales of AFB about one third of the colonies from the new generation eliminated all symptoms of disease (Park *et al.*, 1937). These results made clear that resistance to AFB in honey bees is inherited and responds to artificial selection. It was also shown that no one subspecies has a "corner" on resistance to AFB.

This breeding program was continued for several years and produced a high degree of resistance of AFB. The progress from 1935 until 1949 is shown in Table 1. After 15 selected generations, resistance reached 98%. Table 1 shows setbacks in progress in certain years. Such setbacks are typical for selection programs of all types. In this program artificial insemination was first utilized in 1943 and natural mating was discontinued in 1949 (Roberts, 1967).

Rothenbuhler (1958) changed Park's system of selection and started selecting for resistance in one genetic line and for susceptibility in another. He relied exclusively on instrumental insemination techniques. The resistant material used to form the base population was 10 queens obtained from E. G. Brown of Iowa. The susceptible Van Scoy line was established from a one free-mated queen acquired from H. Van Scoy of New York. Selection accompanied by inbreeding produced two entirely different lines of honey bees which differed tremendously (Rothenbuhler, 1967). Most colonies in the line resistant to AFB did not show signs of disease even after 1000 scales had been placed in their brood nests. Most colonies of the susceptible line were destroyed by disease after they were given just one scale of AFB.

From the beginning of their breeding efforts, Rothenbuhler and his associates searched for mechanisms underlying the remarkable differences between the resistant and susceptible lines. The removal of dead larvae by resistant bees and the failure of the susceptible line to do so was one factor (Woodrow and Holst, 1942; Rothenbuhler, 1964a). Bees ingest dead larvae which have been transformed by the pathogen to a sticky mass. Such larvae cannot be carried by bees in their mandibles (Schulz-Langer, 1956).

Resistant adult workers protect the larvae they nurse from AFB spores by filtering the spores from the liquid food they feed to larvae (Thompson and Rothenbuhler, 1957) and by feeding mandibular secretions having greater antibiotic activity (Rose and Briggs, 1969).

The genetic basis of differential resistance to AFB has been investigated by crosses and backcrosses (Rothenbuhler, 1964b, 1967). When two lines were crossed, the response of the F_1 was very close to that of the susceptible line. This indicated that the gene or genes for hygienic behavior are recessive. Backcrosses of drones from F_1 queens to queens of the resistant line (single-drone matings) gave four kinds of colonies; $\frac{1}{4}$ showed clear hygienic behavior since bees both uncapped and removed dead brood rapidly, $\frac{1}{4}$ uncapped dead brood only, $\frac{1}{4}$ removed dead brood after it was uncapped by a human, and $\frac{1}{4}$ neither uncapped nor removed dead brood. The hypothesis explaining these results suggested that hygienic behavior is controlled by recessive alleles at two loci.

Generally, only minor interest has been shown by commercial breeders in incorporating AFB-resistant characteristics into breeding stocks. However, recently, Taber (1982a,b) began breeding for AFB resistance in his commercial stock. Additionally, Milne (1982) has reported a technique to easily identify bees which rapidly uncap and remove dead brood in a laboratory test.

C. Resistance and Susceptibility to Hairless-Black Syndrome

Virus diseases affect both adult honey bees and their brood. Control of these diseases is very difficult since no known chemical treatment is both effective and economical. As such, resistance breeding is the only practical approach to the control of virus diseases in honey bees.

Kulinčević et al. (1969) and Kulinčević et al. (1970) investigated two cases of an adult honey-bee disease having varied symptoms. The pathogens were not specifically identified, although several different virus-like particles were detected by electron microscopy. Later, Rinderer et al. (1975)

isolated virus particles from infected bees which produced the varied symptoms and were similar to chronic bee paralysis virus (Bailey et al., 1963).

One can usually find some colonies in an apiary having many bees with hairless-black syndrome and other colonies having few such bees. This could be an indication of genetical differences in susceptibility (Rothenbuhler et al., 1979). Drescher (1964) observed that a similar malady in West Germany (Schwarzsucht) was present in one genetic stock but not in another, even though all the colonies were in the same location.

Kulinčević and Rothenbuhler, (1975) reported a successful two-way breeding program designed to produce lines of bees resistant and susceptible to hairless-black syndrome. Selection occurred through four generations. Breeding started with three apparently healthy colonies of different genetic origin chosen from 120 colonies of commercial stock.

The breeding design is similar to the one shown in Chapter 6, Fig. 3. Three free-mated queens (A, B, C) with sperms of various drones in their spermatheca constituted the genetic base stock for this selection experiment. From these queens, after progeny testing, new queens were reared and mated instrumentally with the semen from single drones.

This produced the parental generation from which three of the most resistant and three of the most susceptible queens were chosen to produce the next generation. The same procedure was followed through all four generations of breeding. On average, breeding stock came from the best 22% of the matings tested (Kulinčević and Rothenbuhler, 1975).

Means of the two lines diverged more with each new generation (Fig. 1). In the first generation the response of the two lines was not statistically different. However, in all subsequent generations the differences were highly significant. Both pathogen-fed and control samples from the base stock, the parental, the first, and the second generations showed certain mortality. The pattern of mortality in control samples closely followed the pattern of mortality in pathogen-fed samples. However, overall the mortality in control samples was less. The similar pattern might indicate that the pathogen naturally occurred at equal levels for both control and pathogen-fed groups in the testing system up to the third generation. In the third and fourth generations there was very little mortality in control tests, substantial mortality in pathogen-fed susceptible-line tests, and some mortality in pathogen-fed resistant-line tests.

Rinderer et al. (1975) compared the resistant and susceptible lines produced by Kulinčević and Rothenbuhler (1975) with bees of a nonselected commercial stock. Four doses of virus were tested. The dosage response of all three stocks and recovery of virus particles tied mortality to the virus particles found in sick bees from hairless-black syndrome. The investigation also gave confirming evidence that the lines developed by Kulinčević and

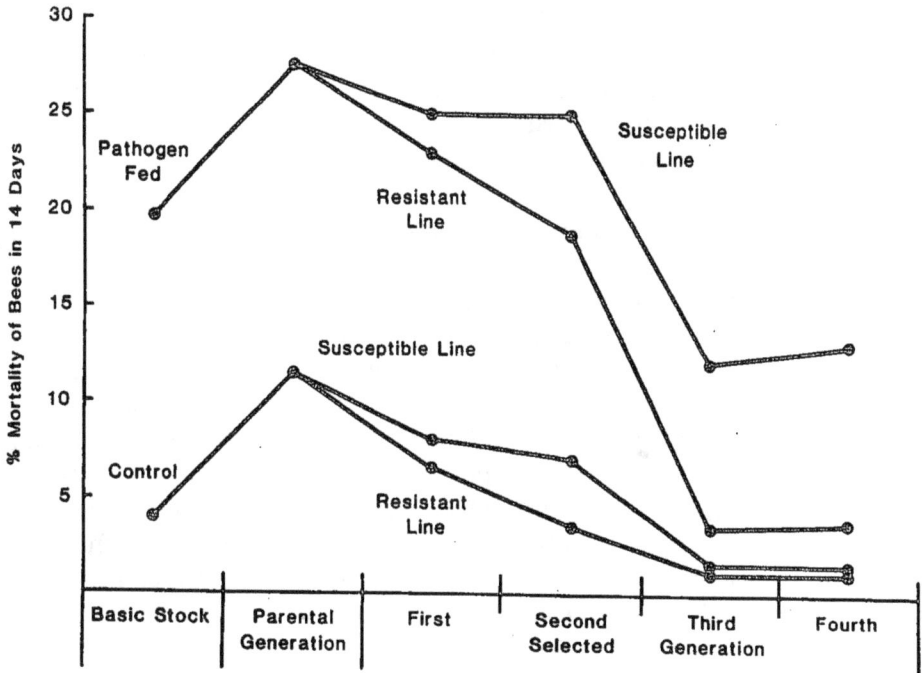

Fig. 1. Results of selection for resistance and susceptability to hairless-black syndrome. [From Kulinčević and Rothenbuhler (1975), with permission. Copyright 1975 by Academic Press.]

Rothenbuhler (1975) were significantly different in their response to the virus.

Of the previous researchers who have reported resistance to virus diseases in insects, all have worked with Lepidoptera, and none have selected for both resistance and susceptibility starting from the same base population.

D. Resistance to Acarosis

The acarine mite (*Acarapis woodi* Rennie; causal agent of acarosis) is a honey-bee parasite present on all continents except Australia. At the beginning of this century the acarine mite destroyed the beekeeping of Great Britain based on the old native British bee (Adam, 1954). Bailey (1964) considers the mite only partially responsible for the demise of this bee.

According to Adam (1962), only imported Italian bees survived the 1905–1919 epizootic because they were more resistant to the mite. He described the appearance of resistant and susceptible colonies in the area of Buckfast Abbey in Southwest England. Selective breeding at Buckfast

Abbey through 20 years has increased resistance to acarosis to such an extent that the mite is no longer a problem. Calvert, cited by Rothenbuhler (1958), confirmed that he overcame acarosis by requeening susceptible colonies with the Buckfast resistant strain of bees. Furthermore, the resistance characteristic was apparently inherited and persisted during the first outcross generation.

Natural variation is also known to exist between stocks of bees in their response to acarosis. Bailey (1967) found that acarosis was detectable in a larger percentage of colonies of "American" bees in Britain than of "British" bees in the same locality. The underlying causes of this difference are unclear since "British" bees added to colonies of "American" bees became equally infested. Alexejenko and Wowk (1971) also reported differences between geographical subspecies of honey bees in acarosis resistance. They found Italian bees were more resistant than the local Russian subspecies.

These experiences with breeding for resistance to acarosis provide hope for programs developed to breed honey bees having resistance to the parasitic mite *Varroa jacobsoni* Oudemans. Recently, this mite has come to pose serious dangers for world apiculture. Indications of natural variation in resistance to varroosis (Piotrowski, 1982) have been found (J. M. Kulinčević, unpublished).

IV. BREEDING AND IMPROVED HONEY PRODUCTION

A. Increased Hoarding

The nectar-hoarding drive in honey bees is essential for the survival of a colony and is the foundation of commercial honey production. However colonies differ in the intensity of their nectar hoarding and consequently in their honey production. Since greater honey production is the main goal of almost every beekeeper, many breeding programs concentrate on this goal. In order to measure and compare the honey production of selected stocks one needs a period of natural nectar availability (nectar flow). These are often uncertain and cannot be prearranged.

Research groups in England and the United States independently showed that testing small groups of worker bees in laboratory cages without reliance on nectar flows could be used efficiently to measure certain aspects of honey production. Bees in cages took sugar syrup and stored it in comb in the cages. This behavior was called hoarding by both groups of investigators. Free and Williams (1972) worked on the effect of environmental factors influencing hoarding. Kulinčević and Rothenbuhler (1973) conducted experiments with bees from several different colonies. While their experiment was designed to test other hypotheses, the range in hoarding they found (4.3 – 13.0 days to hoard 20 ml) indicated substantial genetic variation between colonies. Results in honey gained in the apiary correlated with the laboratory tests. A similar investigation with a different group of colonies (Kulinčević *et al.*, 1974) showed an even higher correlation between hoarding and honey production. This kind of approach to the measurement of hoarding behavior has been accepted and somewhat modified by Rinderer and Elliot (1977), Milne (1977), Rinderer and Sylvester (1978), and Sylvester (1978).

Encouraged by the correlation between hoarding and honey produced in the field and also by the significant genetic variation in hoarding among colonies, Rothenbuhler *et al.* (1979) started a two-way selection to breed both fast- and slow-hoarding lines of bees. This breeding began with 29 colonies having instrumentally inseminated queens (single drone). These colonies were tested for hoarding behavior in laboratory cages (Kulinčević and Rothenbuhler, 1973), and the parents for the first and subsequent generations were chosen solely on the results of laboratory tests. Details of the breeding plan are given by Rothenbuhler *et al.* (1979).

Progress occurred in the breeding for increased hoarding through five generations. The first selected parents differed markedly from the mean of the base population. In the first selected generation, progress was made in increasing the hoarding rate in the fast line but not in decreasing the rate of the slow line. In the second and subsequent three generations, the separation between lines was increased. Fast and slow lines were clearly separated.

When tested in the field, fast-hoarding line colonies of the second generation produced more honey than colonies of the slow line. However, this relationship did not occur in the fourth and fifth generations. These discrepancies were considered to be the result of factors affecting honey yield other than the genetic predispositions of the bees to hoard. Milne (1980a) thought that hoarding tests should be done when worker bees are 7–9 days of age. Rinderer and Baxter (1978) found that increasing the amount of empty comb increases hoarding behavior. The fourth generation was tested during a heavy nectar flow in spring, and the lack of empty comb in the colonies could have resulted in the similar response of colonies from the two lines.

The collection of studies on hoarding behavior done by investigators at several different laboratories shows clearly that hoarding behavior has a genetic basis and is related to the behavioral sequences that lead to honey production. However, environmental influences on both hoarding and honey production do have strong influences. Before hoarding behavior can be used as the testing system in breeding programs designed to improve honey production, more experimental work will be necessary to fully understand the relationship between laboratory hoarding and honey production.

B. Oviposition Rate, Honey Production, and Heterosis

The honey crop produced by a colony depends in part on the number of bees in the colony. Farrar (1937) showed that the honey collected by a colony was correlated with its increase in numbers of colony members. The egg-laying potential of a queen has been considered to be a primary constraint in the development of a productive colony. The relationship between oviposition rate and subsequent honey production was investigated by Cale and Gowen (1956). Their work concentrated on two main points: first, the measurement of the maximum oviposition rates of both some established inbred lines and their F_1 progeny, and second, the measurement of honey production by the progeny of free-mated F_1 queens. The main goals of their study were to look for the presence of heterosis (hybrid vigor) in the F_1 crosses and to explore the relationships between honey production and oviposition rates.

Both oviposition rate and honey production showed substantive heterotic effects. Five out of six F_1s were better than their higher parent in oviposition rate. The progeny of four of the free-mated F_1 queens surpassed the higher parental inbred line in the amount of honey they produced. Also, egg-laying rate and honey production were very highly correlated. Additionally, this study showed that inbred lines do exhibit differences in both general and specific combining abilities.

C. Crossbreeding of Geographical Subspecies

The honey bee evolved in the Near East, Africa, and Europe. A variety of geographical subspecies and ecotypes have been formed by natural selection. Geographical races of the honey bee are defined as subspecies (Ruttner, 1975). Therefore, the word "race" in bee breeding does not have the same meaning as it does in the breeding of other animals.

In many cases, heterosis occurs when two or more geographical races are crossed even if the stock has not been inbred. This occurs as a result of substantial genetical differences existing between honey-bee subspecies. Still, heterosis is not found in all crosses (Ruttner, 1973).

Mârza and Marcovici (1965) crossed *A. m. carpatica* drones with *A. m. caucasica* queens and *A. m. carpatica* queens with *A. m. caucasica* drones. Honey production was increased 26.9 and 10.5%, respectively, and wax production by 28.5 and 7%, respectively. Ruttner (1968) crossed *A. m. mellifera* and *A. m. carnica* and found an increase of 31% in honey production compared to the better parent.

Fresnaye *et al.* (1974) showed that honey production was increased by 40–50% in the first generation of *A. m. mellifera* × *A. m. caucasica* crosses.

However, of the crosses they studied, the most successful were three-way crosses of (*A. m. ligustica* × *A. m. causasica*) × *A. m. mellifera*.

Fresnaye and Lavie (1977) presented the results of a 6-year crossbreeding study using (*A. m. ligustica* × *A. m. caucasica*) × *A. m. mellifera* in France. The average increase in honey production of this three-way cross was 116% better than that of *A. m. mellifera* control stock. Such increases in honey production are very impressive. However, in subsequent generations hybrid effects decreased rapidly.

Raghim-Zade (1977) crossed ecotypes of *A. m. caucasica* (gray × yellow steppe bee) and found increased egg-laying capacity in crossbred queens.

Melnichenko and Trishina (1977) crossed different subspecies of honey bees from Russia with *A. m. ligustica* and *A. m. carnica*. The most productive and prolific stock they tested was the three-way cross. Certain crosses they tested showed resistance to nosematosis, an intestinal disease of adults caused by a microsporidian.

According to Velichkov (1977), in Bulgaria, the most efficient cross-bred stock he tested was the three-way cross of *A. m. caucasica* × "local bees") × *A. m. carnica*. The honey production of this cross-bred stock was 60–70% higher than that of the local stock.

Kurletto (1975) reported that after three generations of backcrossing selecting Africanized bees and *A. m. carnica* he produced stock having good honey-production ability, reduced colony defense, good resistance to disease, and low swarming tendencies.

According to Ruttner (1968), there was no heterosis observed in the offspring of *A. m. cypria* × *A. m. carnica* or *A. m. cypria* × *A. m. mellifera*. Also heterosis was absent in the progeny of an *A. m. mellifera* × *A. m. ligustica* cross (Fresnaye *et al.*, 1974).

D. Breeding Programs in Certain Countries

1. Israel

In Israel after the Second World War the local population of *A. m. lamarkii* bees were substituted by a gentler and more productive "Italian" bee imported from the United States. This alone increased honey production by 30%. In efforts to further improve the introduced stock, a breeding program was initiated. Bar-Cohen *et al.* (1978) presented the results of 13 years of selective breeding (from 1962 to 1974). They continuously selected for both an increased amount of brood and increased honey production. Average annual honey production and number of brood cells per tested colony (Table 2) suggest that selecting breeder queens on the basis of the honey production of colonies produced by their mothers was effective in causing

TABLE 2. Responses to Simultaneous Selection for
Increased Honey Yield and Increased Number of
Brood Cells through 13 Generations[a]

	Honey yield		Brood cells	
Year	Number of colonies	kg	Number of colonies	Thousands of cells
1962	28	33.9	2	29.5
1963	21	27.2	21	26.0
1964	16	43.9	15	25.6
1965	18	40.9	18	21.2
1966	34	49.9	12	17.8
1967	36	53.8	36	22.9
1968	46	41.8	46	27.9
1969	78	39.9	77	29.8
1971	84	48.9	81	28.7
1972	72	56.8	66	25.8
1973	60	57.3	57	31.9
1973	33	65.3	32	33.8
1974	36	43.1	36	32.4

[a] Modified with permission from Bar Cohen et al. (1978).

genetic improvement in honey production. However, the selection for high
oviposition rates was not effective for improvement in either oviposition
rates or honey production. In this study the estimated heritability for honey
yield was 0.54. The selection differential among queens was an average of
17.5 kg honey, and the average annual increase in honey production was
4.7 kg per colony.

This breeding success was achieved with mating selected queens at an
apiary without complete reproductive isolation. The great variety of drones
originating from the selected stock maintained good viability of brood (see
Woyke, Chapter 4).

2. Canada

Szabo (1982) has been developing a bee-breeding program in Alberta,
Canada, using large numbers of colonies. The most important selection
criterion is considered to be the short-term gain (1 day) in weight of a colony.
Short-term gain was significantly correlated with seasonal honey production. The amount of sealed brood and pollen collected during a single day
also correlated well with seasonal honey production. A multiple correlation
of short-term gain both in honey and in pollen and sealed brood with
seasonal honey production ranged from 0.902 to 0.984. Where possible,
other colony characteristics are also considered in the program.

The honey production in the third generation of stock selected on the basis of the multiple characters was 46.3% greater than.that of control colonies in southern test areas and 56.8% greater in northern test areas. These tests involved 267 colonies in 10 apiaries.

This breeding program has involved one-way selection designed to most efficiently produce stock with excellent honey production, improved mating ability, reduced disease problems, and generally having desirable characteristics in the conditions in western Canada.

A second breeding program is ongoing in Canada at the University of Guelph in Ontario (Milne, 1977). Because field tests always contain environmentally induced variation, laboratory testing is used to estimate the breeding value of stock in the program. Laboratory tests of worker hoarding (Milne, 1980a), life-span (Milne, 1980b), and pupae weight (Milne, 1980c) have been developed which correlate with colony honey production. The tests are not themselves correlated, which indicates that each measures a different feature of the complex underpinnings of honey production.

3. Sweden

Ebersten (1979) started a breeding project that is based on data collected directly from beekeepers through a special reporting system. These data concern quantity of brood at different periods, swarming tendency, stinging behavior, inclination to run on the combs, and honey production. A queen pedigree system is being used in the project. In 1977 this project included 25 beekeepers with more than 1000 colonies. According to their long-term plan, the number of colonies will eventually be at least 5000.

V. BREEDING FOR INCREASED POLLEN GATHERING

A. General Pollen Gathering

The production of 90 fruit and seed crop plants depends upon honey bees as pollination agents (McGregor, 1976). Pollen collecting is essential for the survival and reproduction of a honey-bee colony. Cale (1967) found that the pollen gathering of bees was highly correlated with honey production in three different years. These data indicate the possibility of breeding a bee that is excellent both in pollen gathering and in honey production. Differences in pollen collection among colonies and geographical subspecies of honey bees have been investigated and reported by Akerberg and Lesins (1949), Hunkeler (1943), Goetze (1953), and many others.

A two-way selection project for high and low levels of hoarded pollen was recently conducted (R. L. Hellmich, personal communication). Selec-

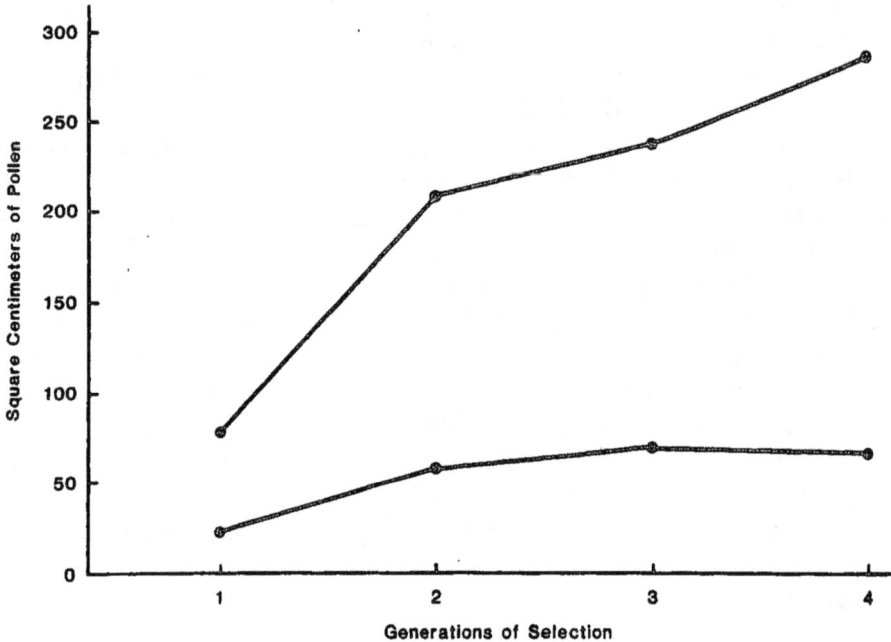

Fig. 2. Results of selection for fast and slow field collection of pollen: two-way selection for pollen hoarding. [From data of R. Hellmich II, personal communication, with permission.]

tions were based on bimonthly measurements of pollen stored in five-frame colonies. The results of four generations of selection, one generation per year, are shown in Fig. 2. The high-pollen-hoarding and the low-pollen-hoarding bees were significantly different during each generation. The lack of appreciable change exhibited by the low pollen hoarders suggests a lower limit to selection of low pollen hoarding. The significant progress exhibited by the high pollen hoarders suggests an upper limit has not been reached for high pollen hoarding.

B. High and Low Alfalfa Pollen Gathering

Alfalfa seed production is a worldwide problem. Modern agricultural practices have eliminated numerous feral insect pollinators. Thus, pollination requirements are often met by moving honey bees to alfalfa fields, even though honey bees are known to prefer other crops which may be in the area (McGregor, 1976). Nye and Mackensen (1965) checked pollen collecting worker honey bees from over 350 colonies in five apiaries and in four locations. They determined the percentage of pollen collectors that gathered alfalfa pollen and found a great variation in the amount of alfalfa pollen collected by different colonies. Because of this variation, they initiated a breeding program aimed at increasing the percentage of alfalfa pollen collected. With their program, Mackensen and Nye (1966, 1969; Nye and Mackensen, 1968, 1970) achieved steady progress in separating high and low alfalfa pollen collecting lines. In generation four the high and low lines no longer overlapped in response, and F_1 crosses of these two lines were intermediate in response. This verified the genetic component of the alfalfa pollen gathering trait.

Selection continued through seven generations. The average percentage of alfalfa pollen reached 87 in the high line and 8 in the low line. It was concluded that alfalfa pollen gathering is probably dependent on several genes having additive effects.

VI. BREEDING FOR LENGTH OF LIFE

Longevity or length of life [see discussion of these terms in Milne (1980b)] in honey bees is of interest for both beekeepers and scientists. Various studies have shown marked differences in the length of life (Free and Spencer-Booth, 1959; Rinderer and Sylvester, 1978; Milne, 1980b).

Kepena (1979) found that the life of worker honey bees becomes shorter with inbreeding. Outcrossing to another stock diminished the effect. The longevity of worker bees in laboratory cages was significantly correlated with colony weight gain in the field. Kulinčević et al. (1984) also observed significant variation in length of life in unselected Italian bees in the United States. The average length of life of worker bees in laboratory cages ranged from 30.5 to 45.5 days.

This observation led to a selection program (Kulinčević and Rothenbuhler, 1982) which bred bees for long and short lengths of life in laboratory cages (Fig. 3). In two selected generations the two lines were significantly different. Both lines improved in length of life but the high selected line improved more. Selection for shorter length of life was not shown to be successful in this experiment. Each generation was produced and tested in a different year, and perhaps the effect of differing environments masked actual genetic changes. This is suggested by the difference shown in Fig. 3 between the two tests of the second generation.

VII. BREEDING FOR MORPHOLOGICAL CHARACTERISTICS

Drescher [reported in Gonçalves and Stort (1978)], using instrumental insemination and applying continuous selection, developed two lines of honey bees with different numbers of hamuli (wing hooks) on the hind

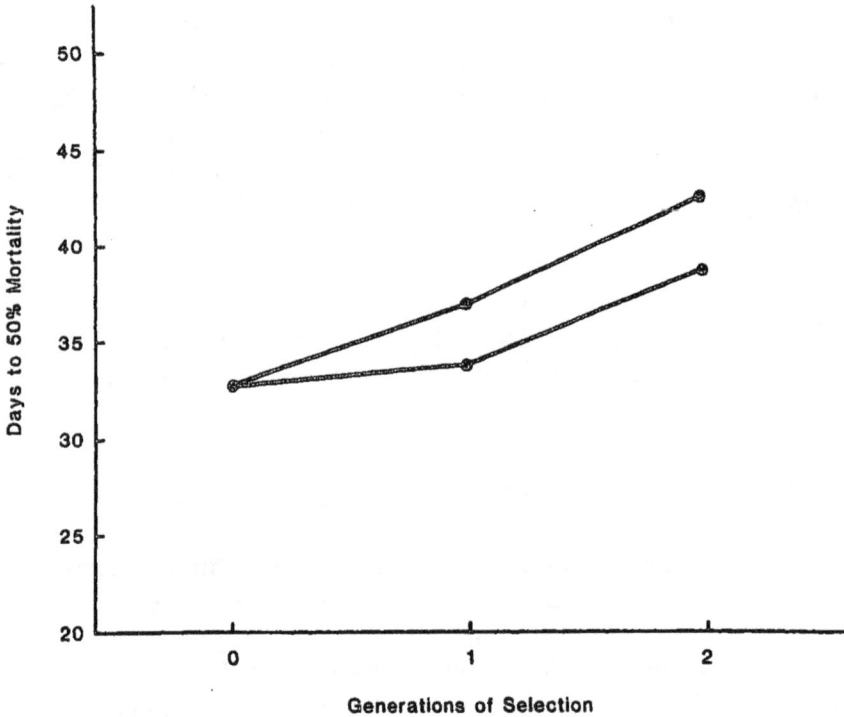

Fig. 3. Results of selection for greater and lesser length of life in honey bees. [Modified from Kulinčević and Rothenbuhler (1982), with permission. Copyright 1982 by Institut National de la Recherche Agronomique.]

wing. After 10 generations, Gonçalves [reported in Gonçalves and Stort (1978)] continued this program through 22 selected generations. The extreme realized values were 28.6 hamuli in bees of the high line and 10.6 in bees of the low line. The F_1 crosses had intermediate numbers of wing hooks. Drescher analyzed worker bees from the selected lines and found a positive correlation between the number of hamuli and the cubital index. He speculated that this correlation resulted from either pleiotropic gene action or accidental parallel selection, since cutibal index was not considered as a criterion for selection.

VIII. COMMERCIAL BEE-BREEDING SUCCESSES

In their work, commercial bee breeders have two choices: the breeding of more or less pure stock or inbred lines for the production of hybrids. With either choice, success has depended upon the very careful selection of parental stock.

A. Line Breeding

This system of breeding refers to the method of selection within a smaller or larger and more or less closed population. The end result is a stock of bees itself suitable for distribution.

Line breeding is practiced by queen producers in both Europe and the United States. There is a long tradition of line breeding of *A. m. carnica* in Austria and Germany. In Austria several strains of *A. m. carnica* are produced. Also, the Austrian Beekeepers Association produces line-bred stock.

Because of the commercial nature of these line-breeding programs there is little published information documenting their success. However, it is known that buyers of these queens often have substantial "stock loyalty." Additionally, it is clear from observing many of these stocks that they often display consistent coloration, good honey production, gentle dispositions, etc. These characteristics could only exist at the high levels of consistency often shown by line-bred stocks where careful selection and breeding were employed in stock development.

Böger (1969) bred within a closed population and concluded that in large populations it is possible to select severely and yet avoid inbreeding. Vinogradova (1977) reported successes in line breeding of *A. m. caucasica* bees for higher honey production, egg-laying capacity, and wintering ability in the USSR.

While several line-breeding programs have been successful, others have not. In New Zealand, Forster (1975) tested three established lines of bees and found that honey production varied little between the lines. The same was true when Kulinčević *et al.* (1984) tested seven stocks of bees from different areas of the United States. However, substantial variation existed between colonies within a stock.

Adam (1954) described his method of line breeding. Among other things he gave information concerning the development of the "Buckfast" bee. This stock has been developed from a cross between dark leather-colored Italian bees and the old native "British" bee. This subspecies outcross was later crossed and backcrossed, in order to improve honey production and other characteristics, with numerous subspecies. Adam (1966) made several investigative trips in search of the best strain of bees to western and central Europe, the Balkans, the Near East, and North Africa.

A large hybrid breeding program using stocks of *A. m. carnica* is ongoing in Czechslovakia (Vésely and Janoušek, 1977). A total of eight unrelated inbred lines are being crossed using instrumental insemination and then tested. This extensive program is being done by 40 separate commercial apicultural groups owning about 10,000 colonies. Regional research stations are testing stock and periodically provide new inbred lines to the program.

Rothenbuhler (1980) suggested that any stock improvement program that is expected to give practical results for beekeeping industries must have three components: field tests under natural conditions, geneticists making genetic decisions, and the commercial production of the improved stock. Field tests could be supplemented by laboratory tests (Rinderer, 1977). This would enhance and speed selection. Methods of testing, data collection, and the number of colonies tested in each generation require constant input from a geneticist. The geneticist makes decisions concerning which colonies are going to produce queens and drones for the next generation, when to use instrumental insemination or natural mating, and how to maintain and further improve new stock. Commercial production of improved stock is the responsibility of queen breeders. Queen quality depends on optimal queen rearing procedures. Poor rearing can produce queens of inferior quality even from improved stock.

B. Hybrid Breeding

The genetic phenomonen of heterosis, which has been successfully applied in the breeding of corn and other crops, can also be used in breeding honey bees (Cale and Gowen, 1956). Hybrids are crosses between previously tested and selected lines. These lines may originate from the same or different subspecies. The main objective in hybrid breeding is higher honey production. However, other characteristics, such as disease resistance, wintering ability, good spring brood-rearing patterns, etc., may also be considered. The creation of inbred lines requires instrumental insemination.

The most widely renown commercial hybrid, named Starline, was released in 1949 by Dadant and Sons, Inc. This hybrid was developed by Dr. G. H. Cale, Jr.. Starline is a four-line hybrid of Italian stock which is continuously improved by combining new inbred lines with preexisting ones. One of the better inbred lines which has been used to produce Starline was recently in its 27th generation (Witherell, 1976). The Midnite is a four-line hybrid of A. m. caucasica and A. m. carnica strains and was first released in 1957. This hybrid is characterized by its gentleness and ability to work in cold conditions. A third four-line hybrid, Cale 876, was released for a short time. With this hybrid, not only were the inbred lines developed and maintained using instrumental insemination, but also instrumental insemination was used in the production of commercial stock.

In hybrids, viability factors are controlled and approach 100%. Usually, line-bred stocks have less viability. Starline is distributed in cooperation with several queen producers in the United States, France, and Chile. Cooperating queen breeders receive grafting and drone-producing stock from the parent company.

According to Witherell (1976), the honey yields of Starline and Midnite are "usually 130 and 200 percent of those of common Italian and Caucasian stock found in the same apiary."

Sugden and Furgala (1982a,b,c) compared Starline, Midnite, Cale 876, Buckfast, Curneen-Black, and Mraz stocks. They found many differences between these stocks in longevity of queens, disposition to sting, use of propolis, propensity to swarm, and honey yield. Certain stocks were considered to have superior value for Minnesota. Perhaps the stocks that did less well were bred to differing standards of acceptability and for different areas. It is interesting to note that the stock they considered unacceptable because of stinging stung about 0.20 times/cm² sec, while "Africanized" bees studied by Collins et al. (1982) stung 0.85 times/cm² sec in similar tests.

IX. PRESERVATION OF SUBSPECIES AND VALUABLE ECOTYPES

The value of certain geographical subspecies and ecotypes of honey bees for breeding has been clearly demonstrated and documented (Ruttner, Chapter 2).

Mass propagation and worldwide distribution of queens is now changing gene frequencies of several geographical subspecies and ecotypes of honey bees. Because of this, several subspecies of honeybees are threatened with extinction. *Apis m. mellifera* in central and northern Europe has almost disappered because of the commercial movement of various bee stocks. The most dramatic case of stock replacement because of imports is the nearly complete "Africanization" of all of South America.

Fresnaye and Lavie (1977) proposed the creation of regional and national reservations for the protection and selection of strains of pure local ecotypes and subspecies.

In the Soviet Union the import of exotic bee stock is prohibited in a number of zones (Kotova and Podolski, 1977). In other areas of the natural habitat of *Apis mellifera*, similar programs are needed.

REFERENCES

Adam, Br. (1954). Bee breeding. *Bee World* 35, 4–13, 21–29, 44–49.
Adam, Br. (1962). Einige Erfarungen mit der Milbenseuche. *Bienenvater, 83,* 35–38, 79–81.
Adam, Br. (1966). "In Search of the Best Strains of Bees." Walmar Verlag, Zell-Weierbach, Federal Republic of Germany.
Akerberg, K., and Lesins, K. (1949). Insects pollinating alfalfa in central Sweden. *K. Lamtbr-högsk. Annir.* 16, 630–643.
Alexejenko, F. M., und Wowk, A. M. (1971). Alters und Rassenwiederstandsfähigkeit der

Bienen gegen Akarioze und deren neue Bekämpfungsmittel. *Proc. Int. Bienenzüchterkongress* (Mosakau, UDSSR) 23, 500–502.

Bailey, L. (1964). The "Isle of Wight Disease": The origin and significance of the myth. *Bee World* 45, 32–37.

Bailey, L. (1967). The incidence of *Acarapis woodi* in North American strains of honey bees in Britain. *J. Apic. Res.* 6, 99–103.

Bailey, L., Gibbs, A. J., and Woods, R. D. (1963). Two viruses from adult honey bees (*Apis mellifera* Linnaeus). *Virology* 21, 390–395.

Bar-Cohen, R., Alpern, G., and Bar-Anan, R. (1978). Progeny testing and selecting Italian queens for brood area and honey production. *Apidologie* 9, 95–100.

Böger, K. (1969). Zur Selektion von Bienenvölkern auf Honigleistung. *Z. Bienenforsch.* 9, 545–571.

Cale, G. H., Jr. (1967). Pollen gathering relationship to honey collection and egg laying in honey bee. *Proc. Inter. Apic. Cong. (Apimondia)* 21, 230–232.

Cale, G. H., Jr., and Gowen, J. W. (1956). Heterosis in the honey bee (*Apis mellifera* L.). *Genetics* 41, 292–303.

Cale, G. H., Jr., and Rothenbuhler, W. C. (1975). Genetics and breeding of the honey bee. *In* "The Hive and the Honey Bee," 4th ed. (Dadant and Sons, eds.), pp. 157–184. Dadant and Sons, Hamilton, Ill.

Collins, A. M., Rinderer, T. E., Harbo, J. R., and Bolten, A. B. (1982). Colony defense by Africanized and European honey bees. *Science* 218, 72–74.

Drescher, W. (1964). Beobachtung zur unterschiedlichen erblichen Disposition von Zuchtlinien von *Apis mellifica* L. für die Schwarzsucht. *Z. Bienenforsch.* 14, 116–124.

Ebersten, K. (1979). Honey bee breeding experimental design of data recording from and by beekeepers in the field. *Proc. 3rd Int. Symp. on Pollination Md. Agric. Exp. Sta.* 1, 245–251.

Eckert, J. E. (1951). The development of resistance to AFB in Hawaii. *Am. Bee J.* 91, 200–201.

Farrar, C. L. (1937). The influence of colony populations on honey production. *J. Agric. Res.* 54, 945–954.

Forster, I. W. (1975). An evaluation of various characteristics in three lines of New Zealand honey bees. *NZ J. Exp. Agr.* 3, 293–296.

Free, J. B., and Spencer-Booth, Y. (1959). The longevity of worker honey bees. *Proc. R. Entomol. Soc. London* A34, 141–150.

Free, J. B., and Williams, I. H. (1972). Hoarding by honey bees (*Apis mellifera* L.). *Anim. Behav.* 20, 327–334.

Fresnaye, J., and Lavie, P. (1977). Selective and cross-breeding of bees in France (*Apis mellifica* L.). *In* "Genetics, Selection and Reproduction of the Honey Bee" (S. Colibaba, ed.), pp. 212–218. Apimondia, Bucharest, Romania.

Fresnaye, J., Lavie, P., and Boesiger, E. (1974). La variabilité de la production des miel chez l'abeille de race noire (*Apis mellifica mellifica*) et chez quelques hybrides interraciaux. *Apidologie* 5, 1–20.

Gochnauer, T. A., Furgala, B., and Shimanuki, H. (1975). Diseases and enemies of the honey bee. *In* "The Hive and the Honey Bee," 4th ed. (Dadant and Sons, eds.), pp. 615–662. Dadant and Sons, Hamilton, Ill.

Goetze, G. (1953). Honigbiene als Bestäuber im Rotklee (*Trifolium pratense*); Rotklee als wichtige Nektarquelle für die Bienenzucht. *Schr. Reihe AID* 66, 5–17.

Gonçalves, L. S., and Stort, C. A. (1978). Honey bee improvement through behavioral genetics. *Annu. Rev. Entomol.* 31, 197–213.

Hunkeler, M. (1943). Die beste Biene. *Schweiz. Bienenztg.* 46, 179–188.

Kepena, L. (1979). Longevity in the laboratory of bees obtained by inbreeding and outbreeding. *In* "Genetics, Selection and Reproduction of the Honey Bee" (S. Colibaba, ed.), pp. 55–58. Apimondia, Bucharest, Romania.

Kotova, G. N., and Podolski, M. S (1977). Bee selection in USSR. *In* "Genetics, Selection and

Reproduction of the Honey Bee" (S. Colibaba, ed.), pp. 130–134. Apimondia, Bucharest, Romania.

Kulinčević, J. M., and Rothenbuhler, W. C. (1973). Laboratory and field measurement of hoarding behaviour in the honeybee. *J. Apic. Res. 12*, 179–182.

Kulinčević, J. M., and Rothenbuhler, W. C. (1975). Selection for resistance and susceptibility to hairless-black syndrome in the honeybee. *J. Invertebr. Pathol. 25*, 289–295.

Kulinčević, J. M., and Rothenbuhler, W. C. (1982). Selection for length of life in the honeybee (*Apis mellifera*). *Apidologie 13*, 347–352.

Kulinčević, J. M., Stairs, G. R., and Rothenbuhler, W. C. (1969). A disease of the honey bee causing behavioral changes and mortality. *J. Invertebr. Pathol. 14*, 13–17.

Kulinčević, J. M., Stairs, G. R., and Rothenbuhler, W. C. (1970). Virus causing paralysis of adult honeybees in Ohio. *J. Invertebr. Pathol. 16*, 423–426.

Kulinčević, J. M., Thompson, V. C, and Rothenbuhler, W. C. (1974). Relationship between laboratory tests of hoarding behavior and weight gained by honey bee colonies in the field. *Am. Bee J. 114*, 93.

Kulinčević, J. M., Rothenbuhler, W. C., and Rinderer, T. E. (1984). Disappearing Disease: III. A Comparison of Seven Different Stocks of the Honey Bee. Res. Bull. 1160, The Ohio State Univ., Ohio Agric. Res. and Dev. Ctr., Wooster, Ohio.

Kurletto, S. (1975). Cruzamento das abelhas Africanizados com as Carnicas. *Ann. Congress Brasileiro de Apicultura 3*, 161–164.

Mackensen, O., and Nye, W. P. (1966). Selecting and breeding honey bees for collecting alfalfa pollen. *J. Apic. Res. 5*, 79–86.

Mackensen, O., and Nye, W. P. (1969). Selective breeding of honey bees for alfalfa pollen collection: sixth generation and outcrosses. *J. Apic. Res. 8*, 9–12.

Mârza, E., and Marcovici, A. (1965). Cercetari privind comportarea albinelor hibride in conditir le din Republica Socialista Romania. *Lucz. Stiiut. Stat. Cent. Seri. Apic. 6*, 5–13.

McGregor, S. E. (1976). "Insect Pollination of Cultivated Crop Plants." Ag. Handbook 496. U.S. Government Printing Office, Washington, D.C.

Melnichenko, A. N., and Trishina, A. S. (1977). Ecological and genetical bases of the heterosis in the honey bee colonies (*Apis mellifera* L.). *In* "Genetics, Selection and Reproduction of the Honey Bee" (S. Colibaba, ed.), pp. 203–209. Apimondia, Bucharest, Romania.

Milne, C. P., Jr. (1977). An improved laboratory measurement of hoarding behavior in the honey bee. *Am. Bee J. 117*, 502, 507.

Milne, C. P., Jr. (1980a). Laboratory measurement of honey production in the honeybee. I. A model for hoarding behavior by caged workers. *J. Apic. Res. 19*, 122–126.

Milne, C. P., Jr. (1980b). Laboratory measurement of honey production in the honey bee 2. Longevity or length of life of caged workers. *J. Apic. Res. 19*, 172–175.

Milne, C. P., Jr. (1980c). Laboratory measurement of honey production in the honey bee. 3. Pupal weight of the worker. *J. Apic. Res. 19*, 176–178.

Milne, C. P., Jr. (1982). Laboratory measurement of brood disease resistance in honey bees. 1. Uncapping and removal of freeze-killed brood by newly emerged workers in laboratory test cages. *J. Apic,. Res. 21*, 111–114.

Nye, W. P., and Mackensen, O. (1965). Preliminary report on selection and breeding of honeybees for alfalfa pollen collection. *J. Apic. Res. 4*, 43–48.

Nye, W. P., and Mackensen, O. (1968). Selective breeding of honeybees for alfalfa pollen: fifth generation and backcrosses. *J. Apic. Res. 7*, 21–27.

Nye, W. P., and Mackensen, O. (1970). Selective breeding of honeybees for alfalfa pollen collection with tests in high and low alfalfa pollen collection regions. *J. Apic. Res. 9*, 61–64.

Park, O. W. (1936). Disease resistance and American foulbrood. *Am. Bee. J. 76*, 12–14.

Park, O. W. (1937). Testing for resistance to American foulbrood in honeybees. *J. Econ. Entomol. 30*, 504–512.

Park, O. W., Pellett, F. C., and Paddock, F. B. (1937). Disease resistance and American foulbrood. *Am. Bee. J.* 77, 20–25, 34.

Piotrowski, F. (1982). Varroosis—the correct term for varroatosis. *Angew. Parasitol.* 23, 49.

Raghim-Zade, M. S. (1977). Prolificness heritability in F_1 and F_2 hybrids of mountain grey × steppe yellow race queens. *In* "Genetics, Selection and Reproduction of the Honey Bee" (S. Colibaba, ed.), pp. 218–220. Apimondia, Bucharest, Romania.

Rinderer, T. E. (1977). A new approach to honey bee breeding at the Baton Rouge USDA Laboratory. *Am. Bee J.* 117, 146–147.

Rinderer, T. E., and Baxter, J. R. (1978). Effect of empty comb on hoarding behavior and honey production of the honey bee. *J. Econ. Entomol.* 71, 757–759.

Rinderer, T. E., and Elliott, K. D. (1977). The effect of a comb on the longevity of caged adult honey bees. *Ann. Entomol. Soc. Am.* 70, 365–366.

Rinderer, T. E., and Sylvester, H. A. (1978). Variation in response to *Nosema apis*, longevity, and hoarding behavior in a free-mating population of the honey bee. *Ann. Entomol. Soc. Am.* 71, 372–374.

Rinderer, T. E., Rothenbuhler, W. C., and Kulinčević, J. M. (1975). Response of three genetically different stocks of the honeybee to a virus from bees with hairless black syndrome. *J. Invertebr. Pathol.* 25, 297–300.

Roberts, W. C. (1967). Development of hybrid bee breeding in the United States. *Proc. Inter. Apic. Cong. (Apimondia)* 21, 226–229.

Rose, R. I., and Briggs, J. D. (1969). Resistance to American foulbrood in the honey bee. IX. Effect of larval food on the growth and viability of *Bacillus larvae. J. Invertebr. Pathol.* 13, 74–80.

Rothenbuhler, W. C. (1958). Genetics and breeding of the honey bee. *Annu. Rev. Entomol.* 3, 161–180.

Rothenbuhler, W. C. (1964a). Behavior genetics of nest cleaning in honey bees. IV. Responses of F_1 and backcross generations to disease-killed brood. *Am. Zool.* 4, 111–123.

Rothenbuhler, W. C. (1964b). Behaviour genetics of nest cleaning in honey bees. I. Response of four inbred lines to disease-killed brood. *Anim. Behav.* 12, 518–583.

Rothenbuhler, W. C. (1967). American foulbrood and bee biology. *Proc. Inter. Apic. Cong. (Apimondia)* 21, 179–188.

Rothenbuhler, W. C. (1980). Necessary links in the chain of honey bee stock improvement. *Am. Bee J.* 120, 223–225, 304–305.

Rothenbuhler, W. C., Kulinčević, J. M., and Kerr, W. E. (1968). Bee genetics. *Annu. Rev. Genet.* 2, 413–438.

Rothenbuhler, W. C., Kulinčević, J. M., and Thompson, V. C. (1979). Successful selection of honeybees for fast and slow hoarding of sugar syrup in the laboratory. *J. Apic. Res.* 18, 272–278.

Ruttner, F. (1968). Methods of breeding honeybees: intra-racial selection or inter-racial hybrids? *Bee World* 49, 66–72.

Ruttner, F. (1973). Zuchttechnik und Zuchtauslese bei der Biene. 3. Auflage. Ehrenwirth, Munich.

Ruttner, F. (1975). Races of bees. *In* "The Hive and the Honey Bee," 4th ed. (Dadant and Sons, eds.), pp. 19–38. Dadant and Sons, Hamilton, Ill.

Schulz-Langner, E. (1956). Ein neues Bild der bösartigen Faulbrut der Honigbiene. *Z. Bienenforsch.* 3, 149–180.

Sturtevant, A. P. (1920). "A Study of the Behavior of Bees in Colonies affected by European Foulbrood." U.S.D.A. Bull. 804. U.S. Government Printing Office, Washington, D.C.

Sugden, M. A., and Furgala, B. (1982a). Evaluation of six commercial honey bee (*Apis mellifera*

L.) stocks used in Minnesota. Part 1. Wintering ability and queen longevity. *Am. Bee J.* 122, 105–109.

Sugden, M. E., and Furgala, B. (1982b). Evaluation of six commercial honey bee (*Apis mellifera* L.) stocks used in Minnesota. Part 2. Aggressiveness and swarming. *Am. Bee J.* 105, 185–188.

Sugden, M. A., and Furgala, B. (1982c). Evaluation of six commercial honey bee (*Apis mellifera* L.) stocks used in Minnesota. Part 3. Productivity. *Am. Bee J.* 122, 283–286.

Sylvester, H. A. (1978). Response of honey bees to different concentrations of sucrose in a hoarding test. *Am Bee J.* 118, 746–747.

Szabo, T. I. (1982). Phenotypic correlations between colony traits in the honey bee. *Am Bee J.* 122, 711–716.

Taber, S. (1982a). Breeding for disease resistance. *Am. Bee J.* 122, 177–179.

Taber, S. (1982b). Determining resistance to brood diseases. *Am. Bee J.* 122, 422–425.

Thompson, V. C., and Rothenbuhler, W. C. (1957). Resistance to American foulbrood in honey bee. II. Differential protection of larvae by adults of different genetic lines. *J. Econ. Entomol.* 50, 731–737.

Velichkov, V. N. (1977). Examination of inter-race hybrids of bees. *In* "Genetics, Selection and Reproduction of the Honey Bee" (S. Colibaba, ed.), pp. 210–212. Apimondia, Bucharest, Romania.

Vésely, V., and Janoušek, J. (1977). Inter-strain cross-breeding of carnica bees in Czechoslovakia. "In Genetics, Selection and Reproduction of the Honey Bee" (S. Colibaba, ed.), pp. 234–235. Apimondia, Bucharest, Romania.

Vinogradova. V. M. (1977). Selection of Caucasian bees. *In* "Genetics, Selection and Reproduction of the Honey Bees" (S. Colibaba, ed.), pp. 229–232. Apimondia, Bucharest, Romania.

Witherell, P. C. (1976). A story of success: the Starline and Midnite hybrid bee breeding programs. *Am. Bee J.* 116, 63–64, 82.

Woodrow, A. W., and Holst, E. C. (1942). The mechanism of colony resistance to American foulbrood. *J. Econ. Entmol.* 35, 327–330.

Index

www.ingramcontent.com/pod-product-compliance
Lightning Source LLC
Chambersburg PA
CBHW080757300326
41914CB00055B/917